MOLECULAR DEVELOPMENTAL BIOLOGY
Second Edition

MOLECULAR DEVELOPMENTAL BIOLOGY
Second Edition

T. Subramoniam

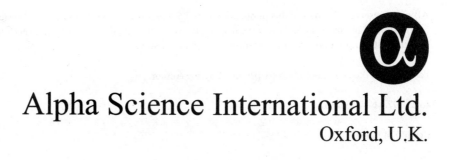

Alpha Science International Ltd.
Oxford, U.K.

Molecular Developmental Biology
Second Edition
364 pgs. | 132 figs. | 15 tbls.

T. Subramoniam
INSA Senior Scientist
Ocean Science and Technology for Islands
National Institute of Ocean Technology
NIOT Campus
Velacherry-Tambaram Main Road
Pallikaranai, Chennai

Copyright © 2008, 2011
Second Edition 2011

ALPHA SCIENCE INTERNATIONAL LTD.
7200 The Quorum, Oxford Business Park North
Garsington Road, Oxford OX4 2JZ, U.K.

www.alphasci.com

All rights reserved. No part of this publication may be reproduced, stored in a retrieval system, or transmitted in any form or by any means, electronic, mechanical, photocopying, recording or otherwise, without prior written permission of the publisher.

Printed from the camera-ready copy provided by the Author.

ISBN 978-1-84265-661-7

Printed in India

For Amruth, Apoorva, Architha and Varshini

PREFACE TO THE SECOND EDITION

The second edition of "Molecular Developmental Biology" is prompted by the phenomenal advancements made in certain aspects such as evolutionary and ecological developmental biology as well as stem cell biology. The earlier edition already gives extensive coverage of these topics. Nevertheless, more recent developments in these fields continue to improve our understanding of developmental mechanisms and their role in morphological evolution as well as phenotypic deviations in response to environmental variables. Increased awareness on the teratogenic effects of certain drugs and environmental chemicals has led to focus attention on the birth defects and congenital diseases. Therefore, a new chapter on teratogenesis and medical embryology has been added. Another chapter to detail the genetic and molecular mechanisms of cancer formation has also been included. Recent research on cancer stem cells has paved way for new therapeutic interventions in arresting cancer spread. Hence, this chapter will be of immense use to students of molecular biology and medical sciences. Stem cell biology is another area in which new discoveries are added almost every day, making the regenerative medicine and bench to bedside therapy, a truly possible dream. In addition to revision and expansion of this chapter, a new section exclusively dealing with the methodology of stem cell harvesting for therapeutic application has been added. As a whole, the revised edition of this book will have a new look covering modern topics in developmental biology. The students and teachers of developmental biology, ecology, evolutionary biology, biotechnology and medical sciences will find this book useful for their course work and for reference. In another perspective, this book will serve as a ready source of current information and hallmark discoveries on developmental biology and related subjects such as ecology, evolution and genetics to any biologist.

In the task of revising this book, I have derived support and help from many of my former students, who painstakingly reviewed and offered many useful suggestions. They are Dr. Thirumalai Rangasamy of Rochester Medical School, USA, Dr. Ganapathi Shanmugasundaram and Dr. K. Rani of Hopkins Medical University, Baltimore, and Dr. Sudha Warrier of Manipal Medical College, Bangalore. Mr. S.Muthukumar of NIOT and R. Naveen Ram of CTS, Chennai helped me in the final formatting of the manuscript. Ms. P. Mathumita of Stella Maris College helped me with the diagrams. I thank all of them whole-heartedly. I am grateful to Indian National Science Academy for offering me Emeritus Scientist positions, which helped me to undertake this book writing work. I also thank Dr. M. A. Athmanand, Director of NIOT and Dr. R. Kirubakaran, Group Head for providing me necessary facilities for preparing the manuscript. Finally, I wish to extend my indebtedness to Gayatri and Nithya, my daughters in law, for taking me to Barnes and Nobles and Rochester central library any number of times for completing most of the revision of the book.

Prof. T. Subramoniam
thanusub@yahoo.com

PREFACE TO THE FIRST EDITION

Developmental biology is one of the fast - growing fields in modern biology. Consequently, the concepts and principles of developmental biology are changing fast. The increased understanding of animal development, even after the completion of embryogenesis, has thrown open the possible extension of this study into adulthood also. In other words, many of the developmental events, independent of normal physiological functions, continue to occur up until the death of the organisms. To cite an example, the haematopoitic stem cells continue to proliferate and differentiate into mature functional blood cells through out the life cycle. In this way, many of the biological systems involved in the replenishment of worn- out tissues and cells utilize a constantly proliferating and differentiating progenitor cells. Such an understanding of animal development has given birth to almost a new discipline called stem cell biology. Similarly, augmentation in our understanding of gene activity together with stimulation of signalling pathways involved in the controlled expression of developmental genes has given recognition to developmental genetics, as an important aspect in the study of animal development. Again, evolutionary principles also derive support from developmental phenomena leading to phenotypic variations and speciation, thus bringing forth a new science of evolutionary developmental biology (Evo-Devo). Furthermore, environmental influence on animal development posed several problems of birth defects both in humans and animals. Understandably, the origin of several disciplines within the broad spectrum of developmental biology has brought this fascinating subject right into the central stage of biological research and teaching.

Recent research on developmental biology has increasingly employed an impressive arsenal of experimental and analytical tools derived from cell and molecular biology. Accordingly, our understanding of animal development at molecular and genetic level is augmented amply. Previously, morphogenetic changes resulting in the phenotypic variation in development have been attributed to interactions of cells and tissues. However, the reductionistic approach employed in molecular studies has yielded a fresh look at molecular and genetic interactions in accomplishing developmental processes too. This book lavishes on genetic regulation of developmental mechanics. Another hallmark of this book is the inclusion of several new chapters to give a new perspective to developmental biology in the context of related subjects like cell and molecular biology, environmental and evolutionary biology. As the principles of plant and animal development have much in common, a detailed chapter on plant development has also been included to give the reader a comparative perspective of development in eukaryotes.

In writing this book, I have derived support from many colleagues, including my former students. Particularly, I wish thank those who reviewed several chapters and offered constructive criticisms. To mention a few are, Dr. Shanmugasundaram of Hopskins Medical University, Baltimore; Dr. P. Mullainathan of Madras University; Prof. J. Khurona of Delhi University; Prof. K.V. Krishnamurthy of Barathidhasan University; Prof. K. Subba Rao of Hyderabad University; Prof. S. Saidapur of Karnataka University and Dr. Sher Ali of National Institute of Immunology. My students Dr. V. Gunamalai, Dr. Bhavanishankar and Mr. Sunil Israel offered considerable assistance in the preparation of the manuscript. I am also grateful to Dr. Kathiroli, Director NIOT and Dr. Venkatesan group head for providing me necessary facilities for preparing the manuscript. Dr. Dharani, Mr. Muthukumar and Mr. Karthikayalu of NIOT, Chennai, offered last minute help in the finalization of the book. My heart-felt thanks go to them. Finally, I am grateful to CSIR, New Delhi and Indian National Science Academy for offering me Emeritus Scientist positions which helped me to undertake this book writing work.

Prof. T. Subramoniam
Chennai

CONTENTS

Preface to the Second Edition   vii

Preface to the First Edition   ix

1. **INTRODUCTION**   1
   Chief events in animal development – History of thoughts and conceptual developments – Experimental embryology – Developmental genetics

2. **GAMETOGENESIS**   5
   Origin of germline cells in mouse- Origin and mechanism of germ cell lineage in other vertebrates-Migration of germ cells to genital ridge - Germ cells in the genital ridge and sex determination-Gamete formation - Significance of gametogenesis

3. **SPERMATOGENESIS**   11
   Sperm structure – Head, midpiece, and sperm flagellum - Types of sperm – Typical and atypical spermatozoa - Sperm cell formation - Spermiogenesis - Primitive sperm type and external fertilization - Modified spermatozoa with internal fertilization - Endocrine regulation of spermatogenesis

4. **OOGENESIS**   19
   Ovary and germ cell development - Stages of Oogenesis - Composition of yolk - Vitellogenesis- Oogenesis in Amphibians - Previtellogenic phase, Vitellogenic phase, Heterosynthetic derivation of yolk, Genetic control of vitellogenin synthesis in amphibians, Endocrine regulation of vitellogenesis and egg maturation - Oogenesis in Insects - Vitellogenesis, Vitellogenins and Vitellins, Vitellogenin uptake into the oocytes, Hormonal regulation of vitellogenesis, Genetic control of vitellogenin synthesis - Types of eggs - Egg envelopes and polarity – Types of eggs – Egg envelopes: Primary egg envelope, Secondary envelope, Tertiary envelopes - Polarity and symmetry

5. **FERTILIZATION**   33
   Sperm aggregation – Sperm activation – Chemotaxis - Sperm maturation and capacitation in mammals - Acrosome reaction - Sperm-egg interaction - Sperm entry into egg - Egg activation – Electrical events and prevention of polyspermy - intracellular calcium release and egg activation - signalling molecules and mechanisms of Ca^{2+} release - Cortical reaction - Physiological polyspermy - Fusion of male and female pronuclei - Post fertilization metabolic activation – Parthenogenesis

6. **CLEAVAGE**   48
   Mechanism of cleavage - Patterns of embryonic cleavage - Radial holoblastic cleavage - Spiral cleavage - Bilateral holoblastic cleavage - Rotational holoblastic cleavage - Amphibians - Meroblastic cleavage - Superficial cleavage

- Determinate and regulatory embryos - Cell lineage - Nuclear transplantation experiments – Cloning – Transgenesis

7. GASTRULATION   63

Molecular mechanisms determining germ layers formation - Factors affecting gastrulation - Cellular and molecular mechanics of gastrulation movements - Types of gastrulation movements - Epiboly, Invagination, Involution, Ingression - Types of gastrulation in representative animal embryos - Sea urchin, Amphioxus, Amphibians - Neuwkoop centre – Exogastrulation - Gastrulation of avian embryos- Gastrulation in mammals - Comparative aspects of gastrulation - Fate maps - Fate map of sea urchin blastula - Fate mapping in amphibian eggs - Fate map of chick embryo

8. EMBRYONIC INDUCTION AND NEURULATION   81

Inductive interaction and neurulation - Neurulation in amphibians - Mechanism of neural tube formation

9. NEURAL CREST CELLS   86

Formation and migration of neural crest cells - Control of neural crest migration - The cranial neural crest - Tooth development - Neural crest cells and axial patterning - cardiac neural crest - Trunk neural crest - Mechanism of trunk neural crest migration

10. CELL DIFFERENTIATION   90

Mechanism of gene action during cell differentiation - Epigenetics - Factors influencing cellular differentiation

11. DEVELOPMENTAL GRADIENTS   94

Developmental gradients in *Hydra*

12. AXIS SPECIFICATION   97

Drosophila - anterior-posterior polarity - Bicoid protein - Nanos and caudal proteins -Formation of terminal body structures - dorsal-ventral polarity - Genetic control of body segmentation - Gap genes - Pair-rule genes - Segment polarity genes - The homeotic genes - Axial gradients in vertebrates (*Xenopus*) - Axis formation in mammalian embryos - *Hox* genes - *Hox* gene regulation of axial morphogenesis in mouse

13. DEVELOPMENTAL CYCLE AND MORPHOGENESIS IN THE SLIME MOULD DICTYOSTELIUM DISCOIDEUM   105

Life cycle - Polarity, Pattern formation and morphogenesis - Role of mitochondria in growth/Differentiation transition - Gene expression during cell growth - Chemotaxis and cAMP signalling - Morphogenetic gradients and pattern formation and gene expression

14. ORGANOGENESIS   110

Formation of vertebrate brain - Formation of brain regions - Placodes - Development of the eye - Heart formation in frog and chick – Frog - Formation of the primitive cardiac tube - Formation of the heart proper - Development of blood vessels and corpuscles – Chick - Urinogenital organogenesis - Pronephros, Mesonephros, Metanephros, Inductive interactions between the ureteric bud and the mesenchymal cells - Sexual differentiation - Freemartin and MIS - Genes involved in gonadal differentiation - Action of sex chromosomes in male and females - origin and evolution of sex chromosome

15. **PATTERN FORMATION AND DEVELOPMENT OF LIMB** 124
Development of vertebrate limb bud - origin of limb bud – Development of wing bud – Role of *Hox* genes in limb patterning - Cellular interaction within the developing wing bud

16. **DEVELOPMENT OF IMMUNE SYSTEM** 130
Origin of the immune system – Organization of the immune system – Primary lymphoid organs – secondary lymphoid organs – T Cells – B Cells – natural killer cells – Macrophages – Dendritic and Langerhans cells – Other accessory cells – Immunoglobulins – Major Histocompatibility Complex – Expression of MHC molecules – MHC loci – Complement system – Cytokines – Neonatal immunity – Maternal immunity – Gene conversion – Disorders in immune system –Allergy – Auto immunity – Acquired immuno deficiency syndrome

17. **EMBRYONIC NUTRITION** 139
Oviparity and yolk utilization – Biochemical composition of yolk – Yolk utilization

18. **EXTRA EMBRYONIC MEMBRANES** 145
Extraembryonic membranes in birds - Yolk sac, Amnion, Chorion, Allantois - Extra embryonic membranes in mammals

19. **PLACENTATION** 150
Placenta of mammals - Implantation - Formation of placenta in Humans – Physiology of placenta - Evolution and classification of placenta - Yolk sac placenta, Allantoic placenta

20. **GROWTH** 157
Types of growth in multicellular organisms – Growth rate – Measurement of growth rate – Growth curve – Mechanism of Cell proliferation – Control of cell growth – Effect of growth hormone and mammalian growth

21. **METAMORPHOSIS** 162
Amphibian metamorphosis – Thyroid hormone receptor - Gene regulation during amphibian metamorphosis – Neoteny - Moulting and metamorphosis in insects – Ecdysteriods action during insect metamorphosis – Role of juvenile hormone in metamorphosis

22. **REGENERATION** 169
Types of regeneration - Regeneration in *Hydra* – Cell proliferation and loss during *Hydra* growth – Role of epithelial cells in *Hydra* regeneration – Axial patterning during regeneration – Morphogenesis during head regeneration -

Regeneration in Planarians - Blastema formation - Urodele limb regeneration - Morphogenesis and redifferentiation - Sources of blastemal cells - Factors stimulating regeneration of limbs - Pattern formation in the blastema – Autotomy and limb regeneration in crab – Hormonal control of regeneration in crab leg

23. **EXPERIMENTAL EMBRYOLOGY** 180
Mammalian reproductive cycle - Hormonal regulation of reproductive cycle - Endocrine changes associated with normal pregnancy - Induced ovulation in humans - Multiple ovulation and embryo transfer in cattle - Embryo splitting - *In vitro* fertilization - IVF in cattle - IVF in humans - Cryopreservation - Methods adopted in cryopreservation - Cryopreservation of human spermatozoa - Human embryo cryopreservation - Ethical issues in cryopreservation

24. **STEM CELL BIOLOGY** 189
Molecular basis of self-renewal and pluripotency - Discovery of stem cells - culture of stem cells - Stem cell niches - Types of stem cells - Adult stem cells - Stem cell division - Types of adult stem cells - Haematopoietic stem cells (HSCS) - Mesenchymal stem cells (MSC) - Multipotent adult progenitor cells (MAPC) - Skeletal muscle stem cells (SMSC) - Cardiac stem cells - Differentiation of embryonic cells into the cardiac lineage - Neural stem cells (NSCS) - Dermal stem cells - gut stem cells - Stem cells in hydracancer stem cells - Ageing of stem cells - embryonic stem cells - Transcriptional control of pluripotency in es cells - Origin of embryonic stem cells in mouse - Surface antigen markers and identification of es cells - In vitro differentiation of Es cells in mouse - Gene expression during eb differentiation - Human embryonic stem cells - Telomerase expression - Gamete production from embryonic stem cells - Therapeutic use of stem cells - Induced pluripotent stem (iPS) cells - First production of viable mice from iPS cells - Regenerative therapy of the heart - Muscular dystrophy - Stem cell therapy of the human central nervous system - Multiple sclerosis - Parkinson disease - Spinal cord injury and disease - Cell based therapy of motor neuron diseases - Stem cells as chaperons - Constraints in stem cell therapy - Ethical considerations of stem cell research

25. **THERAPEUTIC USE OF MESENCHYMAL STEM CELLS** 217
Introduction - What are mesenchymal stem cells - Location of mscs - Use of MSCS in medicine and tissue engineering - Use of MSCS in lung diseases - how are MSCS isolated and characterized - Protocol for isolation and culture of mesenchymal stem cells from mouse bone marrow materials reagents equipment - Reagent setup - harvest buffer - Complete dmem medium - Pocedure - anticipated results

26. **CANCER BIOLOGY** 222
Introduction - Origin of cancer - Characteristics of cancer cells - Genomic alterations as the basis of tumorigenesis - Cancer cell transformation - Chromosomal aberration in cancer cells - Tumor categories - Tumor promotion and tumor initiation - Tumor promotion - Tumor initiation - Genes involved in cancer development - Signal transduction and oncogene activation - Tumor

suppressor genes - p53 tumor suppressor protein - DNA repair genes - Human genetics and cancer - Genes known to be involved in some cancer types - Breast cancer - Hereditary breast and ovarian cancer (hboc) syndrome - Ataxia telangiectasia (a-t) - Li-fraumeni syndrome (lfs) - Cowden syndrome (cs) - Peutz-jeghers syndrome (pjs) - Prostate cancer - Colorectal cancer - Hereditary non-polyposis colorectal cancer (hnpcc) - Familial adenomatous polyposis (fap) - Muir-torre syndrome - gardner syndrome - Turcot syndrome - Myh-associated polyposis (map) - Peutz-jeghers syndrome (pjs) - Juvenile polyposis syndrome (jps) - Kidney cancer - Von hippel-lindau syndrome (vhl) - Hereditary non-Vhl clear cell renal cell carcinoma - Hereditary papillary renal cell carcinoma (hprcc) - Birt-hogg-dubé syndrome (bhd) - Hereditary leiomyomatosis and renal cell carcinoma (hlrcc) - Beckwith-wiedemann syndrome (bws) - Li-fraumeni syndrome (lfs) - Tuberous sclerosis complex (tsc) - Cowden syndrome (cs) - melanomas - Hereditary melanoma - Melanoma-astrocytoma syndrome - Xeroderma pigmentosum (xp) - Retinoblastoma - Li-fraumeni syndrome - Werner syndrome - Hereditary breast and ovarian cancer - Cowden syndrome - pancreatic cancer - Hereditary pancreatitis (hp) - Peutz-jeghers syndrome (pjs) - Familial atypical multiple mole melanoma and pancreatic cancer (fammm-pc) - Hereditary breast and ovarian cancer (hboc) syndrome - Hereditary non-Polyposis colorectal cancer (hnpcc) - Li-fraumeni syndrome (lfs) - Thyroid cancer - Medullary thyroid cancer - Men 2a - Men 2b - Familial medullary thyroid cancer (fmtc) - Papillary and follicular thyroid cancer - Familial papillary thyroid cancer - Familial adenomatous polyposis (fap) - Cowden syndrome (cs) - Cell cycle - Angiogenesis - Viruses and cancer - Environmental factors - Evolutionary and ecological aspect of cancer - cancer stem cells - Discovery and origin of cscs - Cancer stem cells - Tumor migration and metastasis - Role of epithelial to mesenchymal transition in cancer invasion and metastasis- - Cell signaling pathways in cancer stem cells - Identification of cancer stem cells and therapeutic intervention - Current challenges and future directions for cancer therapy

27. **MOLECULAR AND EVOLUTIONARY DEVELOPMENTAL BIOLOGY** 259

Origin of the new science developmental basis of animal evolution - Developmental genes and evolution - Mechanism of gene regulation during development - CRE control of wing pigmentation in insects - Study of homology - Hox genes and body axis formation - Hox genes and the evolution of tetrapod limb - Homology in wnt gene family - Hedgehog gene family - Evolution and spread of hox genes - Hox genes and body segmentation in arthropods - Conserved role of pax6 in eye formation - Evolutionary control of eye development - Developmental basis of morphological variation in the beaks of darwin's finches - Evolutionary pelvic reduction in three-spine sticklebacks - Evo- devo and human evolution - Protein modification and evolution - regulatory RNAs - A conclusion to genetic theory of morphological evolution

28. **ECOLOGICAL DEVELOPMENTAL BIOLOGY** 284

Environmental influence on normal development - Polyphenism - Environmental cues and development of new phenotypes in tadpoles -

Hormonal regulation of temperature-Dependent polyphenism - Temperature and sex determination in reptiles

29. **TERATOGENESIS AND MEDICAL EMBRYOLOGY**   288
Thalidomide and phocomelia - Valproic acid (va) and autism - Retinoic acid - mechanism of drug action - Industrial mercury and minamata disease - teratogenic effects of alcohol - Endocrine disruptors and developmental alterations - Effect of environmetal contaminants on wildlife - Toxicity studies using embryonic stem cells

30. **PROGRAMMED CELL DEATH OR APOPTOSIS**   301
Processes associated with apoptosis – Incidence of Apoptosis - Apoptosis during animal development – Apoptosis in metamorphosis and morphogenesis – Apoptosis in C.elegans ¬ Apoptosis in lens development – Apoptosis in neural cells – Apoptosis during vertebrate limb development – Apoptosis in immune cells – Apoptosis in response to DNA damage and cellular stress – Biochemical and molecular mechanisms involved in apoptosis – Biochemical pathways in mammalian cell apoptosis – Role of BCL-2 family proteins – Macrophage-induced apoptosis – Factors controlling apoptosis – Germ cell apoptosis during spermatogenesis – Apoptosis in female germ cells – Mechanism of oocyte apoptosis – Theories to explain female germline cell death

31. **AGEING AND SENESCENCE**   312
Cellular ageing – Reactive oxygen and cell senescence – Dietary restriction and anti-ageing action – Resistance to oxidative stress - Telomerase theory – Hayflick limit – Ageing of brain cells – Genetic control of longevity – C.elegans – Genetic control of ageing in yeast – Genetic control of ageing in mammals – Age-related diseases: Alzheimer's disease – Protein aggregation as a causative agent of age-related diseases – Aggregation of TAU protein – Cataract – Werner's syndrome

32. **PLANT DEVELOPMENT**   322
Embryonic development in angiosperms – Post–embryonic growth – Meristems – Structure of shoot meristem – Leaf development – Pattern formation in leaf development – Patterning in leaf venation – Patterning in the epidermis – The vegetative to Reproduction transition – Flower development – Genetic pathways regulating floral transition – Homeotic genes and organ identity in the flowers – Root development – Regeneration and somatic embryogenesis – Transgenic plants

References   *338*
Index   *341*

1

INTRODUCTION

CHIEF EVENTS IN ANIMAL DEVELOPMENT

All multicellular organisms arise by a slow process of progressive change called development. This development invariably starts with a single cell, zygote, which is a fusion product of fertilization. The principal features of development from the zygote to adult organism, depicted in Fig.1.1, include the following steps:

i) The zygote undergoes a series of extremely rapid mitotic divisions termed as cleavage. This process culminates in the formation of blastula consisting of several blastomeres, leaving a central hollow region (blastocoel).

ii) The blastomeres undergo dramatic movements within the blastula by changing their position relative to one another. This process is called gastrulation. As a result of these cellular rearrangements, the three germ layers, ectoderm, mesoderm and endoderm are produced.

iii) The cells of the three germ layers undergo further rearrangement and interaction with one another to produce organ systems (organogenesis). During this process, the cells undergo differentiation in order to acquire specific physiological functions (morphogenesis) that are retained throughout its growth phase.

iv) Under ordinary circumstances, the embryo is now fully formed and is functional to lead an independent life.

Embryology or the study of embryo deals with the above narrated steps in embryonic development. However, developmental changes still continue during the growth of the young ones to adult. The best examples for these post-embryonic developmental processes are found in the life of several animals such as the butterfly and frog, wherein the hatched out larvae undergo spectacular change in form before reaching the adult-hood.

Developmental changes do not stop with reaching adult-hood either. The regeneration of the lost parts, as found in many animals like lizards, demonstrates the ability of the differentiated cells to give rise to the embryonic cells which will restore the lost organ completely. Another example is found in some of the organ systems such as the bone marrow, which continuously proliferates different types of blood cells from a group of stem cells throughout the life time. Yet another example is the outstanding ability of the corneal epithelium in the human eye. The corneal epithelium is continuously repairing or renewing itself, shedding the older ones on its surface into the tears of the eyes and replacing it with the younger cells from the next layer below. There are many such examples to demonstrate this ability of continuous cell differentiation that occurs

2   Molecular Developmental Biology

throughout the life span of the organisms. Even the senescence of the cell leading finally to death is an event of programmed cellular differentiation and change.

The developmental history of an individual from the egg to the adult is called ontogenicdevelopment or simply ontogeny. On the other hand, the development of individuals in a population acquiring new characteristics leads to speciation implying evolutionary significance. All these developmental processes at organismal level necessarily incorporates studies from other disciplines such as cell biology, genetics and evolution making developmental biology a multifaceted interdisciplinary subject.

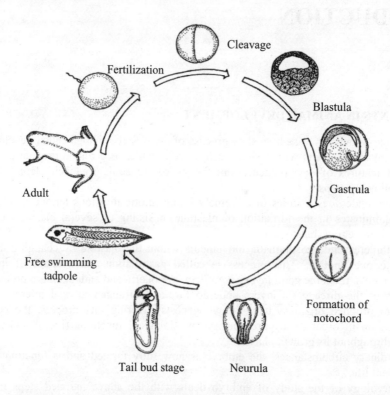

**Fig. 1.1 Important stages in the development and life cycle of the frog *Xenopus laevis*
(Modified from Wolpert, 1998).**

HISTORY OF THOUGHTS AND CONCEPTUAL DEVELOPMENTS

During the seventeenth and eighteenth century most of the theories as well as the conceptual advancements made on animal development were based on theoretical considerations, speculation as well as a few very faulty crude microscopic observations. The initial concept that the embryo emerges gradually during development was proposed by Aristotle. Both Aristotle and Harvey advanced this hypothesis of 'epigenesis' early in the seventeenth century. This explained that most of the embryological structures emerge after fertilization. However, the mechanism by which epigenesis occurs were not speculated upon and it simply said that the early embryo had a forming virtue (vis essentialis). Although the theory of epigenesis found support later in the nineteenth century, this concept of Aristotle and Harvey did not find acceptance by other contemporary embryologists who in turn proposed a new theory that the egg contains its fully formed miniature embryo. Still others believed that a preformed embryo exists inside the sperm as inferred from the microscopic observations that a tiny creature is curled up in the sperm head. These ideas proposed by Malphigini were referred to as the theory of preformation. In short, preformation is the unfolding of existing structures just like a flower. This theory found wide acceptance during that time, mainly because of the fact that the idea of preformation corresponds with the spontaneous generation, advanced by theologians. However, the theory of preformation was rejected by scientists like Caster Wolffe who strengthened the theory of epigenesis proposed already by Aristotle. Wolffe, from his observations on the development of the chick explained that the development is a progressive phenomenon with new parts being formed continuously. By the time this found acceptance in the early part of nineteenth century, the cellular nature of the egg and sperm became apparent. In 1875, Oscar Hertwig made a landmark discovery that fertilization results from the union of male and female gametes, paving the way for the parental inheritance of morphological characters. He himself went on to suggest that the germ cell nuclei are the vehicles of inheritance.

EXPERIMENTAL EMBRYOLOGY

At the beginning of the experimental period in developmental studies, it was proposed that cell differentiation, which accompanies embryonic development, was the result of unequal partitioning of the developmental determinants among the daughter cells of the zygote. The developmental determinants are thought to be associated with the nucleus. In other words, the cell differentiation was thought to involve irreversible differentiation through loss of genetic information. The only cells to maintain their integrity were supposedly the germline cells which were observed to be segregated off from the somatic line early in the development of many animals. In the early days of the germ line theory, the pioneering experiments of Wilhelm Roux provided support for this mosaic theory of development.

Experiments by Roux, Driesch, Morgan and Wilson however showed that the early cleavage cells of different animal species had different developmental potencies. For example, in ascidians, the first cleavage division would produce two half ascidian larvae, if the two blastomeres are separated. This mode of differentiation is referred to as mosaic or determinate development. Conversely, if the first cleavage division blastomeres in sea urchin are separated, these two cells would eventually produce two normal larvae, although of reduced size. This type of differentiation was called regulative development. Roux made a distinction between these two types of development that in determinate, structures develop strictly according to a fixed genetic programme. On the other hand, in the regulative development, one tissue affects the subsequent development of another tissue, through a process known as induction. The new concept of induction led to the famous transplantation experiments of Spemann and Mangold in which a portion of the dorsal blastopore lip induced neural tube tissue in the roof of the primitive gut. Subsequent experiments in this line established the fact that induction plays an important role in the development of organisms with regulative development. The study of the interaction of cells and tissues during ontogeny has now become an independent field of biology, namely, topobiology.

Roux, in particular, made a giant contribution to the method of approach to developmental investigations by seeking mechanisms of development within the embryo itself. Much later, it was discovered that all cells in the somatic line had basically the same chromosome number and DNA content as the zygote. Through the contributions of twentieth century molecular biology, it was realised that all cells undergo a process of differentiation and that, at any particular time, only a small fraction of the genes in the nucleus of a given cell is active. Regulatory mechanisms turn on or turn off a given gene depending on whether its gene product is needed in that cell at that time. Interestingly, the timing of this regulatory activity is in part programmed in the genotype and in part determined by neighbouring cells.

DEVELOPMENTAL GENETICS

Progress in molecular biology and experimental embryology now permits investigations into the genetic mechanisms that control development. Development is progressive and the fate of developing cells is determined at different stages. To achieve this, differential activity of selected sets of genes in different groups of cells plays a key role. Thus, the development of form depends upon the turning on and off of genes at different times and places in the course of development. Differences in form arise from evolutionary changes in where and when genes are used.

With the advent of such genetic studies on development, new explanations and interpretations on the theories of preformation and epigenesis came into force. Essentially, the study of genetics revealed that during development the genotype (genetic constitution of an individual) could control the production of phenotype, which refers to the totality of the observable characteristics of an individual. By providing the information for development, the genotype is the preformed element. In addition, the genetic DNA programme of the zygote can also play a role of vis essentialis of the epigenesists. In short, it may be said that the zygote contains an inherited genetic programme that largely determines the phenotype. Recent molecular studies have now revealed that all cells undergo a process of differentiation and that at any particular time only a small fraction of the genes in the nucleus of a given cell is active. The timing of this regulatory activity is in part programmed in the genotype and in part determined by neighbouring cells.

A new era in developmental genetics was opened when Avery in 1944 demonstrated that DNA was the carrier of the genetic information. DNA controls the production of the proteins, of which an organism is composed. Development, therefore, is the elaboration during ontogeny of different kinds of proteins, which characterize different organ systems. Another land mark discovery in developmental genetics is the discovery of developmental mutants which revealed the contribution of a given gene to development. The analysis of deleterious genes showed that most mutations consisted of a failure to produce a needed gene product, which is usually active during development only in particular time and particular stages of the development. From such analysis, one could describe development as an ordered sequence of gene expression.

Furthermore, it has recently been realized that any individual organism is the product of the interaction during development of all its genes with one another and with the environment. Based on this fact, it has now been established that only a small number genes, termed as "tool kit genes" are responsible for building all animal forms. It has been further argued that any modification in development may influence evolution. After the acceptance of Darwinian Theory, evolutionary implications of animal development have been greatly realized. This gave rise to the recapitulation theory advanced by Meckel-Serres who stated that organisms recapitulate during their ontogeny the phylogenetic stages through which their ancestors have passed. Capitalizing on this law, Ernst Haeckel proclaimed the fundamental biogenetic law as "ontogeny recapitulates phylogeny". Development is intimately connected to evolution because it is through changes in embryos that changes in form arise. Ultimately, this aspect of developmental study has given rise to a new science of Evo-Devo, meaning, evolutionary developmental biology.

2

GAMETOGENESIS

Germ cells may be defined as those cells, whose surviving descendants will become sperm or eggs. In all sexually reproducing animals and plants, these cells play a uniquely important role, namely the transmission of genetic information from one generation to the next. They are segregated early in development, and inherited from the mother's egg a special sort of cytoplasm rich in germ-cell determinants. The origin of germ cell lineage includes the emergence of the first primordial germ cells (PGCs) in the early embryo. In certain organisms, cells were set aside very early in development to segregate the germ cell lineage from the other mortal somatic lineages. More than a centuary ago Boveri (1887-1910) described in the nematode *Ascaris* that the germ cell lineage was segregated during the first few cleavage divisions, retaining its chromatin complement intact, while all somatic lineages suffered "chromatin diminution" in which large terminal regions of the chromosomes were discarded. Similarly, in the developing frog *Xenopus* eggs, aggregates of mitochondria, protein and RNA in the cytoplasm of the vegetal pole could be observed even as early as the unfertilized eggs. These aggregates then segregate into the germ cell lineage as "germ plasm". Later, the presence of germ plasm was confirmed in *Drosophila* in the form of pole plasm. The germ cells of *Drosophila* contained specific cytoplasm (pole plasm) rich in germ cell determinants. Obviously, pole cells are the very first cells to be formed at the posterior end of the fertilized egg. Similarly, in the free- living nematode, *Caenorhabditis elegans*, there are pole granules, called P granules, in the unfertilized eggs which are asymmetrically distributed to daughter cells at each of the first four cleavage divisions, to be concentrated finally in the P4 cells that are the ancestors of the entire *C. elegans* germ- cell lineage. The germ cell lineage in zebrafish and in chick has been followed from the two- cell stage onwards, using germline - specific expression of vertebrate homologues of the *vasa* gene, known to be a germline determinant in *Drosophila* (see below).

ORIGIN OF GERMLINE CELLS IN MOUSE

Cytoplasmic germ cell determinants in eggs or early embryos equivalent of germ plasm are difficult to find out in mammals. However, later in development, primordial germ cells could be identified in the embryos using biochemical markers such as tissue nonspecific alkaline phosphatase (TNAP) activity. These markers were useful in following the migration pathway of mouse PGCs from the base of the allantois to their entry into the genital ridges, the site of the future gonad. At 8 days postcoitum (dpc), the PGCs were found to be embedded in the endoderm, as it began to invaginate to form the hind gut. However, transplantation experiments indicate that PGCs originate in the epiblast cells rather than the endoderm which is a derivative of hypoblast cells. By injecting single epiblast cells at 6.0 and 6.5 dpc with a lineage marker and following the fate of their clonal descendants, it has been shown that the ancestors of the PGCs were derived from the proximal epiblast cells, adjacent to the extraembryonic ectoderm.

Apart from the expression of alkaline phosphatase, other genetic markers such as Oct-4, a Pou transcription factor are widely expressed in epiblast cells as well as germ cell lineage at about 8.0 dpc. Sequential expression of more signalling molecules such as BMP4 are also required to predispose the proximal epiblast cells to give rise to PGCs among their descendants. Specific expression of two other genes, *Fragilis* and *Smad* 1 in proximal epiblast as well as the future allantois also characterize the occurrence of PGCs. Another gene *Stella*, whose expression in the incipient allantois, the location of nascent PGCs, is germ cell specific, since it continues to be expressed in PGCs as they migrate along the hind gut and into the genital ridge. Expression of mesoderm- specific genes (*Brachyury, Fgf8*) is also manifested for a time in nascent germ cell. Evidently, the characteristic feature of germ cell determination relates to two types of genetic expression events. While there is upregulation of all germline – specific genes, there is an obvious transcriptional downregulation of all somatic genes such as the region specific homeobox genes (*HOXB1, HOXa1, Lim1, Evx1*) during germ cells differentiation in mammals.

ORIGIN AND MECHANISM OF GERM CELL LINEAGE IN OTHER VERTEBRATES

It is now well established that in mammals PGCs originate from embryonic ectoderm (epiblast) and not from the endoderm. However, in anuran amphibians (frogs and toads), the PGCs have an early, endodermal origin, with an uninterrupted "germ line" characterised by germ plasm localised to specific cells. On the other hand, in urodeles (newts and salamanders), PGCs arise from presumptive lateral plate mesoderm in midgastrulation, with no indication of germ plasm. Recent experiments on an urodele, Amblystoma, further indicated that the PGCs, along with other mesodermal tissues, originated in ectoderm, under the inductive influence of the ventral yolk mass, during the course of gastrulation. Understandably, this ectodermal origin of PGCs in this urodele is similar to that in mouse described above.

Genes that are highly conserved and expressed in the germ cell lineage of many animals include *DAZ-like* and *vasa*. RNA encoded by the *Xenopus DAZ-like* homolog is a crucial component of germ plasm, associated with the mitochondrial cloud and showing localised expression at all stages of oogenesis and embryogenesis. *Vasa* homologs are expressed in frog germ plasm, *Drosophila* pole plasm, and in the earliest stages of embryogenesis in Zebrafish and chick, but in mice and axolotles, zygotic *vasa* is not expressed until the PGCs begin to colonize the gonad.

MIGRATION OF GERM CELLS TO GENITAL RIDGE

In many animals, germ cells develop at some distance from the gonads, and only later migrate to them, where they differentiate into eggs or sperm. In the amphibians and the mammals, the primordial germ cells migrate by virtue of the filopodial movements through the intervening cells, to reach their final destination namely the gonad. However in the birds and reptiles, these cells are transported by means of blood stream. The primordial germ cells are derived from the epiblastic cells that migrate from the central region of the area pellucida to a crescent shaped zone situated at the anterior border of the area pellucida. From here, these cells enter the blood vessels by 'diapedesis', a type of movement that enables the cells to squeeze in and out of the small blood vessels. These cells finally reach the embryo and then the genital ridges where they are accumulated. This migration of the germ cell progenitors to the gonad is facilitated by the production of chemotactic substances by the gonad that attracts the primordial cells and retains them in the capillaries bordering the gonads.

Migration pathway of PGCs were followed using a green florescent protein (GFP) tagged Oct-4 construct in the mouse embryo. It has been shown by this method that the germ cells show locomotory movements at all times until they enter the genital ridges. After leaving the hind gut, the germ cells tend to contact each other by extending processes, to form a network. Cell adhesion to laminin may also play a role in PGC guidance, involving the regulation of expression of integrins and / or proteoglycans on the cell surface. In addition, genital ridges may release chemotropic signals to attract the germ cells. In zebrafish, PGC migration is guided by a pair of evolutionarily conserved chemotrophic signalling molecules. They are: the chemokine SDF-1 (stroma cell derived factor 1), which is expressed in locations towards which the cells migrate, including the site of the future gonad, and its receptor CXCR4, expressed on germ cells. Knock-down of either SDF-1 or its receptor CXCR4 produced very aberrant PGC migration. During the migratory period, PGCs with two X chromosomes have one X randomly inactivated, like XX somatic cells. However, both X chromosomes are active during oogenesis, since the silent X is reactivated on entry into the genital ridge.

GERM CELLS IN THE GENITAL RIDGE AND SEX DETERMINATION

Once the germ cells are in the genital ridge, they start to express new germ- cell- specific genes such as the highly conserved mouse *vasa* homolog (Mvh), germ cell nuclear antigen 1(Gcna1)and germ cell-less (Gcl).These changes in gene expression are part of the general reprogramming process that the germ cells undergo at this time. In the mouse, the PGCs enter the genital ridges between 10 and 11 dpc. From this time on, they are no longer locomotory and the PGCs are termed as gonocytes. The gonocytes undergo two or three rounds of mitosis, but by 12.5 dpc in both male and female embryos they enter a premeiotic stage and upregulate meiotic genes such as Scp3. However, in the male genital ridge, meiosis proceeds no further. Scp3 is downregulated and the germ cells enter arrest as G0/G1 prospermatogonia. Mitosis in the male genital ridge is not resumed until after birth. The block to meiotic entry in the male genital ridge coincides with the stage at which Sertoli cells have differentiated and testis cords have formed. This suggests that the meiosis inhibitor is a signalling factor produced by Sertoli cells.

On the contrary, the germ cells in the female genital ridge enter meiotic prophase as oocytes, and pass through leptotene, zygotene, and pachytene stages before arresting in diplotene at about the time of birth. The timing of meiotic entry appears to be cell autonomous in the female genital ridge, whereas in the male genital ridge, the entry into the spermatogenic pathway is not cell- autonomous, but is an induced response, as shown above.

GAMETE FORMATION

Gametogenesis is a process by which both male and female gametes are formed. All the gametes originate from the primordial germ cells called gonocytes, which appear early in the development and occupy the gonads. Primordial germ cells multiply by mitosis and produce gonial cells. In the males, the gonial cells are called spermatogonia and in the females, oogonia. The primary gonial cells usually go through a definitive number of mitotic divisions. The resulting cells are called secondary gonial cells, which in males transform into primary spermatocytes and in females the primary oocytes.

Both primary spermatocytes and oocytes undergo nuclear changes during prophase of the I meiotic division. These changes involve duplication, pairing, condensation and partial separation of the homologous chromosomes. These stages are respectively called as leptotene, zygotene, pachytene, diplotene and diakinesis. In spermatogenesis, these changes are followed by the first meiotic division of the primary spermatocytes to form the secondary spermatocytes. The second meiotic division occurs soon after the secondary spermatocytes are formed and produce haploid spermatids (four spermatids for each primary spermatocyte). Differentiation of the spermatids into mature spermatozoa (sperm) is termed as spermiogenesis or spermateleosis.

8   Molecular Developmental Biology

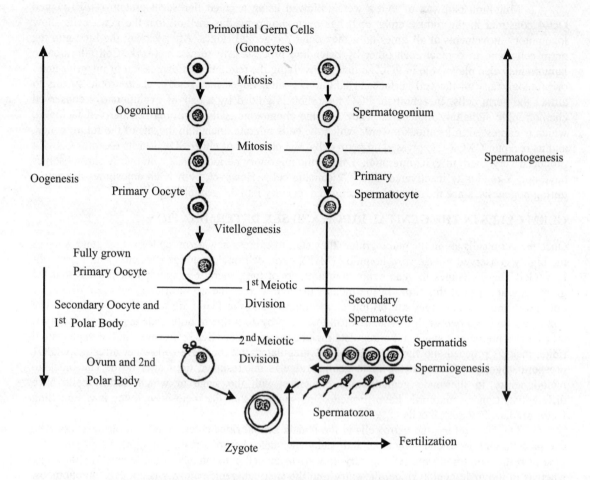

Fig. 2.1 Different stages in Oogenesis and Spermatogenesis of a typical gametogonium (Adapted from Giese & Pearse, 1974).

The primary oocytes undergo two meiotic divisions during oocyte maturation. As a result of the I meiotic division, the I polar body and the secondary oocytes are formed (both having haploid number of chromosomes). Second meiotic division of the secondary oocyte results in the formation of mature oocyte and the second polar body. Details are given in Fig.2.1. This figure also gives a comparative account of spermatogenesis.

The meiotic divisions of the sex cells were first described in the oocyte of *Ascaris* during 1880 by Boveri and his associates. They discovered that two unusual cell divisions (meiotic divisions) occur during gamete formation, reducing the diploid chromosome number to haploid condition. As shown in the Fig.2.2, the egg of *Ascaris*, contains four long chromosomes at the beginning of meiosis. First, these long chromosomes become shortened to form tiny spheres. Homologous chromosomes then form pairs in a process called synapsis. After that, each chromosome of the two chromosome pairs splits. Each pair contains four chromatids, and each group of four is called a tetrad. The tetrads are then aligned on the metaphase plate (Fig.2.2a). During the first meiotic division (2.2b, c) the tetrads split to form dyads. Although the nuclear division is equal, the cytoplasmic division is unequal. During division, the egg nucleus is displaced to the periphery of the egg and hence nearly all the cytoplasm remains with the egg cell. The other cell, namely the polar body receives only a little amount of cytoplasm. During the second meiotic division, the chromosomes do not duplicate themselves. As shown in Fig.2.2 d through h, the dyads separate and, at the completion of the second meiotic division (Fig.2.2.i), the egg nucleus receives two chromosomes, while the other two chromosomes enter the second polar body. In the mean time, the first polar body may also undergo a second meiotic division. Collectively, the potential result of meiosis is one haploid egg and three haploid polar bodies. The unequal cytoplasmic divisions of meiosis in the female gamete ensure that the egg retains the vast majority of the cytoplasm and the yolk is built up subsequently during vitellogenesis. Thus, the essential modification of meiosis that results in production of haploid cells is that there is single duplication of chromosomes for the two divisions. On the contrary, each mitotic division is accompanied by chromosomal divisions, maintaining a constant chromosome number at each division. In contrast to the egg cell division, during the first meiotic division in the male gametes, the nucleus is centrally placed so that two equal-sized daughter cells are produced, each receiving two dyads. Similarly, in the second meiotic division, there is separation of the dyads and there is distribution of two chromosomes to each of the four cells (spermatids).

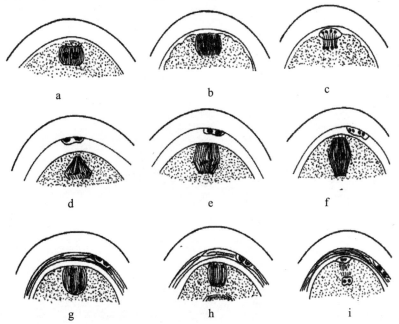

Fig. 2.2 Meiosis in the ovary of Ascaris following the description by Boveri (for details see text)
(Redrawn from Browder, 1980).

SIGNIFICANCE OF GAMETOGENESIS

In sexual reproduction, the gametes form interlinks between two generations. The male and female gametes carry haploid number of chromosomes so that their fusion products during fertilization viz., the zygote, will possess the diploid number of chromosomes. The presence of diploid number of chromosomes is a prerequisite for their further division. The production of haploid number of chromosomes is made possible only through meiotic divisions that happen during gametogenesis. During meiotic divisions, there is a possibility of chromosomal crossing over and recombination resulting in genotypic variation even at the time of gamete production. Genotypic variation thus produced will enhance speciation during evolution. During the growth phase of female gamete, the accumulation of macromolecules as well as the organelles during oogenesis is used to regulate and sustain early embryogenesis. The yolk molecules that are presynthesised from maternal tissues and then carried to the ovary for deposition in the egg will even serve as a molecular memory of their past association with the maternal tissues.

3

SPERMATOGENESIS

Differentiation and maturation of male and female gametes make them competent to take part in fertilization, although their respective role in fertilization is different. Whereas the egg is a passive partner, the sperm actively fertilizes the egg. Various structures present in the sperm cells have evolved in such a way to accomplish their function during fertilization with the egg cell. Essentially, the function of the spermatozoa is to transmit the paternal genome to the egg, in addition to activating it.

SPERM STRUCTURE

A typical mammalian spermatozoa is an intricate and highly polarised cell which is organised into three distinct regions namely, a head, a midpiece, and a long filamentous tail (Fig. 3.1).

HEAD

Head is the anterior most part of the sperm. This is the organ with which the sperm establishes contact with the egg during fertilization. The size and shape of the head vary significantly among different animal species. The head consists of a highly condensed nucleus with an overlying acrosome.

Nucleus: The nucleus, housed within the head, is highly streamlined and elongated due to the extreme condensation of the chromatin. The tight packing of the DNA makes it less susceptible to physical damage or mutation during storage and transport to the site of fertilization. The unique basic chromosomal protein namely, protamine is suggested to be responsible for chromosomal condensation by its interaction with the DNA molecules. According to one theory, protamine is in an extended form, winding along the minor groove of the DNA double helix, with successive arginyl groups interacting alternately with the phosphate groups of the two DNA strands. The interaction between the protamines and DNA also results in the transcriptional inactivation. Close alignment of cysteinyl groups between the protamine molecules also result in disulphide linkages, which render the chromatin resistant to dispersion by chemical treatment. The chromosomal condensation and the resulting changes in the shape of the nucleus is also brought about by the pressure applied by microtubules outside the nucleus. The nucleus is additionally supported by an extra envelope called perinuclear theca.

Acrosome: Acrosome is a membrane bound sperm structure capping the anterior pole of the nucleus. The near- universal presence of acrosome on sperm head has an important role to play in sperm activation and fertilization. Acrosome is a derivative of Golgi complex and its material content isalways PAS- positive. The shape of the acrosome is quite variable and it, along with nucleus, determines the shape of the sperm head. The acrosomal vesicle has functional homology with the lysosome inasmuch as it contains a variety of hydrolytic enzymes as well as sharing a common origin from the Golgi complex. The hydrolytic enzymes contained inside the acrosome are necessary for the digestion of proteinaceous and high carbohydrate containing egg tertiary membrane at the time of sperm penetration into the egg. A few examples of the hydrolytic enzymes present in the mammalian acrosome are, acrosin, which is a trypsin-like serine proteinase, hyaluronidase, acid phosphatase, aryl sulphatase, collagenase-like proteins, calpain II, α-cathepsin D-like protease, ß-N-acetylglucosaminidase, ß-glucuronidase, neuraminidases, various esterases and phospholipases. The acrosomal enzymes are packed in an organised pattern and are released during fertilization. The acrosomal vesicle also contains non- enzymatic proteins, which serve mainly as egg recognition proteins. For example, in the sea urchin sperm, the acrosome releases a protein called bindin which is species specific in its binding with vitelline envelop of eggs during fertilization.

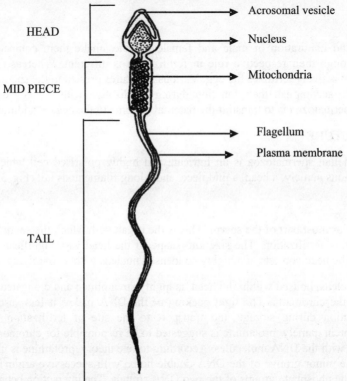

Fig. 3.1 External morphology of the human spermatozoa.

In the spermatozoa of several marine invertebrates such as sea urchin and molluscs, there is a subacrosomal vesicle lying in between the anterior acrosomal vesicle and the nucleus. The subacrosomal vesicle contains globular actin, which participates in the formation of finger-like acrosomal processes during sperm activation.

MIDPIECE

Immediately posterior to the sperm head is a short neck which connects the sperm head distally to the thickened midpiece. The midpiece is characterized by the presence of helically arranged mitochondria, surrounding the inner axoneme as found in mammalian sperm head. Sperm mitochondria have typical metabolic capabilities. Thus, sperm contain enzymes capable of aerobically metabolizing glycolysable sugars, glycerol, sorbitol, lactate, pyruvate, acetate, other fatty acids and several amino acids. The sperm mitochondria also contain lactate dehydrogenase (LDH - C_4 isozymes) which can oxidize lactate to pyruvate, thereby allowing further energy production via the Krebs cycle.

The formation of the midpiece is highly variable, since the mitochondria in different species assume different shape and structure. For example, in the sea urchin sperm, the midpiece is very short and simple, enclosing clusters of a few enlarged mitochondria. In many insects, the mitochondria in the midpiece are highly modified to give rise to a spherical structure called nebenkern. The mitochondrial cristae in the nebenkern are completely lost. The nebenkern is closely associated with the axonemal complex, suggesting their role in energy transfer and mobility of the sperm. In general, the midpiece containing mitochondria also enters the egg cytoplasm during fertilization. But in ascidians, as the sperm penetrates the chorion, the mitochondria slides posteriorly out of the head and along the axoneme to the tip of the tail, where it is discarded.

SPERM FLAGELLUM

The differentiation of sperm flagellum is basically the process of axoneme formation. Axoneme gives the cytoskeletal support to the sperm tail. The axoneme, the main motor portion of the sperm tail, is formed by the microtubules originating from the distal centriole, found at the base of the sperm nucleus. Sperm axonemes closely resemble those of other flagellate organisms with microtubules comprising a 9+2 array occupying the central portion of the flagellum through out its length. The core of the axoneme essentially consists of two central microtubules surrounded by a row of nine doublet microtubules. The microtubules are formed by the protofilaments which are made by the contractile proteins, α- and ß- tubulin. The peripheral tubules are interconnected by nexin bridges and are joined to the central sheath by nine spokes (Fig.3.2).

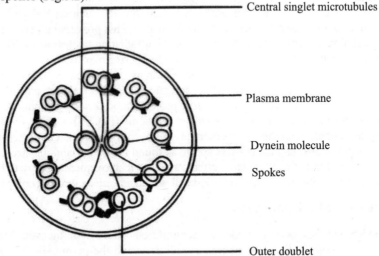

Fig. 3.2 Schematic diagram of the sperm flagellum in cross section. (Redrawn from Afzelius 1961)

A class of flagellar ATPases called dynein is involved in the generation of axonemal movements to effect the lashing movement of the flagellum. Dynein molecules are attached to the peripheral microtubules in the form of two arms, which by forming transient cross bridges between adjacent microtubule doublets, effect sliding movements in the microtubules. People suffering from a genetic syndrome (Karagener's triads) in which the sperm tail specifically lacks the dynein molecule, produce only sterile immotile spermatozoa.

The mammalian sperm, which is required to swim in viscous genital fluid, gets additional structural support for the axoneme in the form of peripheral dense fibrous sheath. There are nine outer dense fibres located peripheral to each peripheral doublet of the axoneme, making the axonemal formula of 9+9+2, characterising the mammalian sperm. The fibres are thickest in the proximal half of the tail and progressively decrease in diameter towards the tip.

TYPES OF SPERM

The main function of the spermatozoa is to actively fertilize an egg which is invariably non-motile. Hence the sperm morphology varies depending upon the type of fertilization that the sperm cells effect in a particular animal species. In general, the spermatozoa can be broadly classified into typical and atypical forms.

TYPICAL SPERMATOZOA

In the typical spermatozoa, all the complementary structures found in the motile spermatozoa described above are found. There is a distinctive anterior head containing nucleus and acrosome, a midpiece containing the mitochondria and a long filamentous tail with axonemal complexes for sperm motility.

ATYPICAL SPERMATOZOA

The atypical spermatozoa generally lack a typical tail with its axonemal complexes. They also exhibit bizarre shapes as found in many crustacean species such as the crabs, shrimp and lobsters. The spermatozoa in true crabs are spherical in shape with a proximal anterior acrosome overlying the distal posterior nucleus provided with several slender radiating nuclear arms. The acrosome is providing with an apical concavity or the acrosomal cap. The spermatozoa of the lobsters resemble those of the crabs, but the nucleus is provided with fewer numbers of extensions in the form of spikes. In the case of marine shrimp, *Sicyonia ingentis* the spermatozoa assume a thumb nail-like structure, with a prominent head or nucleus anteriorly and a narrow projection posteriorly forming the acrosomal complex, with a single spike seen as the extension of the acrosome (Fig.3.3). During fertilization, the spike makes the first attachment with the egg membrane, followed by a characteristic acrosomal reaction. The atypical spermatozoa also include tail-less sperm as found in *Ascaris* and mesozoa. In *Ascaris*, the spermatozoa though lacking flagellar tail, effect amoeboid movements to reach the egg for fertilization. Nevertheless, it is a strongly modified cell which has arisen through complicated spermiogenic events.

SPERM CELL FORMATION

Spermatogenesis occurs inside the seminiferous tubules of the vertebrate testis. The seminiferous tubules contain radially distributed Sertoli cells and the germ cells namely the spermatogonial cells. The Sertoli cells are columnar with broad bases and narrow tips that extend into the lumen of the tubules. Spermatogonia are found in between the Sertoli cells. By repeated mitotic multiplication, these primary spermatogonial cells give rise to a definitive number of secondary spermatogonial cells which transform into primary spermatocytes. The primary spermatocytes are pushed from the outer periphery to the inner region of the seminiferous tubules. They then undergo two successive meiotic maturation divisions to give rise to secondary spermatocytes and then, the spermatids. The spermatid contains haploid number of chromosomes and assumes a spherical shape. The spermatids undergo extensive nuclear and cytoplasmic differentiation to give rise to the typical flagellate sperm cells. The whole process of spermatid differentiation into sperm is called spermiogenesis or spermateleosis.

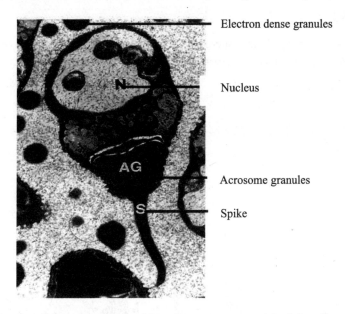

Fig. 3.3 Electron micrographic pictures of the mature spermatozoa of the shrimp *Sicyonia ingentis* (From Subramoniam, 1995).

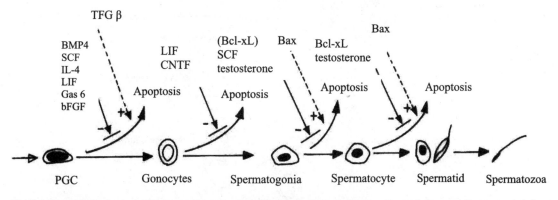

Fig. 3.4 Gonadal development during embryogenesis and in the first wave of spermatogenesis is normally accompanied by extensive apoptosis of all germ cell types. Signals that may regulate germ cell apoptosis are indicated by arrows and named in black above the cells they influence, with "+" indicating promotion of apoptosis, and "-" indicating inhibition of apoptosis. Abbreviations: BMP, bone morphogenetic protein; SCF, stem cell factor, IL, interleukin; LIF, leukaemia inhibitory factor; FGF, fibroblast growth factor (Redrawn from Print *et al.*, 2000).

In the adult mouse, spermatogenesis is characterized by continuous maturation towards the centre of the seminiferous tubules: mitotic proliferation of spermatogonia, meiotic division of spermatocytes, differentiation of spermatids and finally release of spermatozoa into the tubule lumen. To achieve precise homeostasis of each germ cell type in the adult, germ cell renewal, proliferation, export and apoptosis must be finely balanced (see Fig.3.4). This appears to occur at the cost of substantial germ cell wastage, however, since it is estimated that up to 75% of potential spermatozoa degenerate in the testes of adult mammals.

SPERMIOGENESIS

The primary events occurring during spermiogenesis include: 1. modification of the nucleus, 2. formation of the anterior acrosome, 3. reorganization of the mitochondria, 4. formation of the tail with its axonemal complex

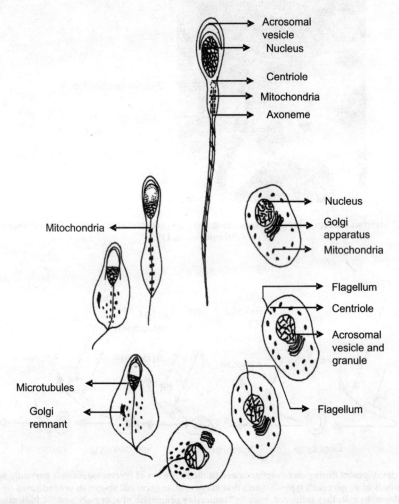

Fig. 3.5 Different stages in the spermiogenesis of a mammalian sperm (Redrawn from Gilbert 1997)

The nuclear modification involves extensive condensation of the chromatin due to lateral compression of the DNA molecule. During this period, the DNA binding basic proteins, the histones, are slowly getting replaced by the unique highly basic DNA binding proteins, protamines. This protein by virtue of its interaction with the DNA double helix produces condensation of the chromatids. During this period of nuclear condensation, the acrosomal vesicle is formed by the aggregation of Golgi complex anterior to the nucleus. The Golgi complex produces several vesicles, known as proacrosomal vesicles, which ultimately fuse together to form the acrosome proper. The characteristic feature of the acrosome is its specific staining with the carbohydrate stain periodic-acid Schiff's (PAS) reagent. The spermatid is now becoming elongated in shape. The proximal and distal centrioles are now displaced to occupy a posterior position to the nucleus. The distal centriole gives rise to the 9+2 axonemal complexes which begin to elongate into a long flagellum. Concurrently, the mitochondria which are originally spread over the entire area of the cytoplasm aggregate around the base of the flagellum and become incorporated into the midpiece of the sperm. Following these events, the nucleus undergoes transformation into a pear-shape with most of the redundant cytoplasm being discarded off. The flagellate spermatozoa can again be divided into two major types depending upon the type of fertilization that they effect.

1. THE PRIMITIVE SPERM TYPE AND EXTERNAL FERTILIZATION

The primitive sperm is a small cell with a sharp, rounded or conical head with a midpiece containing a few mitochondria (often four) and a tail consisting of a long flagellum (Fig.3.6). They exhibit very vigorous movements and effect external fertilization as in the case of sea urchin.

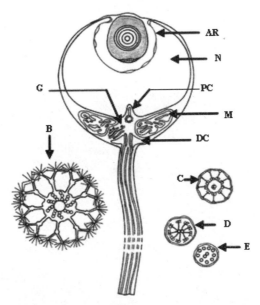

Fig.3.6 Showing the head region of the holothurian Leptosynapta spermatozoon (Sea cucumber, Echinoderm) to show the acrosomal, nuclear and mitochondrial arrangement. The acrosomal region, found in a depression anterior to the nucleus consists of concentric layers of acrosome surrounded by a granular subacrosomal region. (Redrawn from Atwood, 1974). (A) longitudinal diagram of intact sperm, (B) cross-sectional diagram through the distal centriole and satellite projections, (C) section through proximal region of tail containing Y-shaped membrane connectives, (D) section through mid region of tail containing arms and spokes, (E) section through distal region of tail, AR acrosomal region, DC distal centriole, G golgi complex, M mitochondrion, N nucleus, PC proximal centriole

2. MODIFIED SPERMATOZOA WITH INTERNAL FERTILIZATION

This type of spermatozoa is characterised by modifications in the midpiece and head. The mitochondria are reorganised and spread down the flagellum. The entire spermatozoa are enlarged and elongated. In these thread-shaped spermatozoa, the divisions between the acrosome, nucleus and the midpiece may be difficult to discern with the light microscope. Example: Phoronida, Acanthocephala and Ctenophora.

ENDOCRINE REGULATION OF SPERMATOGENESIS

In many animal species, spermatogenesis is a cyclical event and hence is controlled by hormones. The production of spermatozoa in the vertebrate testis is regulated by two gonadotropic hormones, namely, follicle-stimulating hormone (FSH) and luteinizing hormone (LH), both produced from the pituitary gland. The pituitary gland in turn is controlled by gonadotropin-releasing hormone from the hypothalamous. Within the testis, the initiation, maintenance and restoration of spermatogenesis is accomplished by the testicular hormone, testosterone. The Leydig cells found in the interstitial space of the seminiferous tubules of the testis secrete testostone in response to luteinizing hormone. FSH is also involved in the regulation of spermatogenesis by acting on Sertoli cells which produce a peptide called androgen-binding protein. This protein binds to testosterone, thereby sustaining its effect on spermiogenesis.

4

OOGENESIS

Oogenesis can be considered as the first phase of embryogenesis, as it refers to the period during which the egg acquires its developmental potential by storing not only the nutritionally important yolk, but also a variety of regulatory macromolecules to direct early developmental processes. Oogenesis comprises the origin, multiplication, and differentiation of the female gametes. There are two distinct phases occurring during oogenesis: a proliferative phase and a differentiative phase. The proliferative phase that occurs in the germinal zone of the ovary increases the number of oogonial cells by mitotic multiplication. During this phase, the proliferating primary oogonia undergo rapid multiplication by mitosis to give rise to secondary or terminal oogonia. The secondary oogonia then transforms to primary oocytes which is destined to grow and mature into typical egg cells by the differentiative process. Following this, the primary oocyte undergoes two meiotic maturational divisions to possess a haploid set of chromosomes that signify the completion of egg maturation. Meanwhile, the volume of the egg increases several folds accumulating enormous amounts of macromolecules, which are generally termed as the yolk.

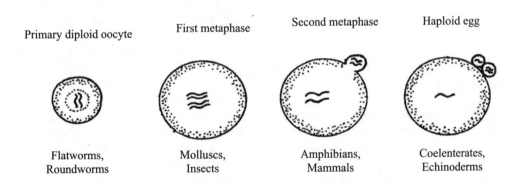

Fig. 4.1 Meiotic maturational status of eggs of different animals at the time of fertilization

In the humans, the oocytes remain arrested in the I meiotic prophase at the time of birth. The resumption of meiosis thereafter occurs by hormonal stimulation. Under hormonal influence, which happens at the time of puberty, the oocytes undergo nuclear progression from the I meiotic prophase (with four times the haploid DNA complement) to metaphase II (with two times the haploid DNA complement). The egg at this stage is characterised by the presence of a polar body. During this progression from the I meiotic prophase to II metaphase stage, the nucleus exhibits dramatic changes. The germinal vesicle undergoes dissolution followed by the condensation of the diffused chromatin into the distinct bivalents, separation of homologous chromosomes and emission of the first polar body, and arrest of meiosis at metaphase II. The whole process is called meiotic maturation, which is a prerequisite for the eggs to get fertilized. Further resumption of meiosis occurs at the time of fertilization, when the second polar body will be produced, with the egg possessing a haploid set of chromosomes. The state of meiosis at the time of ovulation and insemination may vary among different animal species (Fig.4.1). For example, in the echinoderms and the coelenterates the eggs have completed all the meiotic divisions and attained the pronuclear stage at the time of insemination. On the contrary, in mammals, only the first meiotic division is completed and the second polar body is formed only after fertilization.

OVARY AND GERM CELL DEVELOPMENT

The female gonad is a complex, multi-compartmental structure with diverse biological functions that include production of fertilization-competent germ cells (oocytes) and synthesis of a variety of hormones that influence the function of many other organs and tissues of the body. The current concept is that germ cell production in most mammalian females is restricted to fetal or early neonatal life. During the perinatal period, the ovary contains a reserve of oocytes arrested in meiosis-I and enclosed by a specialized population of somatic epithelial cells referred to as granulosa cells. This germ cell-somatic cell unit, termed a follicle, then progresses through a series of maturational steps involving growth of the oocytes and proliferation of the granulosa cells. Concurrently, during the final stage of follicle maturation, there is addition of another somatic cell population, namely, theca-interstitial cells. Under appropriate hormonal conditions, the fully matured follicle releases the oocyte into the reproductive tract for fertilization and subsequent embryonic development. However, only a small fraction of the pool of oocytes generated during gametogenesis survives to be ovulated for fertilization. For example, in the human female nearly 7×10^6 germ cells are produced in the fetal ovaries by mid-gestation, but only 1×10^6 or so remain in the ovaries at birth. There is a continuous decline in the number of oocytes throughout young and adult life, such that only a few hundred mostly degenerative oocytes are found in the ovary at menopause. Apparently, more than 99.9% of the potential germ cell population generated in the human females degenerate rather than survive by a process of apoptosis (Details, see section on apoptosis).

STAGES OF OOGENESIS

The period of oocyte differentiation is divided into premeiotic, previtellogenic, and vitellogenic phases. The premeiotic phase is characterized by a regular sequence of nuclear events during prophase of the first meiotic division. During previtellogenic phase, the nuclear events are arrested in the diplotene stage of the first meiotic division and the chromosomes remain dispersed in the nucleoplasm. The nucleus swells up tremendously to reach the germinal vesicle stage. Previtellogenic stage is often protracted and the oocyte grows slowly with increase in ooplasm and its organelles such as mitochondria. This is the period of intense RNA synthesis. Previtellogenic period is followed by the vitellogenic phase during which the yolk is elaborated. In general, the yolk can be obtained by two methods: (1) by autosynthesis within the ooplasm of growing oocytes, and (2) by heterosynthesis in which the yolk precursor materials are synthesised in the extraovarian somatic organs such as liver (oviparous vertebrates) and fat body (insects), and are transported through blood for final incorporation into the oocytes. Specific uptake of this yolk precursor protein (vitellogenin) from the blood is mediated by the follicle cells.

COMPOSITION OF YOLK

Yolk is the main nutritive material, accumulated in substantial quantities in the egg cytoplasm to meet the basic requirements of embryonic development in all oviparous animals, where embryo develops independent of the maternal organism. The composition of yolk could vary from species to species depending on diet and their requirements during embryogenesis. Yolk is comprised of three chemical components, viz., proteins, lipids and carbohydrates. Proteid yolk is mainly found in the form of lipoproteins or phosphoproteins, which are referred to as lipovitellin or simply vitellins. Lipid reserves include fatty yolk globules, phospholipids and triglycerides. Carbohydrate yolk reserves include glycogen, galactogen and various polysaccharide-protein complexes. Yolk protein, present in the form of droplets or platelets, is a major constituent of the ooplasm of many eggs. The precursor yolk protein after their synthesis in the extraovarial organ undergoes posttranslational changes such as phosphorylation, glycosylation and sulphation in addition to lipid binding. In the vertebrates, two major classes of yolk proteins, lipovitellins and phosvitins are found. Whereas the vertebrate yolk is highly phosphorylated, invertebrate yolk protein is less phosphorylated. The primary function of yolk is that it is used as a source of free amino acids for protein synthesis during embryogenesis. It also acts as a carrier protein during vitellogenesis for ions, lipids, sugars and vitamins, in addition to transporting several steroid hormones for storage along with the yolk. The hormones thus transported through yolk protein serve as morphogenetic hormones during embryogenesis.

VITELLOGENESIS

Vitellogenesis or the formation of yolk is the central event in oogenesis of oviparous animals. The evolutionary origin of vitellogenic system is primarily autosynthetic in nature, since all early metazoans synthesized most of the yolk products within the growing oocytes, without or with minimal assistance from extra oocytic sources. However, with increasing demand from embryonic nutritional requirements, several animal species expanded the vitellogenic activities to a heterosynthetic process. This method of yolk formation involves an extraovarian organ to synthesize yolk precursor protein and an efficient transport system to carry the released yolk precursor molecules into the ovary, where they are selectively sequestered to form the storage yolk protein of the mature eggs. Yet another form of vitellogenesis, commonly found among invertebrates, is heterosynthetic mode of yolk formation. Thus, in certain crustaceans such as crayfish, elaboration of cell organelles such as endoplasmic reticulum and Golgi complex during primary vitellogenesis heralds the biosynthesis of yolk protein within the oocytes. During secondary vitellogenesis this active phase of protein synthesis is followed by a peroid of micropinocytotic activity involving structural alteration in the oocyte cortex.

Internalization of yolk precursor protein, vitellogenin, by the developing oocyte is accomplished via receptor-mediated endocytosis. Receptor-mediated endocytosis, an essential process in all eukaryotes, is required for general cellular functions, including uptake of nutrients (e.g., low density lipoproteins or transferrin) and recycling of membranes and membrane proteins. The receptor-mediated endocytosis of yolk protein has been revealed by several electron microscopic investigations as well as molecular studies in many animal species. Vitellogenin receptor, an oocytic membrane glycoprotein has been characterized extensively in the amphibian, *Xenopus laevis*, birds, reptiles, and in insects such as *Drosophila* and *Aedes ageyptii*. The vitellogenin receptor protein binds to the circulating serum vitellogenin at the membrane level and the complex is internalized through micropinocytosis. In birds, in addition to vitellogenin, large quantities of very-low density lipoprotein, a carrier of energy in the form of triacylglycerol, are incorporated and sequestered in oocytes during vitellogenesis. Interestingly, the uptake of both vitellogenin and very low-density lipoprotein are mediated by the same vitellogenin receptor. A contrasting condition however occurs in insects in respect of vitellogenin and non- vitellogenin yolk precursor molecules. In insects like *Locusta migratoria* and *Drosophila melanogaster* a class of lipoprotein, namely lipophorin, is internalised along with yolk precursor proteins during vitellogenesis. Their main function is lipid transport into the ovary. In the mosquito *Aedes ageyptii* and *Locusta migratoria*, it is internalised via receptor- mediated endocytosis, but utilizing an ovarian lipophorin receptor which is distinctly different from insect vitellogenin receptor.

It is now known that vitellogenin receptor of several oviparous animals including the invertebrates, such as the nematode, *Caenorhabditis elegans*, insects and crustaceans and vertebrates such as the chicken belongs to the low- density lipoprotein receptor superfamily. In addition, both vitellogenin and apolipoprotein belong to the large lipid transfer protein, and they are also found to be evolutionarily related not only in function but also in amino acid sequence. Again, the binding region of vitellogenin showed a pattern similar to the binding sites of apoB and apoE. Positively charged basic residues such as lysine or arginine in the vitellogenins and apoB and apoE bind to the acidic residues found on the ligand binding sites of receptors by electrostatic interactions (Fig.4.2).

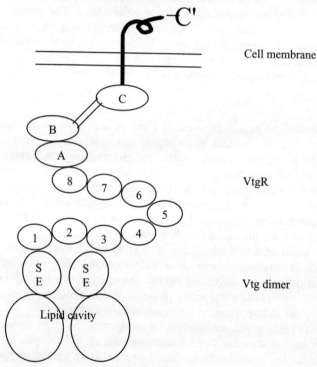

Fig.4.2 The model of Vtg-VtgR interaction. VtgR-Vitellogenin receptor; Vtg dimmer-Vitellogenin dimer. Redrawn From Li A, Murali S, and Ding J.L. (2003)

During vitellogenesis, the developing oocytes are characterized by the presence of an enormous quantity of coated vesicles in the cortical cytoplasm. Extracellular vitellogenin molecules bind receptors anchored in the plasma membrane of the oocyte and the receptor-ligand complexes then cluster in specialised domines called coated pits. Coated pits invaginate into the cytoplasm and pinch off, forming coated vesicles. Several coated vesicles fuse to form the endosomes. Within the endosomes, the vitellogenin is processed to yield the final storage yolk product, called vitellin. Thus, the endosomes are the storage compartment for the yolk protein. In chicken, these giant endosomes are called yolk spheres reaching a diameter of 140μm.

The process of oogenesis with special reference to the synthesis and accumulation of RNAs and yolk, together with the hormonal regulation of the whole process has received extensive studies in two animal groups, namely amphibians and the insects. Hence these two animal groups have been chosen as case studies to illustrate the female gametogenesis.

OOGENESIS IN AMPHIBIANS

In the frog, *Rana pipiens*, oogenesis is a prolonged process lasting for three years. The young oocytes start growing after the tadpoles metamorphose into young froglets. The oocyte growth continues for two years and by the third year the eggs grow faster and fully mature for spawning. Every year a new batch of oocytes is produced as a result of oogonial division, but these do not mature till three years later, so that oocytes of three generations may be contained in the ovary at the same time.

PREVITELLOGENIC PHASE

Premeiotic changes are arrested when the oocyte reaches the diplotene stage of the first meiotic phase. The subsequent stages of meiosis are postponed and instead, the nucleus becomes greatly swollen to assume the characteristic germinal vesicle stage. At this time, the chromosomes become greatly distended, with their threads or loops developing perpendicular to the long axis of the chromosome proper. This gives the chromosomes the appearance of lampbrush. The loops represent actual sections of the chromosomes, which have become completely despiralized and this is favourable for the gene activity, namely, the synthesis of messenger RNA (mRNA). By using a radioactively labelled RNA precursor (^{3}H Uridine), RNA synthesis in the loops has been demonstrated. Another important activity of the oocytes in this stage is the formation of one or more nucleoli inside the germinal vesicle. They are concerned with the synthesis of ribosomal RNA. The formation of numerous nucleoli is the result of outward expression of the increase in the number of genes coding for the ribosomal nucleic acids.

In the oocytes of *Xenopus laevis*, the genes encoding for two main rRNA molecules (18S and 28S RNAs) are multiplied by a factor of several hundreds. This increase in the number of genes without mitosis is referred to as 'amplification of genes'. The amplification of ribosomal genes produces numerous nucleolus organisers and is responsible for the large number of nucleoli in the amphibian oocytes. These nucleic acids remain dormant until after fertilization and take part in protein synthesis in the early part of development.

In addition to the large accumulation of RNA materials, the amphibian oocytes also characteristically accumulate DNA in the cytoplasm. These DNA are contained in the mitochondria. These cytoplasmic DNA materials are utilised during the cleavage, when there is great need for DNA synthesis to meet the successive quick mitotic divisions of the blastomeres.

VITELLOGENIC PHASE

In amphibians, the highly condensed yolk is contained in yolk platelets. These are flattened ovoid structures consisting primarily of two components, phosvitin and lipovitellin, both organised into a crystalline lattice. Phosvitin is a highly phosphorylated phosphoprotein and the lipovitellin is a large lipophosphoprotein always found as a dimer. Two monomers of phosvitin molecules are fitted on to the lateral sides of the lipovitellin dimer as shown in the Figure 4.3. The whole molecular complex is defined as yolk, which undergoes crystallization at the end of vitellogenesis. The crystalline yolk can be dissociated into phosvitin monomers and lipovitellin dimers, which can be fully dissociated into lipovitellin monomers after treatment with an alkali. A characteristic feature of amphibian

24 Molecular Developmental Biology

vitellogenesis is that the crystalline yolk platelets are also found inside the mitochondria. A mitochondrial enzyme called protein kinase may play an important role in yolk platelet formation. This enzyme adds a phosphate group to phosvitin molecule causing the yolk to become insoluble. This may directly cause the yolk to crystallize out of the solution to form yolk platelets.

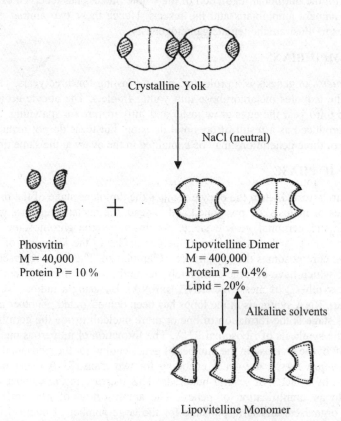

**Fig. 4.3 Schematic models representing sub-units of the crystalline yolk in *Xenopus laevis*
(Redrawn from R.A. Wallace & M.J. Dumont, 1968)**

In addition to yolk, many other molecules needed for early embryogenesis accumulate in the oocytes. They include glycogen, lipids and a variety of enzymes. Furthermore, the amphibian eggs also accumulate cortical granules, which are produced by the Golgi complex. These are carbohydrate rich molecules enclosed in a vesicle. The cortical vesicle moves to the periphery of the oocytes and plays a major role in fertilization.

HETEROSYNTHETIC DERIVATION OF YOLK1

As in other oviparous vertebrates, vitellogenesis in amphibians is a heterosynthetic process, in which an extraovarian tissue produces yolk protein precursors. The yolk protein precursor or vitellogenins are secreted into blood circulation for transportation into the ovary, where they are selectively taken up by the growing oocytes for final processing and storage. In all amphibians, liver secretes the yolk precursor protein and secretes into the serum in the form of a single polypeptide, which is called serum lipophosphoprotein. This precursor molecule consists of about 12% lipid, 1.5% phosphate and 1%

carbohydrates, in addition to calcium attached to protein phosphate groups. The vitellogenin reaches the oocyte periphery through interfollicular channels and internalized by micropinocytosis. After entering into the oocyte, the vitellogenin is converted into lipovitellin and phosvitin by specific proteolytic cleavages. Although trypsin and chymotrypsin were reported to be involved in such proteolytic cleavages in *Xenopus* eggs, recent evidence indicates that cathepsin D participates in such cleavages to give rise to lipovitellin I and II and the phosvitin yolk molecules. Cathepsin D is an aspartic protease and is usually packaged in lysosomes. They are found in the nascent primordial yolk platelets, taking part in the processing of yolk precursor proteins into the final product, namely, the crystalline yolk.

GENETIC CONTROL OF VITELLOGENIN SYNTHESIS IN AMPHIBIANS

Technical advancements in molecular biology and DNA cloning technology have made it possible to identify the genes responsible for the vitellogenin synthesis. In the amphibians and birds, the gonadal steroid hormone, estrogen, controls the vitellogenin gene functions. In the frog, *Xenopus laevis*, vitellogenin (Vtg) is encoded in a family of at least four expressed genes, as deduced from the analysis of a number of cDNA clones. The four *Xenopus laevis* Vtg genes have been designated as A1, A2, B1, and B2. The Vtg gene family in *Xenopus*, encoding distinct, but related proteins, has apparently arisen by gene duplication.

Hormonal induction of Vtg gene activity and vitellogenin synthesis in the frog has been extensively investigated by the facility that the liver of male *Xenopus* can be induced by estrogen to synthesize and secrete large amounts of Vtg into the blood. This facilitated the study of differential gene expression as regulated by a hormone in the frog. The estrogen binds to nuclear receptors, which bind to chromatin at specific acceptor sites, and consequently influences gene activity. Recent studies have revealed the occurrence of the estrogen response elements in the 5' region of the chicken Vtg1 and the four *Xenopus laevis* Vtg genes. The estrogen response element is the target DNA sequence for the hormone receptor and is responsible for the initiation of Vtg gene expression (Fig.4.4) That the estrogen acts at the transcriptional level of Vtg genes is indicated by the fact that Vtg mRNA accumulates during hormonal stimulation of liver in the male frogs. Complementary DNA (cDNA) synthesized on purified Vtg mRNA has been used as a probe to titrate the accumulation of the mRNA during hormonal stimulation.

ENDOCRINE REGULATION OF VITELLOGENESIS AND EGG MATURATION

In amphibians, the oocyte growth and differentiation on one hand and the meiotic maturation and ovulation on the other hand are under definite hormonal control. Gonadotropic hormones originate from the pituitary, which is in turn regulated by the neuroendocrine factors released by the hypothalamus. The gonadotrophic hormone FSH (follicle stimulating hormone) influences the ovary in a dual way. First it stimulates the follicle cells to produce the steroid hormone, estrogen, which on release into the blood stream induces the liver, to produce the yolk precursor protein, vitellogenin. This is released into the blood as serum lipophosphoprotein. The FSH again induces the vitellogenin uptake into the ovary, by activating the follicular epithelial cells surrounding the oocyte. The follicle cells produce numerous villi-like processes which interdigitate with those of the oocyte periphery. The yolk precursor protein is selectively sequestered through pinocytosis, through the interdigitating villi-like processes which interdigitate with those of the oocyte periphery. The yolk precursor protein is selectively sequestered through pinocytosis, through the interdigitating villi-like processes. The serum lipophosphoprotein, once gaining entry into oocyte undergoes proteolytic cleavage to give rise to the phosvitin and lipoprotein molecules that finally conjugate to form the crystalline yolk platelets. Once the yolk platelet formation is completed, meiotic maturation commences, taking the hormonal cue from the pituitary gland to resume meiosis. The pituitary secretes another hormone, namely, lutinizing hormone which stimulates the follicle cells to synthesize yet another steroid, progesterone. Progesterone induces the synthesis of a cytoplasmic factor called maturation promotion factor, which resumes and completes the meiotic maturation. Progesterone also influences the ovulation, by causing the follicle cells to release lytic substances that rupture the epithelium.

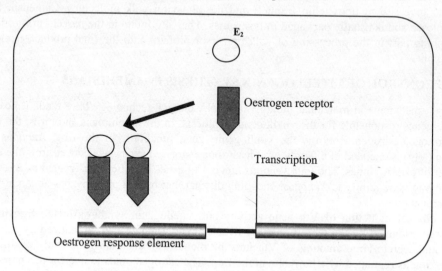

Fig. 4.4 Oestrogens such as 17β-oestradiol (E2) regulate cellular processes by binding to oestrogen receptors in the nucleus. The activated receptors dimerize and bind to specific oestrogen response elements in the promoter regions of target genes (Vtg gene), thereby triggering gene transcription.

OOGENESIS IN INSECTS

Insect oogenesis has been studied very intensively with reference to vitellogenesis and endocrine mechanism regulating it. The insectan ovarioles are divided into three zones. The anterior most zone is the terminal filament, which is followed by the germinal zone or germarium. This zone is the site of active mitotic multiplication of both oogonial cells and prefollicular cells. Posterior to the germinal zone is the growth zone or vitellarium where the primary oocytes derived from the germarium undergo growth and differentiation to attain final maturity. The follicle cells originating from the germinal zone also move into the vitellarium to surround the oocyte.

As in other animal species, RNA synthesis and yolk formation form the primary event of the insect oogenesis. However, insect ovary differs from that of the amphibians in reference to the origin of RNA materials. In general, insect ovary is divisible into two categories (Fig.4.5) based on the sources of RNA for the growing oocytes.

VITELLOGENESIS

In most insects, vitellogenesis occurs through heterosynthetic means, by which the presynthesized yolk precursor protein is sequestered from hemolymph into the oocyte. In almost all the insects vitellogenin is synthesized by the fat body and released into the hemolymph. From hemolymph, the vitellogenin is selectively sequestered by the growing oocyte. The follicular epithelial cells undergo morphological modifications during this process aiding in the uptake of blood-borne vitellogenin into the oocyte, as in the case of amphibians.

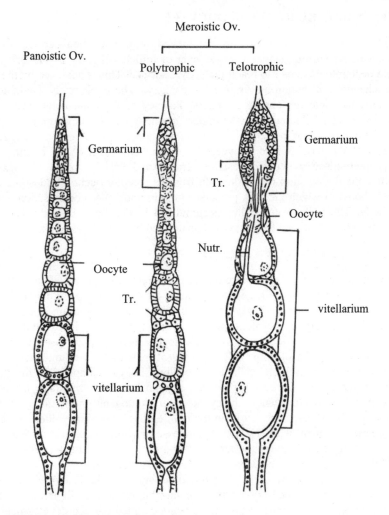

Fig. 4.5 Three types of ovarioles in insect species. Nutr- nutritive cord, Tr- trophic tissues. (Modified and redrawn from Franz Engelmann, 1970)

VITELLOGENINS AND VITELLINS

In most insects, vitellogenins are large oligomeric glycolipophosphoproteins consisting of two or more subunits. They are synthesized in the fat body as single or multiple precursors before being processed for secretion into the hemolymph. The vitellogenin also undergoes posttranslational modification such as glycosylation and sulphation. A characteristic feature of insect vitellogenesis is that, in addition to the uptake of vitellogenin, the oocyte also accumulates nonvitellogenin yolk proteins. The transportation of lipids to the ovary is mainly achieved by a class of hemolymph lipoproteins called, lipophorins. These high or very high-density lipoproteins carry lipids to ovary and deliver the lipids to the oocytes and get dissociated to transport more lipid molecules from the hemolymph. The oocytes also accumulate large quantities of glycogen granules.

VITELLOGENIN UPTAKE INTO THE OOCYTES

The sequestration of yolk precursor proteins is achieved by receptor-mediated endocytosis. Since the vitellogenic oocytes are surrounded by a single layer of follicle cells, the extraovarian yolk precursor proteins should penetrate this layer to reach the oocyte surface. This is achieved by the enlargement of interfollicular channels, a condition known as patency. These channels facilitate the entry of vitellogenin molecules onto the oocyte periphery. Patency of follicular cells occurs at the onset of vitellogenesis. Cytoskeletal elements play an important role in patency development, although $Na+/K+$ ATPase is known to participate in the process. In the mosquito, *Aedes aegypti*, vitellogenin binds to its receptor, located at the base of and between the oocyte microvilli. After binding to its receptor, vitellogenin is internalised by coated vesicles. The coated vesicles reach the next compartment, endosome, where the dissociation of vitellogenin from its receptor occurs. Endosomes coalesce into a transitional yolk body, in which the yolk precursor protein undergoes condensation until it becomes a mature yolk body. The yolk proteins thus accumulated inside the yolk body do not undergo any degradation until embryogenesis when, they will be utilized by hydrolytic enzyme processing.

HORMONAL REGULATION OF VITELLOGENESIS

In the majority of the insects, juvenile hormone (JH), secreted by corpora allata is the gonadotrophic hormone regulating oogenic activities in the female. However, in the dipteran flies, the molting hormone, 20-hydroxy ecdysone also plays a role in the control of vitellogenin synthesis by the fat body. In addition, neurosecretory hormones from the brain may also be involved directly in regulating vitellogenesis in some insects. The JH control of vitellogenesis in a typical insect is given in figure 4.6.

JH has multiple gonadotrophic actions in that it induces the vitellogenin synthesis by the fat body, promotes patency in follicle cells and develops oocyte competence for the yolk protein internalization. In the locust, *Locusta migratoria*, a brain hormone stimulates the corpora allata to secrete JH. Juvenile hormone influences the synthesis of vitellogenin in the fat body at transcriptional and translational levels. Allatectomy results in the failure of vitellogenin synthesis by the fat body, but it can be restored by treatment with JH. After the release of vitellogenin into the hemolymph, JH also activates the ovarian follicle to sequester vitellogenin into the oocyte. The receptor-mediated endocytosis of the vitellogenin molecule is also influenced by JH activity.

However, in the mosquito, *Aedes aegypti*, vitellogenin synthesis and secretion by the fat body are regulated by 20- hydroxy ecdysone (20E). In this blood sucking dipteran fly, the blood meal stimulates brain neurosecretory cells to produce a peptide called the egg development neurosecretory hormone (EDNH). EDNH stimulates the ovary to secrete ecdysone, which in concert with JH, stimulated the fat body to synthesize vitellogenin. The uptake of vitellogenin into the oocytes is, however, mediated by the JH influence, as in other insects (Fig.4.6).

Recent studies have indicated that blood feeding triggers a 20-E hormonal cascade, which activates yolk protein precursor genes in the fat body of the female *Aedes aegypti*. In this insect, there is a previtellogenic arrest preventing the activation of the yolk protein genes. Similarly, there is termination of their expression, so that another arrest is achieved after a batch of eggs are laid. 20-E manifests the gene regulatory effect through its receptor that is a heterodimer of two members of the nuclear receptor gene family, ecdysone receptor (EcR) and a retinoid X receptor (RXR) homologue, ultraspiracle (USP). The EcR-USP complex recognizes sequence-specific DNA motifs, called ecdysteroids response elements (EcREs). EcRE is shown to be required for the activation of the Vg gene.

After binding to 20E, the EcR-USP complex also induces several early response genes, which are necessary for a high level Vg gene expression. USP exerts its functions by associating with distinct partners at different stages of vitellogenesis. During inactive stage of previtellogenesis, the USP exists as a heterodimer with the orphan nuclear receptor AHR38. During vitellogenesis (after a blood meal) in the presence of 20E, EcR can efficiently displace AHR38 and form an active heterodimer with USP. When vitellogenesis proceeds to the terminal stage, falling 20E titres shifts USP heterodimerization towards SVP, another co-repressor molecule found in the nucleus (see Fig. 4.7).

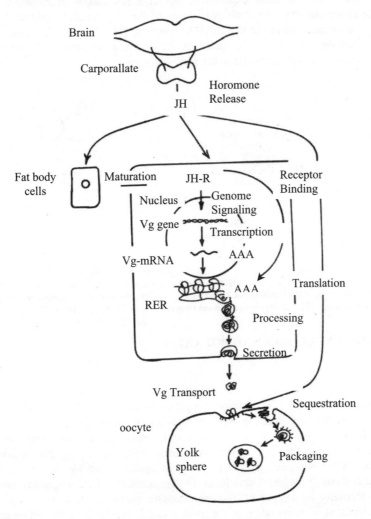

Fig. 4.6 JH control of vitellogenin synthesis and uptake in insects (Adapted from Chen and Hillen, 1983). JH-juvenile hormone, JH-R- juvenile hormone receptor, Vg- vitellogenin, RER- rough endoplasmic reticulum.

GENETIC CONTROL OF VITELLOGENIN SYNTHESIS

Genetic control of Vtg synthesis has been studied mainly in locusts and the fruit fly *Drosophila melanogaster*. In the migratory locust, *Locusta migratoria*, there are two coordinately expressed Vtg genes, A and B, which are situated on the X chromosome. The locust Vtg genes specify a single transcript, which codes for a single previtellin molecule that undergoes proteolytic processing to generate the large and small subunits. On the other hand, in *Drosophila* the yolk proteins are small and are not products of larger precursors. They are synthesized in the fat body and follicle cells and regulated by both JH and 20- hydroxyecdysone. In *Drosophila melanogastor*, three yolk polypeptides (Yp1, Yp2 andYp3) have been characterized and the genes encoding the mRNA for these polypeptides have been found to be linked to the X chromosome. Thus, unlike in amphibians, the Vtg genes in insects are not only sex limited in their expression, but also sex linked in location, suggesting functional significance to the sex chromosomal localization of the Vtg genes. In insects, both JH and ecdysone are shown to cause increase in the Vtg mRNA present in the fat body. This indicates that the hormones exert their effect directly on the Vtg genes. In the locust, a cytoplasmic factor serving as a JH receptor may also mediate the transfer of JH to the genomic sites.

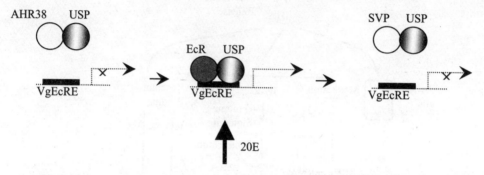

Fig. 4.7 Working model of modulation of 20E response via different protein heterodimerisations during activation and termination of mosquito Vtg gene transcription. AHR38 - Orphan nuclear receptor; USP - retinoid X receptor homologue Ultraspiracle; VgEcRE - Vitellogenin Ecdysteroid response element; EcR - Ecdysone receptor; SVP - Seven-up; 20E - 20-hydroxyecdysone (Redrawn from Raikhel et al, 2002)

TYPES OF EGGS, EGG ENVELOPES AND POLARITY

TYPES OF EGGS

It is known that the major component of any egg cytoplasm is the yolk. The amount of yolk in the ooplasm has a definitive relationship with the types of fertilization and embryogenesis that the egg is going to have. In all the true viviparous animals (mammals), where the embryogenesis occurs within the female reproductive system, the egg carries very little yolk for the simple reason that the embryonic development is not sustained by the stored yolk materials but by the supply of nutrient material from the mother through placental structures. On the contrary, in the oviparous animals (eggs which are incubated externally by the parent) the eggs contain enormous quantity of yolk. Here the entire embryonic development is independent of the mother and hence the yolk should supply nutrient material for the entire embryonic development (example: hen's egg). Nevertheless, the type of embryonic development in releasing a larva or a miniature adult will also have a bearing on the yolk distribution within the egg. For example, the eggs of sea urchins contain very little yolk and hence, the embryonic development is very quick and the larvae hatch out of the egg within few hours.

With all these exceptions a classification of the egg can be conveniently made depending on the amount of yolk that they contain. The first type of egg has very small amount of evenly distributed yolk. They are called isolecithal or homolecithal. The examples are sea urchin, *Amphioxus* and tunicates. Another type of egg with apparently no yolk is called alecithal eggs. Examples are eutherian mammals. Macrolecithal and megalecithal eggs are a category of the egg containing enormous quantity of yolk materials, which almost fill the entire egg displacing the egg cytoplasm to the animal pole to form a cytoplasmic cap that contains the nucleus. The best examples are bony fish, birds and reptiles. These eggs are also called as telolecithal. In the amphibians, there is a gradient of yolk distribution with the vegetal hemisphere having more of yolk than the animal hemisphere. This category of yolk is also called mesolecithal. In arthropods such as insects, the yolk assumes a central position and the cytoplasm is surrounding the centrally placed yolk as a thin coat. In the centre of the egg, there is an island of cytoplasm, which contains the nucleus. These eggs are called as centrolecithal.

Most of the eggs of the terrestrial oviparous animals are protected from the environment by a calcareous outer shell. Such eggs are classified as cledoic eggs. Examples are reptiles and birds. These eggs have the facility of storage excretion within the shell. The noncledoic eggs, on the contrary, are not having thick shells. In general, these eggs are found in viviparous animals, where the embryonic development is internal (e.g. mammals).

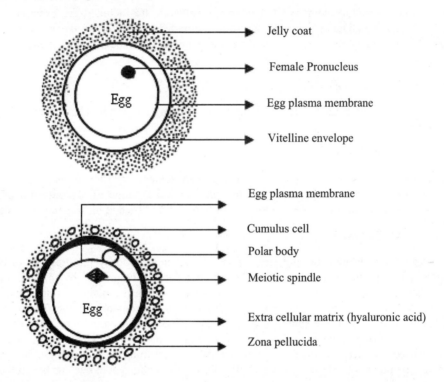

Fig. 4.8 Schematic diagram of eggs and their extra cellular layers. A. Mouse eggs. B. Sea urchin eggs (Redrawn from Evans, 2000)

EGG ENVELOPES

Egg itself is a cell comparable in all respects to the somatic cells and possesses a cell membrane or plasmalemma. In addition to the cell membrane, the eggs are protected externally by several layers of envelopes depending upon the type of fertilization, nature of embryogenesis and the relationship with the environmental condition (Fig 4.8). For example, an egg laid on the land should contain many protective layers to safeguard the soft egg constituents from environmental hazards such as temperature and light. In addition, some of these egg membranes also contain sperm receptors in order to facilitate fertilization with homologous sperm cells. The best example is the zona pellucida glycoprotein ZP3 that is specifically responsible for sperm binding to initiate fertilization in the mammals. Other important functions of the egg membranes would include the release of the substances responsible for activation of male gametes to undergo acrosome reaction. As an example, the jelly coat of the sea urchin eggs exudes substances, which will induce natural acrosomal reactions in the homologous spermatozoa. The egg membrane in some cases also provides nutrients such as albumen for the embryonic development as in the case of molluscs and birds. A generalised function of all the egg membranes is however to hold the dividing blastomeres in their respective position till such changes as gastrulation and morphogenesis starts. Based on the origin of the egg membrane the following classification can be made.

PRIMARY EGG ENVELOPE

It is formed mainly by the ovum itself. Under this category are included the examples of vitellin membrane, the zona radiata which represents the degraded microvilli of the growing oocytes in mammals, the structure-less jelly surrounding the eggs of echinoderm and many other marine invertebrates.

SECONDARY EGG ENVELOPE

They are secreted by the follicular epithelial cells surrounding the ovum. The chorion of insects and ascidians are examples.

TERTIARY EGG ENVELOPES

These are formed by the oviduct or other female accessory sex organs. They are normally secreted and deposited outside the fertilized egg when they descend down the oviduct. The albumen of amphibians, reptiles and birds as well as the protective shell membrane of reptiles and birds is the examples.

POLARITY AND SYMMETRY

All animal eggs have an invisible polarity, which can be understood by the presence of an anterior animal pole and a posterior vegetal pole. These two poles are connected by the axis of the egg. Assuming that the egg is like a globe all the planes that contain the main axis are called meridian planes. The plane that bisects the main axis at right angle is the equatorial plane. The polarity of the egg is difficult to discern in the spherical shaped eggs. However, in the oblong eggs of insects and cuttle fish, the polarity is revealed in the shape of the egg itself. In other cases, the egg polarity can be readily visualised. Further, the polarity can be readily inferred from the fact that the polar bodies are given off at the animal pole of the egg. Again after fertilization, the zygote nucleus lies in the eccentric position nearer to the animal pole of the egg.

The arrangement of the yolk granules in the egg cytoplasm also reveals the polarity of the egg. For example, the yolk density increases from the animal pole towards the vegetal pole as in the amphibian eggs. In addition, the egg pigmentation also shows gradients in that the animal region of the egg surface is darkly pigmented whereas the vegetal pole is unpigmented.

Apart from polarity, many eggs possess an obvious bilateral symmetry. In amphibian eggs, a zone of lighter pigmentation, the grey crescent, develops on one side of the egg along the boundary of the dark animal half in the first few hours after fertilization. This marks the dorsal side of the egg and also of the embryo. The opposite side becomes the ventral side. Once the grey crescent has been formed, we can divide the egg into symmetrical half only in one way namely through the meridian plane, which bisects the grey crescent. This plane contains the main axis and also the dorsoventral axis, which is at the right angle to the main axis and connects the two opposite points on the equator of the egg.

5

FERTILIZATION

Fertilization is the central event in sexual reproduction to perpetuate the species. It is the process in which the mature male and female gametes fuse to form the zygote, which will initiate development. But in many metazoans, the spermatozoa must penetrate one or more egg investments before the actual process of fertilization occurs. In most animals, these egg coats are glycoprotein in nature and are penetrable by the spermatozoa with the aid of acrosome. In other animals, such as some insects and teleost fish, these egg coats are impenetrable layers, like chorion which are necessarily provided with one or more portals or holes called micropyles to permit sperm entry. Sperm belonging to this category do not possess acrosome (eg, teleost sperm). Furthermore, as a prerequisite to fertilization, the spermatozoa must undergo several activational processes before reaching the egg. In many marine invertebrates, fertilization occurs externally in seawater medium and hence, recognition of sperm and egg belonging to the same species is achieved by some of the prefertilization events by which the sperm and egg are attracted towards each other by chemotactic and behavioural means (Fig.5.1). These events include the following steps.

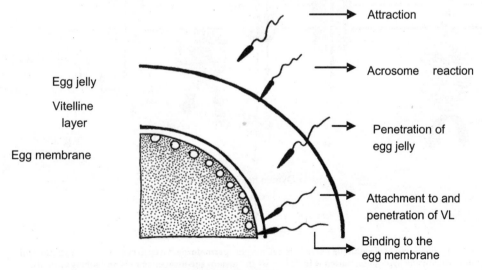

Fig. 5.1 Figure showing different steps in the sperm-egg interactions of sea urchin (Redrawn from K.R. Foltz, 1995).

SPERM AGGREGATION

In the sea urchin, the spermatozoa, on contact with the egg jelly substances, undergo a kind of aggregation. This behaviour of the spermatozoa provides the basis for the fertilizin-antifertilizin theory of Lillee. He preferred to call this aggregation as isoagglutination, comparable to the interaction of the antibody with antigen. The agglutinating substance (fertilizin) is provided by the jelly coat. Multivalent antifertilizin is thought to be present in the sperm head to effect its interaction with fertilizin. However, recent investigations have shown that the agglutination is a reversible process caused by sperm behaviour. Whatever be the reason for such aggregation of the spermatozoa, this process is not a prerequisite for fertilization, because successful fertilization can occur by individual spermatozoa without undergoing such swarming.

SPERM ACTIVATION

The first sperm chemoattractant isolated and characterized was the sea urchin egg peptides, speract (GFDLGGGVG) and resact (CVGAPGCVGGGRL-NH2), from *Strongylocentrotus purpuratus* and *Arbacia punctulata*, respectively. These peptides stimulate sperm motility and respiration at subnanomolar concentrations in a species-specific manner. This bioactivity is found in the carboxy terminal amino acids in speract, whereas, in the resact this activity is potentiated by the amino-terminal half of the peptide.

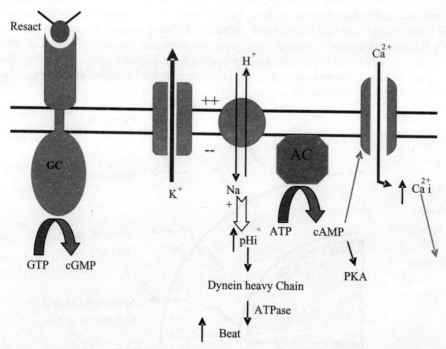

Fig. 5.2 The egg peptide-activated signaling cascade in sea urchin spermatozoa. Grey arrows indicate hypothetical interactions. Membrane hyperpolarization is indicated by the membrane-associated plus and minus symbols. GC – Guanylyl cyclase; AC – Adenylyl cyclase. (Redrawn from: Quill, T.A. and Garbers, D.L. 2002).

Using the radioiodinated peptide analogs, the receptor in the sperm of S. purpuratus was found to be a protein of Mr.77000. In A. punctulata, a similar approach using a radiolabeled resact analog identified a membrane guanylyl cyclase as the sperm receptor. A model of the signal transduction pathway activated by the sea urchin egg peptides is shown in the Fig 5.2. In this model, the peptide binding with the receptor stimulates a transient increase in the cGMP production and an associated K+ dependent membrane hyperpolarization, involving the opening up of K+ channel, allowing K+ efflux. As a consequence of K+ dependent hyperpolarization, a Na+/ H+ exchanger is activated, causing the alkalinization of the sperm. This elevation of internal pH (pHi) stimulates both motility and respiration in the sperm by ATPase activation in the dynein heavy chain of the axonemal complex. Another effect of increase in internal pH is the activation of the sperm adenylyl cyclase leading to the opening of a calcium channel. The consequent increase in cAMP and Cai 2+ regulates sperm motility through the actions of these second messengers on the axoneme proteins (See fig.5.2).

CHEMOTAXIS

Among the marine invertebrates, the most common method of fertilization is the external or broadcast fertilization. A method to recognise the homologous gametes is, therefore, necessary to achieve successful fertilization especially when many different animals discharge their gametes in a common place during peak breeding season. Chemotaxis is a mechanism by which the sperm cells are attracted towards the eggs of their own species by a species-specific factor originating from their egg membranes. Several invertebrates such as coelenterates, tunicates and chitons exploit this method of sperm attraction to effect normal fertilization between homologous gametes. Some chemotactic molecules like a 14- amino acid peptide, resact, also act as a sperm- activating factor. Resact has been isolated from the egg jelly of the sea urchin *Arbacia punctulata* and it exhibits species specific sperm attraction and activation. This peptide, on contact with the sperm cells, causes dramatic increase in sperm motility and oxygen consumption. Concurrently, resact also brings about an increase in sperm cAMP and cGMP that appear to activate the dynein ATPase to stimulate tail beating in the sperm (Fig. 5.3).

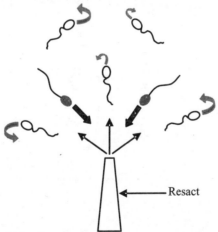

Fig. 5.3 A sperm chemoattraction model for sea urchin. Spermatozoa with grey heads, indicating low intracellular calcium, are shown moving up the resact concentration gradient. Spermatozoa with white heads, indicating elevated intracellular calcium, are not moving up the resact concentration gradient and consequently turn. Linear black arrows indicate straight movement. Curved grey arrows indicate turning
(Redrawn from; Quill T. A. and Garbers, D. L. 2002).

SPERM MATURATION AND CAPACITATION IN MAMMALS

A phenomenon similar to sperm activation of sea urchin also occurs in mammalian sperm. Mammalian spermatozoa undergo a process of maturation and become motile while being transported to the proximal caudal region of the epididymis. Yet, this maturation is not enough for fertilization. The spermatozoa should undergo an additional maturation process, termed capacitation, after ejaculation

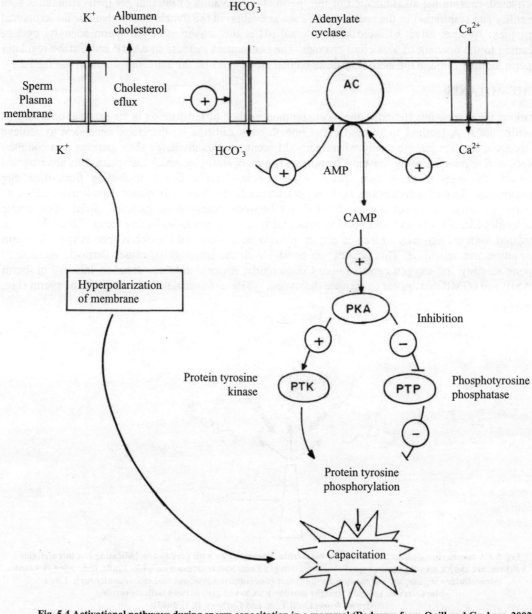

Fig. 5.4 Activational pathways during sperm capacitation in a mammal (Redrawn from Quill and Garbers, 2002)

into the female reproductive tract, to acquire fertilizing potential. Sperm maturation and sperm capacitation is unique to mammals. Sperm capacitation is also a prerequisite for the acrosome reaction. Initiation of capacitation begins with removal of seminal fluid, because it contains decapacitating factors that inhibit capacitation. Cholesterol present in the seminal fluid is the chief decapacitating factor. Substances present in the female genital tract, including progesterone, may exert an influence in regulating capacitation. Removal of cholesterol and a variety of phospholipids on sperm surface is the first step in capacitation. Cholesterol loss from the sperm plasma membrane brings about membrane fluidity, destabilization of the plasma membrane and increased fusion ability. Cholesterol efflux also results in the increase in internal pH and protein tyrosine phosphorylation. More importantly, it causes the exposure of mannose receptor on to the sperm surface, enabling acrosome reaction, after binding to egg surface components. Before capacitation, the receptor resided beneath the plasma membrane.

During capacitation, the intracellular Ca^{2+} level is also increased. This brings about a cascade of signalling activities starting with activation of adenylate cyclase, followed by rise in cAMP and protein kinase A activation (Fig. 5.4). Capacitation of sperm thus results in hyperactivation of sperm, making it competent for acrosome reaction and subsequent events of sperm-egg interaction and fusion.

ACROSOME REACTION

Acrosome reaction is an absolute prerequisite to successful fertilization under physiological conditions. This is the activational process, wherein the acrosome undergoes structural changes. Essentially, acrosome reaction involves the fusion and vesiculation of the plasma membrane overlying the acrosome with the outer acrosomal membrane to release the acrosomal content to the environment. In sea urchin, the egg jelly exudes a carbohydrate factor, namely a fucose sulphate to stimulate acrosome reaction. The stimulatory activity of this polysaccharide is species- specific, engaging specific receptors on the sperm surface. The acrosome reaction occurs as the sperm penetrates the jelly coat and it is a biphasic event in sea urchin sperm. Firstly, the acrosome vesicle undergoes exocytosis by membrane fusion between the acrosomal membrane and the over-lying sperm plasma membrane. This results in the release of the acrosomal materials including the sperm lytic enzymes such as proteases which are necessary to digest the jelly coat substances to pave the way for the sperm entry into the egg.

Following this, the globular protein, actin, present in the subacrosomal fossa undergoes polymerization to give rise to the acrosomal or fertilization filaments. The intact inner acrosomal membrane ensheaths the extending acrosomal filament. Acrosomal filament formation is a characteristic feature of many marine invertebrate sperm, whereas they are absent in the mammals (Fig.5.5). The acrosome reaction in sea urchin also exposes a protein called bindin on the tip of the acrosomal filament and that is responsible for sperm adhesion to the vitelline envelop of the egg. The mammalian spermatozoa also undergo acrosome reaction, but with a difference. The acrosome in the hamster sperm envelopes the anterior half of the inner sperm nucleus giving a conical shape. During acrosome reaction, there is a fusion between the plasma membrane and the outer acrosomal membrane at different sites giving rise to vesiculation. This leads to the complete disruption of the acrosomal vesicle resulting in the release of the acrosomal materials. The sperm at the anterior tip is now naked with the anterior nuclear region enveloped by the intact inner acrosomal membrane. It is this part of the acrosome reacted anterior region of the spermatozoa which makes contact with the egg during fertilization.

SPERM EGG INTERACTION

A fundamental difference between the acrosome reaction of sea urchin and mammalian sperm is that in sea urchin, only acrosome reacted sperm attaches onto the vitelline envelop, whereas, in the mammals, acrosome reaction is initiated only after the sperm adhesion to the zona pellucida. But in both the cases, acrosome reaction is a prerequisite to sperm membrane fusion with egg plasma membrane. In the sea urchin, acrosome reaction releases a protein called bindin, which confers a species-specific recognition of the sperm with the egg. Bindin is a 30,500 Da protein, isolated from the acrosome of the sea urchin *Strongylocentrotus purpuratus*. It is capable of agglutinating its own dejellied eggs, but not those of other species. Sperm bindin has a glycoprotein receptor found on the outer surface of the vitelline envelope. Their interaction results in the sperm adhesion to egg surface.

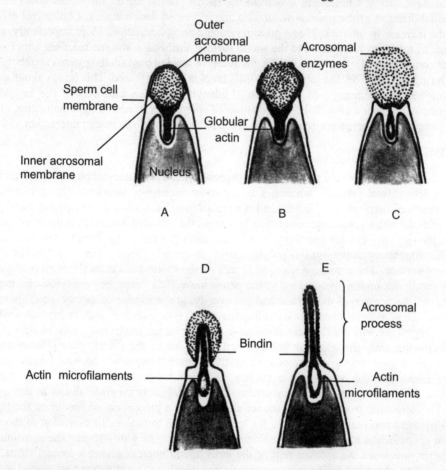

Fig. 5.5 Acrosome reaction of the sea urchin sperm A to C – Exocytosis of acrosomal granular material and D, E – Formation and extension of acrosomal process (Redrawn from Gilbert, 1997).

The mammalian spermatozoa have to penetrate through two egg barriers, namely the outer cumulus cell layer and the inner glycoproteinous zona pellucida. Hyaluronidase activity found on the sperm head is responsible for the dispersal of cumulus cells which are cemented by hyaluronic acid. After crossing the cumulus layer, the sperm makes its contact with the inner zona pellucida layer. A zona pellucida glycoprotein named ZP3 serves as the sperm receptor molecule. A sugar component, N-acetylglucosamine on the ZP3 interacts and binds to the sperm surface protein, namely, glycotransferase to effect interaction between sperm and egg. Sperm binding to ZP3 induces exocytosis in the form of acrosome reaction. In the mammals, the attachment of sperm with zona pellucida is not strictly species specific, as the fertilization is always internal. Carbohydrate-protein interactions appear to provide the molecular basis for species specific binding of sperm to the egg zona pellucida, a process homologous to that found in cellular adhesion. The sequence of sperm-egg interaction leading to fertilization in mammals is depicted in Figure 5.7.

$$
\begin{array}{ccccc}
Ca^{++} & Na^{++} & & Stage\ II & \\
Egg\ jelly\ + & \longrightarrow\ Stage\ 1 & \longrightarrow & Acid\ release & \longrightarrow\ Late\ changes \\
sperm & \downarrow & (Na+\ influx/H+\ efflux) & & K^{+}\ efflux \\
\downarrow & \downarrow & \downarrow & & \\
Exocytosis\ of & Actin\ polymerization & Ca^{++}\ uptake & & \\
acrosome & Formation\ of\ acrosomal\ process & & &
\end{array}
$$

Fig. 5.6 Ionic changes associated with the acrosome reaction in the sea urchin sperm

SPERM ENTRY INTO THE EGG

The site of sperm entry and sperm incorporation in many eggs is restricted to specific parts of the ovum. In fish and insects, the egg with a thick chorion is provided with an opening called micropyle. This acts as the portal for the sperm entry, as this directly leads to the soft plasmalemma region of the egg for easy entry and fusion. In the majority of the other cases, the sperm fuses with the animal pole region where normally the polar bodies are extruded. In mammals such as mouse and hamster, the sperm fuses in the region containing microvilli. However, in the sea urchins, fertilization can occur anywhere on the surface of the egg.

As mentioned in the preceding sections, the eggs are protected by several investment layers which have to be penetrated by the fertilizing sperm. In the sea urchin, the jelly coat layer dissipates easily in the seawater. Even when intact, the acrosomal region can easily penetrate through this layer both by the release of certain proteases and by mechanical force of the extending acrosomal filaments. Bindin molecules present on the sperm tip act as a fusigenic protein to initiate membrane fusion.

Once the acrosomal tip touches the vitelline envelope, the membrane fusion occurs between the egg plasma membrane and the acrosomal filament. This fusion results in the formation of a cytoplasmic bridge which widens to engulf the spermatozoan which is actually peeled off from the plasma membrane. The fusion between the sperm and the egg always occurs within the microvilli. The internalization of the sperm is a process comparable to engulfing of the prey by an amoeba. In the area of the sperm fusion with the egg, there is an elevation of egg surface to give rise to a 'fertilization cone'. This is made possible by the clustering and elongation of the egg membrane microvilli by the extension of their inner microtubules.

In the mammals, acrosome reaction releases a protease, called acrosin which digests the zona pellucida layer enabling the sperm to cross this barrier. Acrosomal reaction exposes another protein, fertilin which binds to an integrin-like receptor on the egg plasma membrane to initiate sperm-egg fusion. The acrosomal process, otherwise called percutor organ, attaches onto the egg in a tangential position to allow membrane fusion between the lateral side of the acrosomal process and the microvilli of the egg membrane for internalizing the sperm into the egg proper. This is in striking contrast to that found in sea urchin, where the acrosomal tip makes the contact with the egg membrane in a perpendicular manner, like an arrow piercing through any surface.

EGG ACTIVATION

The union of sperm and egg triggers an egg activation process to initiate the developmental program that leads to embryogenesis. Sperm-induced egg activation is accompanied by specialised events such as (1) a change in membrane potential, (2) a transient increase in intracellular calcium, (3) meiotic resumption of the egg nuclei and (4) fusion of the male and female pronuclei. A subtle difference between egg activation in sea urchin and mammals is that egg activation includes completion of meiotic division in mammals, whereas in the sea urchin, female pronucleus is already formed prior to fertilization.

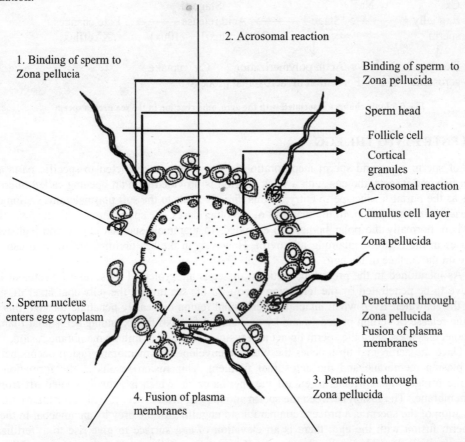

Fig. 5.7 Various stages in the fertilization of mammalian egg (Redrawn from Wolpert, 1998).

ELECTRICAL EVENTS AND PREVENTION OF POLYSPERMY

In the sea urchin, the first step in egg activation involves changes in the membrane potential caused by ionic exchange across the egg membrane. Before fertilization, the egg membrane is at a resting potential of about −70 mV. Shortly after sperm- egg contact, the sea urchin sperm induces an inward current, resulting in a transient depolarization of the egg membrane. This depolarization results in an action potential that takes the membrane potential to positive values such as +20 mV (Fig.5.8). This action potential is generated mainly by an influx from the extracellular medium of Na^+ and Ca^{2+} ions. As a result of this change in electrical potential of the egg plasma membrane a fast block to polyspermy is achieved. The entry of extra sperm is prevented by the positive potentials established during the depolarizing phase of the action potential. Therefore, in sea urchins at least, the membrane potential has to be at negative potentials in order to allow a sperm to initiate the fusion process. However, this transient block lasts only for a minute, after which a permanent block to polyspermy should be achieved by a process called cortical reaction. Jaffe and her co-workers found that polyspermy can be induced in eggs if the eggs are artificially supplied with electrical current that keeps their membrane potential negative. Conversely, fertilization can be prevented entirely by artificially keeping the membrane potential of eggs positive. Action potentials, otherwise known as fertilization potentials, are also seen at fertilization in other eggs such as those of starfish, frogs, annelids, tunicates and *Urechis*. However, in mammalian eggs there is no evidence for an action potential at fertilization.

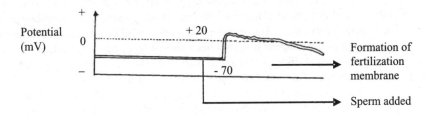

Fig. 5.8 Activation potential of a monospermic sea urchin, *Strongylocentrotus* (Redrawn from F.J. Longo, 1987).

INTRACELLULAR CALCIUM RELEASE AND EGG ACTIVATION

Changes in membrane potential could only bring about a fast block to polyspermy; but, the permanent block to polyspermy by way of cortical reaction as well as the actual egg metabolic activation will be achieved after the intracellular calcium release. Thus, the release of Ca^{2+} ions from the endoplasmic reticulum inside the egg is essential for the development of the egg into an embryo. The resulting rise in the free cytoplasmic Ca^{2+} concentration within the egg seems to be a general feature of egg activation at fertilization in a wide variety of animal species. The role of Ca^{2+} as an intracellular signalling element is even found in eggs from the plant kingdom. Egg activation involves a number of morphological and biochemical changes; the most obvious ones are those caused by exocytosis, such as the formation of fertilization envelope in the sea urchin. In addition, an important aspect of activation is the completion of meiotic stages, which was arrested at metaphase of the second meiotic division as found in the amphibian, *Xenopus* eggs. The rise in the intracellular free calcium is oscillatory or wave-like. The calcium wave starts at the point of sperm entry and thus, this region becomes the starting point of egg activation in general. An increase in the intracellular Ca^{2+} at fertilization was observed for the first time in the eggs of medaka fish. The proposal that Ca^{2+} is the most important signal for development in eggs is supported by the finding that artificially inducing a Ca^{2+} increase can trigger many of the early events of egg activation. Furthermore, the introduction of Ca^{2+} chelators into the egg cytoplasm, in order to prevent a sperm induced- rise in Ca^{2+}, abolishes all events associated with activation. Thus, the signal transduction at fertilization starts from sperm- egg interaction at plasma membrane to the opening of the Ca^{2+} release channels in the endoplamic reticulum.

SIGNALLING MOLECULES AND MECHANISMS OF CA^{2+} RELEASE

There are several models for the mechanism of Ca^{2+} release at fertilization following sperm-egg fusion. A widely held hypothesis for signal transduction at fertilization is that sperm acts as a honorary hormone and triggers internal Ca^{2+} release via interaction between receptors at the plasma membrane.

The sperm-egg adhesion at the level of plasma membrane generates inositol 1,4,5,-trisphosphate (IP$_3$), which then opens its specific Ca^{2+} channel on the endoplasmic reticulum. When an inhibitor of the IP$_3$ receptor, heparin, is used, Ca^{2+} oscillation at fertilization is arrested. Similarly, an antibody to the IP$_3$ receptor pore region also blocks Ca^{2+} oscillation in hamster and mouse eggs, implicating IP$_3$ and its receptors production in generating Ca^{2+} release from the endoplasmic reticulum at fertilization. Once released, the calcium ions can facilitate the release of more calcium ions by binding to calcium-sensitive receptors located in the cortical endoplasmic reticulum. The resulting wave of calcium release is propagated throughout the cell, starting at the point of sperm entry. The release of calcium from intracellular storage can be demonstrated by calcium-activated luminiscent dyes such as aequorin or fluorescent dyes like fura-2 which emit light when they bind free calcium ions. IP$_3$, as second messenger of fertilization to effect intracellular Ca^{2+} release is now known in a number of eggs. In mammalian eggs, sustained injection of IP$_3$ can trigger Ca^{2+} oscillations. Increased turnover of phospho-inositide-containing phospholipids has also been shown to occur at fertilization of frog and sea urchin eggs. In analogy with signalling pathways in many somatic cells, the sperm is considered to stimulate phosphatidylinositol-specific phospholipase C (PI-PLC) to generate IP$_3$.

Several models have been proposed to explain the sperm-activated IP$_3$ production during fertilization by way of activating the enzyme phospholipase C. In the first model, the sperm receptor (bindin receptor in sea urchin) at the egg plasma membrane has a protein tyrosine kinase activity in its cytoplasmic domine. The binding of bindin with its receptor brings about phosphorylation of this receptor tyrosine kinase which in turn activates PLC. In another model, the bindin receptor is linked to a protein tyrosine kinase in the cytoplasmic side. This tyrosine kinase is activated by receptor cross linking by sperm during fertilization.

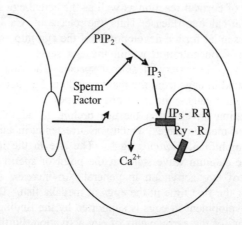

Fig. 5.9 Sperm factor hypothesis for signalling at fertilization in vertebrate eggs. The sperm fuses with the egg plasma membrane and introduces a soluble protein into the egg cytoplasm. The sperm factor is a protein complex that contains an essential phospholipase C (PLC) activity. This generates inositol 1,4,5-triphosphate (IP$_3$) from phosphatidylinositol biphosphate (PIP$_2$) within the egg. This leads to the generation of Ca^{2+} release via the IP$_3$ receptor (IP$_3$-R) within the endoplasmic reticulum, as in sea urchin eggs. The ER of mammalian eggs also contains ryanodine-sensitive Ca^{2+} release channels. Ry-R; Ryanodine receptor.
(Redrawn from: Swann, K. and Jones, K. T. 2002).

In yet another hypothesis, a sperm factor is introduced into the egg cytoplasm by way of sperm – egg fusion during fertilization. This sperm factor is a protein termed "oscillin" which is responsible for the generation of IP$_3$ (Fig.5.9).This hypothesis gains additional support from the fact that the intracytoplasmic injection of sperm in an *in vitro* fertilization technique results in the formation of male pronucleus and the subsequent zygote formation and normal embryo development (See section on *in vitro* fertilization under chapter, experimental embryology).

The release of calcium from the endoplasmic reticulum results in the rise of calcium ions in the cytoplasm of the eggs, thereby increasing the intracellular pH. The rise in the intracellular pH is also due to the second influx of sodium ions which causes a 1:1 exchange between sodium ions from the seawater and hydrogen ions from the egg. The sodium/ hydrogen exchange pump in the membrane is also activated by diacylglycerol (DAG), released from the membrane phospholipid phosphatidylinositol 4,5-bisphosphate by the action of phospholipase C. This loss of hydrogen ion causes the intracellular pH to rise. The increase in pH and the calcium ion elevation act together to stimulate egg metabolism, new protein synthesis and DNA synthesis (Fig.5.10). Another important consequence of calcium ion increase is the initiation of cortical reaction.

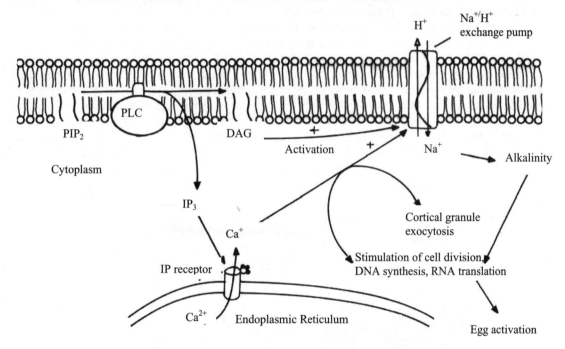

Fig.5.10 IP3 pathway in initiating calcium release from the endoplasmic reticulum and the initiation of egg activation. The combined action of DAG and Ca^{2+} activates the Na^+/H^+ exchange pump. The resulting alkalination of egg cytoplasm stimulates biochemical changes related to egg activation and zygote division. Ca^{2+} also induces cortical reaction bringing about permanent block to polyspermy. (Redrawn from Gilbert, 2000).

CORTICAL REACTION

Many marine invertebrate eggs contain distinctive cortical granules in the cortex of the egg. The cortical granules are manufactured by the Golgi complex and are closely apposed to the egg membrane. The cortical reaction is the second step in the egg activation process. This is brought about by the release of free calcium from its bound form or its fresh entry from the external medium. The importance of calcium ions in bringing about the cortical reaction has been experimentally proved by the calcium ionophore A23187, which transports the calcium ions across the membranes, by opening up the calcium channels.

The egg is surrounded by a vitelline membrane, which lies outside the plasma membrane. The membrane-bound cortical granules lie just beneath the egg plasma membrane. When the cortical reaction is initiated, the cortical vesicular membrane fuses with adjoining egg plasma membrane, leading to the release of its contents into the perivitellin space created as a result of the sperm entry (Fig. 5.11). The perivitelline space is the region between the outer vitelline envelope (zona pellucida in the case of mammals) and the plasma membrane proper. The cortical granule contents include mainly mucopolysaccharide substances which, after release into the perivitelline space, adhere to the overlying vitelline envelope to make it thicker and tougher. This vitelline envelope again undergoes a hardening by di- and tri- tyrosine linkages. Such linkages are promoted by an enzyme called peroxidase which is again released from the cortical vesicle exudates. As a result of the hardening of the vitelline envelope, t is transformed into a tough fertilization envelope.

Another product of the cortical granule of the sea urchin egg is a protein called hyaline. This product forms a layer between the fertilization membrane and the egg plasma membrane. Hyaline layers in the sea urchin embryos have important functions during cleavage, by holding up the dividing blastomeres in their respective positions. In mammals, a separate hyaline layer is absent but the zona pellucida undergoes hardening as a result of their interaction with the cortical granules. Such a transformation of the outer egg membranes (VE & ZP) creates a barrier for the entry of the supernumerary spermatozoa thereby preventing polyspermy. Yet another consequence of cortical reaction is the abolition of the sperm receptor molecules on the surface of the vitelline envelope aided by the release of an enzyme which knocks off the receptors to prevent further sperm binding

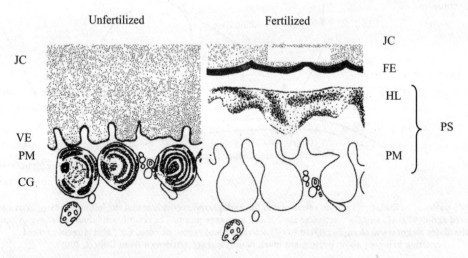

JC – Jelly coat ; VE – Vitelline envelope ; PM – Plasma membrane ; CG – Cortical granule ; FE – Fertilization envelope; HL – Hyaline layer ; PS – Perivitelline space

Fig. 5.11 Cortical reaction in the sea urchin (Redrawn from Hartman J.F. 1983)

PHYSIOLOGICAL POLYSPERMY

Although the prevention of polyspermy is an important prerequisite for normal zygote formation, in some cases polyspermy is also permitted. However, there is suppression of supernumerary male pronuclei formation. For example, in birds and reptiles, where the egg cytoplasm is segregated from the yolk mass, several spermatozoa enter into the cytoplasmic mass. However, only one sperm is fused with the nucleus of the egg. All other spermatozoa get disintegrated in the cytoplasm. This process is called physiological polyspermy, a condition which is supposed to increase the cytoplasmic volume for the cleavage to occur. The mechanism of a polyspermy block is well understood in urodele amphibians. In *Cynops*, as many as 20 sperm enter the egg at normal fertilization. Most of the sperm cells enter the egg at the animal hemisphere and equatorial regions. However, some of the spermatozoa enter through the vegetal hemisphere also. All incorporated sperm undergo nuclear decondensation and form sperm pronuclei with functional centrosomes. However, only one sperm pronucleus, nearest to the egg pronucleus fuses with the latter and forms the zygotic nucleus in the animal hemisphere.

The sperm pronucleus that fuses with female pronucleus is called principal sperm nucleus, whereas all the other sperm nuclei are called accessory sperm nuclei. When the zygote nucleus enters the prometaphase, its single centrosome divides and forms a bipolar spindle with a diploid set of condensed chromosomes. All the accessory sperm nuclei do not produce spindle and ultimately degenerate in the egg cytoplasm by the process of pycnosis. Thus, in the physiologically polyspermic eggs of urodeles, the block to polyspermy occurs in the egg cytoplasm. In all the monospermic eggs, the polyspermy is prevented at the egg membrane level by activational process, such as plasma membrane depolarization as well as the formation of fertilization envelopes by cortical reaction.

FUSION OF MALE AND FEMALE PRONUCLEI

The nucleus after completing the first and second meiotic divisions becomes haploid and is called the female pronucleus. After entry into the egg cytoplasm, the sperm nuclear envelope vesiculates exposing the compact sperm chromatin to the egg cytoplasm. The condensed sperm chromatin now undergoes decondensation by an exchange of DNA binding proteins with the egg proteins. In sea urchins decondensation appears to be initiated by the phosphorylation of two sperm specific histones (protamines) that bind tightly to DNA. Increased cAMP-dependent protein kinase activity phosphorylates several of the basic residues of the sperm specific histones and thereby interacting with their binding to DNA. This loosening is thought to facilitate the replacement of the sperm-specific histones by other histones stored in the egg cytoplasm. In the mammalian sperm, DNA bound protamines are tightly compacted through disulfide bonds. Once in the egg, glutathione reduces these disulfide bonds and allows the uncoiling of the sperm chromatin. Once decondensed, the DNA can begin transcription and replication. The decondensed chromatin now acquires a nuclear envelope to form the male pronucleus. The male pronucleus rotates 180°, so that the sperm centriole is positioned between the sperm pronucleus and the egg pronucleus. The microtubules of the centriole of the male pronucleus extend and make contact with the female pronucleus, thus facilitating the migration of the pronuclei towards each other. The membrane fusion between the two pronuclei results in the diploid zygote nucleus. DNA synthesis follows in order for the zygote to enter into cleavage divisions.

POST-FERTILIZATION METABOLIC ACTIVATION

The metabolic activation of the egg starts soon after the cortical reaction. In the sea urchin, there is an activation of NAD kinase which results in the increase of NADP and NADPH. The production of NADPH has important role as a co-enzyme for lipid biosynthesis. Synthesis of lipids is related to the construction of many new cell membranes required during cleavage. Concurrently, there is an increase in the oxygen consumption by the fertilized egg. In sea urchin eggs, this oxygen burst is related to the production of peroxide, which is the substrate for the enzyme ovoperoxidase. Ovoperoxidase joins tyrosyl residues in dityrosyl linkages which serve to cross-link polypeptide chains of the fertilization membrane (Fig.5.12).

An acid efflux (H^+ ion release) increases the internal pH which activates the amino acid transport and protein synthesis. This protein synthesis does not depend on new mRNA, but rather utilizes mRNAs already present in the oocyte cytoplasm. Such mRNAs encode proteins such as histones, tubulins, actins and morphogenetic factors that are utilized during early development. These changes in egg physiology brought about by increase in internal pH, are evidenced by the fact that such burst in protein synthesis could be induced by artificially raising the pH of the cytoplasm by ammonium salts.

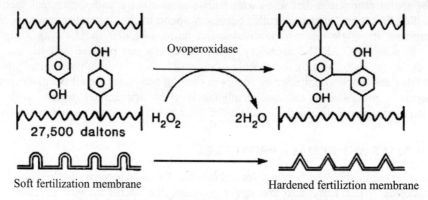

Fig. 5.12 Mechanism of hardening of fertilization membrane in the sea urchin egg

As a late stage metabolic activity, there is an initiation of DNA synthesis leading to mitosis that occurs during the first cleavage. The increase in calcium and pH levels is required for the initiation of DNA synthesis. The chronology of the chief fertilization events in the sea urchin eggs is given in the table no.5.1.

PARTHENOGENESIS

Two important contributions of the sperm that fertilize the egg are (i) egg activation and (ii) donation of haploid set of chromosome to restore diploidy in the zygote. These two events are essential for the initiation of normal embryogenesis. Many animal species often achieve these phenomena, without the participation of the sperm. Such a mechanism of initiating egg development without requiring spermatozoa is called as parthenogenesis. In the Amazon black molly, there are no males in the population and the eggs are activated by the spermatozoa of another fish species without their fusion with the egg. Similarly, restoration of the somatic chromosomes is achieved by modified meiotic mechanisms. In the parthenogenetic lizard, *Cnemidophorus uniparens*, the oocyte chromosome is doubled prior to meiosis, but the two meiotic divisions reduce the chromosome number in the egg back to the somatic level. The eggs go on to develop into normal embryos. Alternatively, in other animals, diploidy is reconstituted either by fusion of pairs of first cleavage nuclei or by suppression of the second polar body extrusion, followed by its fusion with the female pronucleus. Parthenogenesis once again underscores the importance of oogenesis in that it provides all developmental information for normal embryogenesis and animal development without requiring any obvious contributions from the male gamete.

Recently, mammalian eggs, such as of mouse, rabbit as well as humans, have also been found to activate on their own without sperm. Mouse and rabbit parthenotes (eggs developed by parthenogenesis) appear morphologically indistinguishable from fertilized eggs for the first few days of development, including the formation of the small group of pluripotent stem cells. However, these parthenotes are not viable and fail to develop into offspring. Hence, stem cells derived from these mammalian parthenotes including those of humans, can be used in stem cell therapy, without getting into moral and ethical controversies, as the embryos do not develop to offspring.

Table 5.1 Chronology of the chief fertilization events in the sea urchin eggs

Event	Approximate time postinsemination based on the data from sea urchin
Sperm egg binding	0 sec
Fertilization potential rise (fast block to polyspermy)	within 1 sec
Sperm-egg membrane fusion	within 6 sec
Calcium increase first detected	6 sec
Cortical vesicle exocytosis (slow block to polyspermy)	15-60 sec
Activation of NAD kinase	starts at 1 min
Increase in NADH and NADPH	starts at 1 min
Increase in O_2 consumption	starts at 1 min
Sperm entry	1-2 min
Acid efflux	1-5 min
Increase in pH (remains high)	1-5 min
Sperm chromatin decondensation	2-12 min
Sperm nucleus migration to egg centre	2-12 min
Egg nucleus migration to sperm nucleus	5-10 min
Activation of protein synthesis	starts at 5-10 min
Activation of amino acid transport	starts at 5-10 min
Initiation of DNA synthesis	20-40 min
Mitosis	60-80 min
First cleavage	85-90 min

Parthenogenetic reproduction often occurs by environmental influence. But in many animals, parthenogenesis is a part of the normal life cycle. For example, in aphids, a parthenogenetic phase alternates with sexual phase. Females that hatch out in the spring reproduce parthenogenetically throughout the summer, and at the end of the summer a single sexual generation develops. Parthenogenesis is prevalent in animal species such as flat worms, rotifers, arthropods, fish and lizards.

6

CLEAVAGE

Becoming multicellular is the first step in the development of any individual through sexual reproduction. The zygote formed by the fusion of male and female pronuclei acquires the genetic potential to undergo a series of quick mitotic divisions, termed as cleavage.

The cell cycle during cleavage is characteristically shortened by the absence of cell growth inbetween two mitotic divisions. In a typical somatic cell cycle, mitosis (M) is followed by an extended interphase which includes G1 (growth phase I), S (DNA synthesis) and G2 (growth phase II) phases. However, in the cell cycle of embryonic cleavage, only S and M stages are present and the two interphase stages namely, G1 and G2 are absent. Hence, the cytoplasmic volume does not increase during embryonic cleavage. On the other hand, the enormous volume of zygote cytoplasm is divided into an increasing number of smaller cells, thus bringing the ratio of cytoplasmic volume to nuclear volume down, as cleavage progresses. In *Xenopus laevis,* eggs, this decrease in the cytoplasmic to nuclear volume ratio is crucial in timing the activation of certain genes. In *Xenopus,* the G1 and G2 phases are inserted only in the blastula stage, when new transcripts of zygotic genes are needed. This stage is known as the midblastula transition. Thereafter, the cell cycles become longer and the divisions of the various cells become asynchronous.

MECHANISM OF CLEAVAGE

The control mechanism of cleavage division involves the activation of mitosis promoting factor (MPF) in the egg cytoplasm. MPF is also responsible for the resumption of meiotic cell divisions in the ovulated frog egg. MPF continues to play a role during cleavage regulating the biphasic cell cycle of early blastomeres. It undergoes cyclical changes in its level of activity in mitotic cells. The MPF activity of early blastomeres is highest during M phase and undetectable during S. The MPF contains two subunits such as a large cyclin B subunit and a small subunit called cyclin- dependent kinase. Cyclin B component of MPF shows a periodic behaviour, accumulating during S and then being degraded after the cells have reached M. Interestingly, cyclin B is encoded by mRNAs stored in the oocyte cytoplasm. Cyclin B regulates cyclin- dependent kinase which activates mitosis by phosphorylating several target proteins, including histones, the nuclear envelop lamin proteins, and the regulatory subunit of cytoplasmic myosin. This brings about chromatin condensation, nuclear envelop depolymerization, and the organisation of mitotic spindle. Thus cleavage divisions are regulated by genome apparently mediated by the egg cytoplasm. Evidently, the egg cytoplasm has been programmed by the genome prior to fertilization to divide at a certain rate.

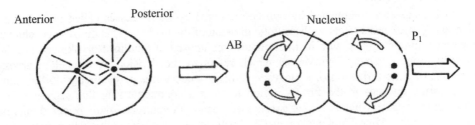

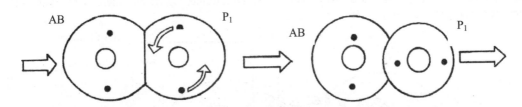

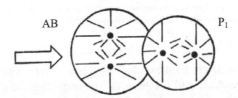

Fig. 6.1. Centrosomal influence on the plane of cleavage. (Redrawn from Lewis Wolpert, 1998)

The plane of cleavage in animal eggs is determined by the orientation of the mitotic spindle, which in turn is due to the action of the asters (Fig. 6.1). The asters are microtubules that are radiating out from a centrosome, which acts as an organizing centre for microtubule growth. Before mitosis, the centrosome becomes duplicated and the two daughter centrosomes move to the opposite sides of the nucleus and form asters. The orientation of the asters determines the plane of cleavages. The plane of cleavage is also important in distributing the cytoplasmic determinants into each dividing blastomere. In the nematode eggs, the cytoplasmic determinants are asymmetrically distributed and hence unequal divisions produce cells with differently specified fates. The first cleavage of the nematode zygote results in two daughter cells, namely, an anterior AB and a posterior P1 cell. At the next division, the centrosome of the two cells moves in different directions, as shown in Fig.6.1. In the AB cell, the duplicated centrosome move in such a way that the next cleavage is at right angles to the first cleavage. In the P1 cell, the nucleus and duplicated centrosomes rotate so that the cleavage of P1 is in the same plane as the first cleavage.

Cleavage is accomplished by two co-ordinated cyclic processes called as karyokinesis and cytokinesis. Karyokinesis refers to the mitotic division of the nucleus. This division is achieved by the formation of mitotic spindle with its microtubules, composed of tubulin. Karyokinesis is followed by the second process namely, cytokinesis which refers to the division of the cell. Karyokinesis is accomplished by the formation of a cleavage furrow, which forms perpendicular to the spindle axis. The egg cortex thickens in the furrow region to form a contractile ring that separates the cell by constricting the zygote. Contracile ring consists of a band of circumferentially oriented microfilaments of 30 - 70 Å length. Microfilaments are made of actin which interacts with another contractile protein in the cortex namely, myosin to bring about the constriction of the furrow. During cell division, the mitotic spindle and contractile ring are perpendicular to each other. The mitotic spindle occurs in the central cytoplasm, whereas, the contractile ring is formed in the cortical cytoplasm. The contractile ring exists only during cleavage. This splits the zygote into blastomeres, just like an intercellular purse string, tightening about the egg as cleavage continues. This tightening of the microfilamentous ring creates the cleavage furrow (Fig. 6.2).

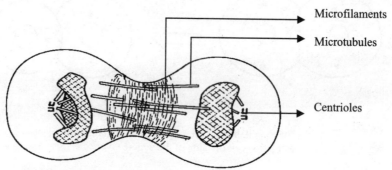

Fig. 6.2 Diagram showing the mechanism of cleavage furrow formation and the role of microtubules and microfilaments (Redrawn from Bodemer, 2000)

PATTERNS OF EMBRYONIC CLEAVAGE

The pattern of cleavage in a species is determined by two major parameters. They are (1) the amount and distribution of yolk within the cytoplasm and (2) those factors in the egg cytoplasm influencing the angle of the mitotic spindle and the timing of its formation. In general, the rate of mitotic multiplication is faster in eggs containing less quantity of yolk than in those with high quantity of yolk. The high concentration of yolk inhibits cleavage. The embryonic development is clearly longer in the heavily yolked eggs than in the eggs with very little yolk, which complete the development fast and release the young ones as a larval form which has to grow, develop and metamorphose into adult. Depending on the lecithality of the eggs, the embryonic cleavage can be broadly categorised into two patterns, namely, holoblastic and meroblastic. Holoblastic means 'complete cleavage'; whereas; meroblastic refers to 'incomplete cleavage'. Holoblastic cleavage occurs mostly in eggs with low yolk content; whereas, meroblastic cleavage occurs in heavily yolk laden eggs. Holoblastic cleavage is further divided into radial, spiral, bilateral and rotational types of cleavages.

RADIAL HOLOBLASTIC CLEAVAGE

Radial cleavage is characteristic of echinoderms and the protochordate, *Amphioxus*. In the sea cucumber, *Synapta digita*, the first cleavage furrow passes directly through the animal and vegetal poles, creating two equal sized daughter cells. This cleavage is said to be meridional because it passes through the two poles, like a meridian on globe. The mitotic spindle of the second cleavage is at right angles to the first, and the cleavage furrow appears simultaneously in both the blastomeres and also passes through the two poles. Thus, the first two divisions are both meridional and perpendicular to each other. The third division is equatorial. The mitotic spindle of each blastomere is parallel to the animal-vegetal axis and the resulting cleavage furrow divides the embryo into eight equal blastomeres. Thereafter, the divisions are meridional alternating with equatorial, as a result of which the successive divisions produce embryos consisting of blastomeres arranged in horizontal rows along the central cavity which is referred to as blastocoel (Fig.6.3).

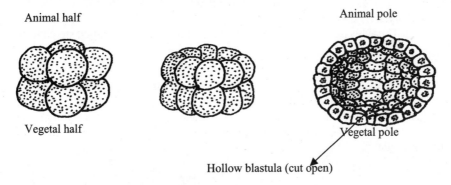

Fig. 6.3 Radial holoblastic cleavage in the sea cucumber *Synapta digita* to show the blastocoel
(Modified from Gilbert, 1997)

The embryo at this stage is called as blastula. The characteristic feature of radial cleavage is that any time during cleavage, an embryo bisected through any meridional plane will produce two embryos, which are mirror images of each other. Since this type of cleavage is characteristic of a sphere or a cylinder, it is called radial symmetry. In other echinoderm species, the deviation from this regular pattern appears after the third division. Such a deviation results in three types of blastomeres namely micromeres, mesomeres and macromeres, as it occurs in the sea urchin, at the 16 cell stage.

SPIRAL CLEAVAGE

The spiral cleavage pattern is characterized by an oblique orientation of the spindles with respect to the embryonic animal-vegetal axis. This type of cleavage is found in annelids, flatworms, nemertenes and the non-cephalopod molluscs. It differs from radial cleavage in many ways. The eggs do not divide in parallel or perpendicular orientations to the animal-vegetal axis of the egg. The cleavage is at oblique angle, forming the spiral arrangement of daughter blastomeres. The first division is typical in that, it passes through the animal-vegetal pole axis although it is slightly inclined. The spindles at the second division are perpendicular to the first, but they are not perpendicular to the main axis. The spindles are inclined opposite to each other. This causes one daughter blastomere from each cleavage pair to lie obliquely above the other pair, while the cells opposite to one another are at the same level. The two higher cells are traditionally designated as A and C and the two lower cells are designated, B and D. The identities of these cells are easy to ascertain, since the polar bodies are located at the site where A and C are in contact. The third cleavage is equatorial (i.e. horizontal), but the spindles are again oblique rather than parallel to the main axis. Consequently, the cleavage planes are shifted, causing the cells of the animal pole to lie in the spaces between the lower cells. In most cases of spiral cleavage, the spindles of the third cleavage are shifted to the right, when the embryo is observed from the animal pole. This causes the clockwise displacement of the upper tier of cells and is known as dextral cleavage (Fig. 6.4).

Left handed or sinistral cleavage is rare. The third cleavage is frequently unequal, so that the cells of the upper tier (micromeres) are smaller than those of the lower tier (macromeres). Thereafter, the micromeres divide equally, while the macromeres continue to divide unequally producing additional quartets of micromeres which are arranged in tiers stacked below the pre-existing ones. The rigid pattern of spiral cleavage produces a highly ordered embryo with each cell occupying a specific position. In the spiralian cleavage, the fate of the individual blastomeres is found to be identical among different spiralian phyla. The blastulas so produced have only a few numbers of blastomeres at the time of gastrulation. These blastulas generally lack blastocoel and hence are called stereoblastula.

BILATERAL HOLOBLASTIC CLEAVAGE

Bilateral holoblastic cleavage is found in ascidians. The most striking phenomenon in this type of cleavage is that the first cleavage plane establishes the only plane of symmetry in the embryo separating the embryo into its future right and left sides. Each successive division orients itself to this plane and the half embryo formed on one side of the first cleavage is the mirror image of the half embryo on the other side.

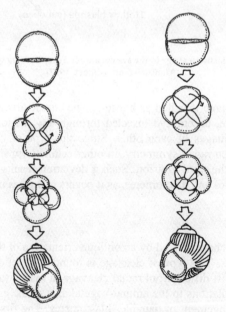

Fig. 6.4 Spiral cleavage in the snails, showing dextral and sinistral cleavage pattern. Arrow indicates the direction of shifting of blastomeres in the upper tier (Modified from Wolpert, 1998)

The second cleavage plane is perpendicular to the first, but is displaced towards the posterior side of the embryo. The resultant four-cell stage embryo has two unlike cells on either side of the median plane, but each cell is mirrored by an identical cell on the side opposite. The embryo has now recognisable right and left side as well as an anterior and posterior region. As in spirally cleaving eggs, the pattern of bilateral cleavage is important in embryonic organization. This is evidenced even in the first cleavage, which parcels the presumptive right and left halves of the embryo into daughter blastomeres.

ROTATIONAL HOLOBLASTIC CLEAVAGE

The mammalian eggs contain very little yolk and hence undergo complete cleavage. But the pattern of cleavage and the resulting blastula-like embryo is totally different from all other embryos with holoblastic cleavage. This is because of the fact that the mammalian embryonic development occurs within the female reproductive tract and the entire embryonic nutrition is supplied by the maternal tissue through the placental connections. Hence, the cells derived from the cleavage of mammalian embryos should provide facility for embryonic development in the new environment of uterine endometrium.

The mammalian cleavage differs from other cleavage types in (1) the relative slowness of the divisions, and (2) the unique orientation of the blastomeres with one another. The first cleavage is a normal meridional division. However, in the second cleavage one of the two blastomeres divides meridionally and the other divides equatorially. Hence, this type of cleavage is called rotational cleavage (Fig.6.5). Unlike the other types, mammalian blastomeres do not divide all at the same time. Again, the mammalian genome is activated during early cleavage and the genome produces the proteins necessary for cleavage to occur.

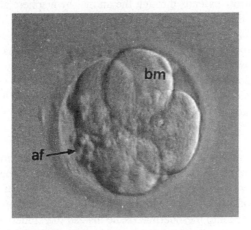

bm – blastomere ; af – anucleated fragments

Fig. 6.5. 4 cell stage of human embryo cleavage (Photo : Courtesy Dr. Balaji Prasath, K.K. Women's and children's Hospital, Singapore

Up to eight cell stage, the mammalian blastomeres are loosely arranged with plenty of space in-between them. However, at the eight cell stage, the blastomeres undergo a spectacular change in their behaviour. The cells suddenly huddle together maximising their contact with other blastomeres and form a compact ball of cells. This process is called 'compacting', again a unique feature of mammalian cleavage. Compaction is mediated by cell surface interactions between the adjacent blastomeres. This results in membrane polarization of the interacting blastomeres. Specific cell surface proteins such as E-cadherin are also involved in such cellular interaction. E- cadherin, an adhesive glycoprotein becomes restricted to these sites on cell membranes that are in contact with adjacent blastomeres. Activation of protein kinase C initiates compaction by localising E- cadherin in between the interacting blastomeres. Compaction is also mediated through the microvilli by their contractile action of actin microfilaments.

The cells of the compacted embryo divide to produce a sixteen cell morula which consists of a small group of internal cells surrounded by a large group of external cells. The descendants of the external cells become the trophoblast (or trophectoderm cells). These cells do not produce any embryonic structures but produce the chorion, the embryonic portion of the placenta. Chorion plays an important role for the growing embryo in getting the nutrients from the mother, secreting hormones to stabilise the implantation and providing immunity for the embryo from rejection by the mother.

The inner cells of the sixteen cell stage embryo generate the inner cell mass that give rise to the embryo as well as the extraembryonic structures such as the amnion and yolk sac. Initially, the morula does not have an internal cavity. However the trophoblast cells secrete fluid into the morula to create a cavity by a process of 'cavitation'. The inner cell mass is positioned on one side of the ring of the trophoblast cells. This structure is called blastocyst.

AMPHIBIANS

Cleavage in amphibian eggs is holoblastic although they contain considerable yolk accumulated in the vegetal half (telolecithal). However the yolk does exert an impediment to cleavage. Thus the first division begins at the animal pole and slowly extends down into the vegetal half. Before the first cleavage division is completed the second division is initiated from the animal pole. In the axolotl salamander, the cleavage furrow extends through the animal hemisphere at a rate of 1mm per minute. The cleavage furrow bisects the grey crescent and slows down to 0.02 - 0.303 mm/min as it approaches the vegetal pole. The third cleavage is equatorial, as in other holoblastic cleavages but because of the vegetally placed yolk and location of the nuclei in the four blastomeres (displaced towards animal pole), the cleavage furrow is much closer to the animal pole. This results in the formation of four small micromeres in the animal blastomeres and four large blastomeres (macromeres) in the vegetal region. The successive divisions are faster in the micromeres and slower in the macromeres due to the impact of yolk on cytokinesis. After passing through the morula stage (at 16 to 64 cell stage), the embryo attains the blastula stage at 128 cell stage. The blastocoel is also displaced towards the animal pole accordingly (Fig.6.6).

In the amphibian embryos, the blastocoel permits cell migration during gastrulation. The blastocoel appears to prevent the contact of the vegetal cells destined to become endoderm with those cells that will in future give rise to ectoderm. If cells from the roof of the blastocoel (animal hemisphere) are removed and placed next to the yolky vegetal cells from the base of the blastocoel, these animal cells differentiated into mesodermal tissue instead of ectoderm in the newt embryos.

During cleavage, cell adhesion molecules such as EP-cadherin keep the blastomeres together. The mRNA for this protein is already present in the egg cytoplasm. If this mRNA is destroyed by injecting anti sense oligonucleotides complementary to this mRNA, the adhesion between the blastomeres is dramatically reduced.In all the above holoblastic examples of embryonic cleavages, the multiplied blastomeres first form a morula, representing a solid cluster of cells. Very soon a fluid-filled central cavity namely blastocoel is formed surrounded by layers of blastomeres. This epithelial layer of blastomeres is called blastoderm and the embryo at this stage is termed as blastula.

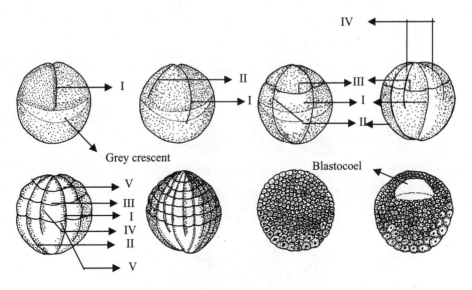

Fig. 6.6 Cleavage of a frog egg ; the cleavage furrows are numbered in the order of appearance (Redrawn from Gilbert, 1997)

MEROBLASTIC CLEAVAGE

The prohibitory effect of heavy lecithality on cleavage division is highlighted in the egg of bony fish, reptiles and birds. The concentration of yolk in these eggs is so heavy that the highly complexed yolk is segregated from the egg cytoplasm to occupy the major portion of the egg. Consequently, the egg cytoplasm is restricted to a thin peripheral layer at the animal pole. This cytoplasmic area is called blastodisc. Interestingly, cleavage in these eggs is restricted to the blastodisc cytoplasm only, as the yolk is impenetrable for cytokinesis to occur. By virtue of this restricted cell division, this cleavage pattern is also called discoidal meroblastic cleavage.

The first cleavage is initiated near the blastodisc, as a short slit in the surface. The second cleavage furrow follows again as a slit at right angles to the first one. The four blastomeres remain continuous with the yolk both below and at their outer margins. At the periphery, cleavage furrows radiate for a short distance in all directions. These blastomeres retain their continuity both with the underlying yolk and the peripheral cytoplasm. After the formation of the single layered blastoderm, equatorial cleavages divide this layer into a tissue, three to four cell layers thick. Between the blastoderm and the yolk is a space called subgerminal cavity. This space is created when the blastoderm cells absorb fluid from the albumin and secrete it between themselves and the yolk. At this stage two distinctive regions in the blastodisc can be identified, the area pellucida, composed of the cells above the subgerminal cavity and the area opaca consisting of the darker cells at the margin of the blastodisc and yolk. The marginal cells eventually lose the cell membrane and become syncytial, and are called the periblast, which breaks down the yolk to feed the growing embryo.

The blastoderm after acquiring adequate numbers of cells by the discoidal cleavage, delaminate or shed some of their blastomeres into the subgerminal cavity to form a second layer of cells. The posterior marginal cells of the area pellucida also migrate forwards to join the cells already shed from the blastoderm above. This cellular layer is now called as the hypoblast and the upper intact germinal layer is termed epiblast. The cavity created between the epiblast and hypoblast cells corresponds to the blastocoel of other holoblastic eggs.

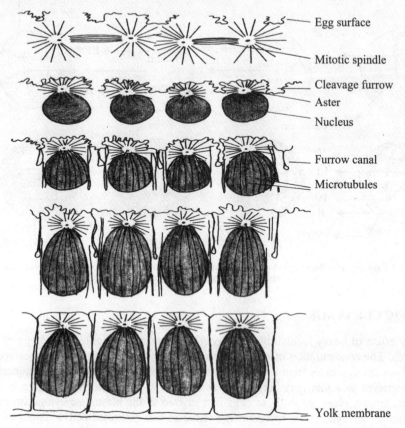

Fig. 6.7 Superficial cleavage in the centrolecithal eggs of insects (Redrawn from Gilbert, 2000).

SUPERFICIAL CLEAVAGE

Another example of meroblastic cleavage, where the enormity of yolk has significantly affected the cleavage pattern, is found in insects. The insect eggs are endowed with centrally placed yolk granules (centrolecithal) surrounded by a thin layer of cytoplasm termed as periplasm. The nucleus is however embedded in an island of cytoplasm found in the centre of yolk granules. The periplasm is formed as a result of reorganization of the egg cytoplasm, occurring after fertilization.

The cellularization of the syncytial blastoderm is a dramatic event in insects. In *Drosophila*, after 12th or 13th nuclear divisions, the nucleus begins to divide and elongate in a plane perpendicular to the plasmalemma. Bundles of microtubules surrounding the nuclei help in the perpendicularly oriented nuclear elongation. As the nuclei elongate, cleavage furrows form between them pushing in from the surface and segregating the nuclei into distinct cells. However, the cell membranes at the base of the cleavage furrows do not fuse below the nuclei; instead the cells remain connected with the inner yolk sac by a cytoplasmic stalk. At the end of cleavage, embryo consists of a single layer of columnar epithelial cells viz., the cellular blastoderm, a syncytial core of yolk surrounded by a yolk membrane and a cluster of pole cells at the posterior end.

DETERMINATE AND REGULATORY EMBRYOS

Animal development could be divided into two distinct types, regulative and mosaic. In the mosaic development the pattern of cleavage segregates localised morphogenetic determinants into specific blastomeres. Cleavage in mosaic embryo is also said to be determinate, since the intervention of cleavage furrows determines the fate of different regions of the zygote. In other words, each part of the embryo retains its determination or individuality. Mosaic developmental patterns are found mainly in bilaterally and spirally cleaving embryos (e.g. Ascidians, annelids and molluscs).

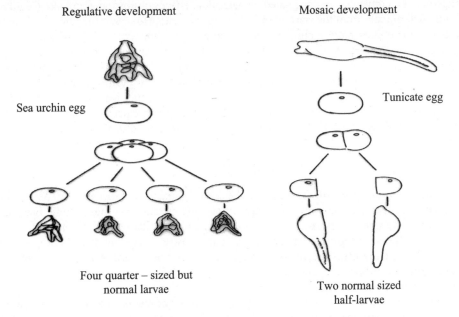

Fig. 6.8 A comparison of mosaic and regulatory development by embryo splitting method at 2 cell and 4 cell stage (Redrawn from Davenport, 1979)

In tunicate embryos, blastomeres separated at the two cell stage produced exact half larvae either a right half or left half (Fig.6.8). Similarly, defects produced in older tunicate embryos indicated that if a single blastomere is removed the larvae will completely lack the organ or the part that the excised blastomere would normally produce. Thus, such embryos have no capacity to regulate the loss of parts.

In tunicates, the mosaic nature of the embryo arises at fertilization. Eggs cut into fragments prior to fertilization can regulate to form normal but smaller larvae, when fertilised, if they are of sufficient size. However, after fertilization, similar fragments produced partial larvae. Evidently, a radical shift in the symmetry of cytoplasmic heterogeneity occurs at fertilization. The fixation of cleavage pattern is also a contributory factor for mosaicism. On the contrary, cleavage in echinoderm and amphibian embryos is described as being indeterminate, since these embryos can be subjected to experimental abuse and still regulates to form normal organisms. Such a developmental pattern is also called regulatory.

A German scientist Han Driesch conducted a series of 'defect' studies as well as 'isolation' studies in sea urchin embryos to prove the existence of regulative developments. In this he removed certain blastomeres but still he could see that remaining blastomeres would develop into a full embryo. Similarly, when blastomeres were separated from 2-, 4-, and 8 celled embryos, each blastomere developed into a complete larva. A comparison of mosaic and regulative development as seen in the sea urchin egg and tunicate egg is given in Figure 6.8.

CELL LINEAGE

Although eggs may be conveniently designated as determinate (mosaic) or indeterminate (regulative), no egg is rigidly determinate from the start or entirely indeterminate. However, the remarkable patterns seen in the determinate eggs have facilitated the so called 'cells lineage' studies, which consists of painstaking tracing of the sequence in each developmental type until cells become too small and numerous to follow any further. Alternatively, cell lineage studies were combined with extirpation or defect experiments in which one or more blastomeres were picked off or destroyed in order to see what might consequently be missing from the embryo. Similarly, in the isolation experiments, the development of isolated blastomeres was followed to determine their respective potentials. Vital dyes like Janus green, neutral red and Nile blue sulphate have also been used to follow the developmental destiny of each blastomere. More details in this regard are given under the section on fate map.

NUCLEAR TRANSPLANTATION EXPERIMENTS

One of the major consequences of cleavage is the establishment of intrinsic differences in the cytoplasm of the blastomeres by segregating unique ooplasmic constituents into specific blastomeres. On the other hand, the genome is distributed equally among the daughter cells at each cleavage division. Hence, cellular determination occurring either very early in the cleavage (mosaic development) or late in the cleavage phase (regulative development) arises from the influence of cytoplasmic constituents on specific gene expression.

The importance of nucleus in initiating and accomplishing the cleavage divisions has been realised for a long time. Spemann conducted some ingenious experiments to adduce evidence on this fact. Using a fine human hair he constricted the just fertilized amphibian egg so that it was divided into two portions of equal size, which were connected by a very narrow cytoplasmic bridge. One side of the constriction contained the zygote nucleus and the other was non-nucleated. He found that the cleavage was initiated only in the nucleated portion and continued at a normal rate. Meanwhile, the nonnucleated portion remained uncleaved. However, when the blastomeres became very small, cell divisions near the constriction allowed one of the daughter nuclei to pass through the narrow space into the non-nucleated half. This initiated cleavage in the other half ultimately leading to the formation of twin larvae connected centrally by a cytoplasmic bridge. This experiment indicates not only the importance of the nucleus for cleavage to take place, but also points to the absence of restriction of nuclear potency in forming the normal larvae even after participation of the nucleus in several cleavage divisions.

In some exceptional cases, such as in the case of the embryos of *Ascaris* and some insects, all the cells except the potential germ cells eliminate a large number of chromosomes during the progression of their cleavage. However, in all the other animal forms, the genotype of the cell is unaffected by the differentiation. This has been elegantly revealed by the nuclear transplantation experiments which showed that a complete set of genes capable of directing the development of a whole organism is retained by differentiated cells. Briggs and King, first activated the frogs eggs by pricking with a glass needle, when the egg's nucleus approached the surface of the egg (Fig.6.9). They then removed the nucleus by again pricking the egg in the spot where the peripheral nucleus was located. Next, a cell of an advance embryo was separated by sucking into the tube of an injection pipette. Due to the smaller diameter of the pipette the injected cell wall was damaged and the nucleus got separated. This nucleus was then separated and injected into the enucleated egg. Up to 80% of the injected embryos started cleaving and developing suggesting that the nucleus of an advanced embryo did not diminish in its potentiality to develop into an adult. Similarly, nuclei from neural plate as well as the ciliary cells of the alimentary tract in the tadpoles also gave rise to tadpoles which went through metamorphosis.

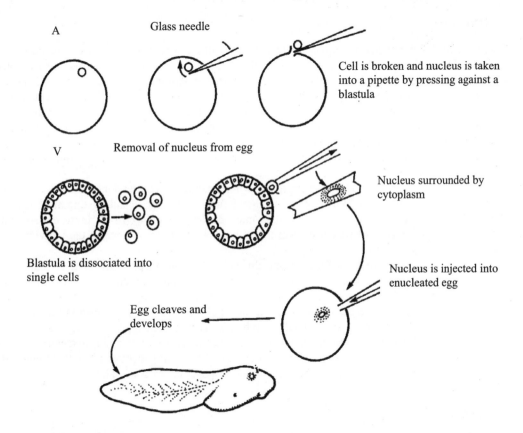

Fig. 6.9 Experimental procedures for nuclear transplantation in the frog em`bryo (Adapted from Davenport, 1979)

These experiments again demonstrated that the nucleus of even well differentiated cells provide the necessary genetic information for the differentiation into an adult frog. However, when the nucleus of an adult frog was transplanted into an enucleated egg they could not support development. Nevertheless, if the adult cells are cultured *in vitro,* they lose certain properties of the differentiated cell and start dividing mitotically. If the nucleus from the defective embryo was injected into the eggs, the second generation developed into much better froglets, after completing metamorphosis. These nuclear transplantation experiments of Gurdon and associates clearly demonstrate the fact that although the nuclei of differentiated cells possess all potentialities to develop into adults, their chromosomes were found to have defects such as deletions and translocations. These defects are due to the inability of the chromosomes taken from the adult cells to adapt themselves to the rapid reproductive rhythm of early development. The nuclei of the newly fertilized eggs, although small in proportion to the egg cytoplasm, are much larger than the nuclei at later stages. Accordingly, transplanted nuclei increase upto 30 fold in volume during the first 40 minutes after transplantation. They may also revert their synthetic activity to those of embryonic condition. Thus, a transplanted nucleus reverts in every respect to the condition in which nuclei would be after normal fertilization. In conclusion, the above experiments demonstrate that the nuclei well advanced in development possess the full potentialities for producing all parts of the adult body.

CLONING

Retention of developmental potentialities in the nuclei of adult animal cells was revealed recently in the animal cloning experiments. In 1997, Ian Wilmut and his colleagues at a Scottish research institute reported the first cloning of a sheep, called Dolly. In this experiment, an enucleated egg from a sheep of one breed was fused with a mammary gland cell from a female of another breed. The activated egg developed into a healthy lamb. Since all the genes in the newborn lamb were derived from the transplanted nucleus, it possessed all phenotypical characteristics of the nuclear donor sheep. It is evident from these experiments that nuclei of specialized cells of the udder contain the genetic information required for differentiation into many different types of cells. The cloning procedure of Dolly is depicted in Fig.6.10.

An alternative method of nuclear transfer procedure involves the donor nucleus obtained from somatic cell of the individual to be cloned. As shown in Fig.6.11, successful cloning has been done for the mouse with the transfer of donor nucleus into enucleated oocyte. In all these experiments, the reconstituted egg generates a new organism that is a genetic copy or a clone of the individual nuclear donor.

The restoration of totipotency in the differentiated somatic nucleus, after its transplantation into the oocyte is by a process of dedifferentiation. When transplanted into an oocyte, a somatic nucleus may respond to the cytoplasmic factors and be reprogrammed back to totipotency. These cytoplasmic factors must be capable of erasing the 'molecular memory' (imprints) that give somatic cells their characteristic properties. It would also be necessary for the reprogrammed nucleus to switch off specific genes that are expressed by the somatic nucleus and initiate embryo-specific genes at the two -cell stage in the mouse embryo. The reprogrammed genome generates pluripotent epiblast cells, and undergoes rapid transdifferentiation to generate trophectoderm cells. The epiblast cells then give rise to the embryo proper.

TRANSGENESIS

Transgenesis could better be referred to as genetically modified animals and plants. This means putting a gene into an animal so that it and its offspring are permanently altered. The technique involved the cloning of a known gene of interest and infects into a single celled embryo (say mouse) and introduce carefully inside one of the cell's two nuclei. The result is a 'transgenic' mouse with the gene incorporated in a random position on one of its chromosomes. The inserted gene need not be derived from a mouse, but it could be from a person.

For instance, a mouse that is abnormally susceptible to cancer can be made normal again by the introduction of a human chromosome 18, containing a tumour-supressor gene. But rather than inserting whole chromosomes, it is more usual to add a single gene.

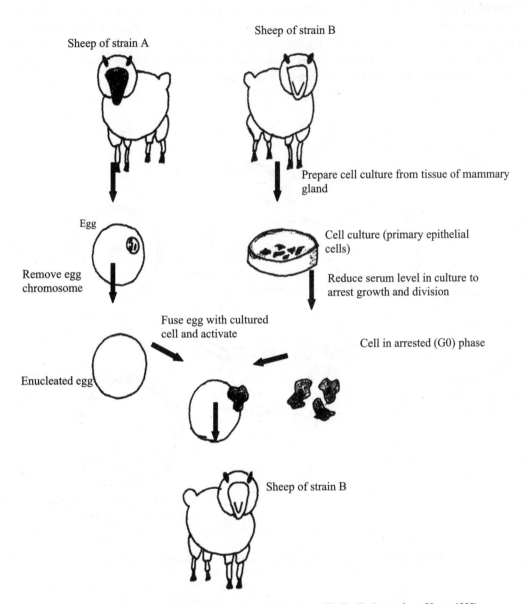

Fig. 6.10 Experimental procedures adopted in cloning of Dolly (Redrawn from Karp, 1999)

Microinjection is giving way to a more advanced technique which enables the gene to be inserted in a precise location. On the 2nd day of embryogenesis, the mouse embryo contains cells known as embryonic stem cells or ES cells. If one of these is extracted and injected with a gene, the cell will splice that gene at precisely the point where the gene belongs, replacing the existing version of the gene. Mario Capecchi, the discoverer of this technique, took a cloned mouse cell by briefly opening the cell's pores in an electric field, and then observed as the new gene found the faulty gene and replaced it. This procedure is called homologous recombination. The technique exploits the fact that the mechanism that repairs broken DNA often uses the spare gene on the counterpart chromosome as a template. It mistakes the new gene for the template and corrects its existing gene accordingly. This genetically altered ES cell can then be placed back inside an embryo and grown into a 'chimeric' mouse-a mouse in which some of the cells contain the new gene.

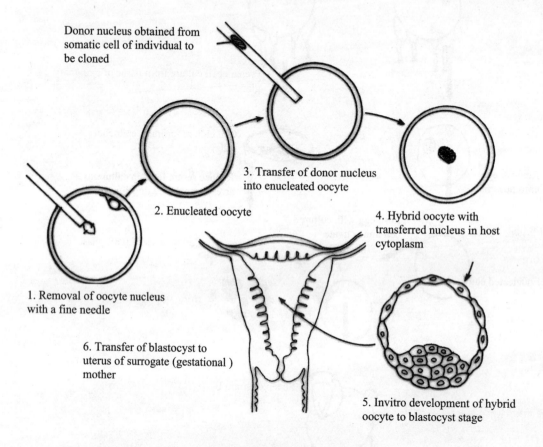

Fig. 6.11 Schematic illustration of the somatic nuclear transfer protocol developed for mouse (Redrawn form Romon Pinon, 2002).

7

GASTRULATION

In the development of the embryo, cell differentiation is the key event. Morphogenesis, a process of cellular and tissue reorganization is the precursor to cell differentiation. Cellular rearrangement is, therefore, a necessary prerequisite before organogenesis. The events that bring such cellular reorganization is termed as gastrulation. The term gastrulation literally means the formation of a primitive archenteron. However, the formation of three germ layers by the rearrangement of the blastodermal cells is the central event of gastrulation. The germ layers, which are concentrically arranged in the gastrula, include: 1. external ectoderm, 2. the interstitial mesoderm, and 3. the internal endoderm. This orderly arrangement of the three layers is a prerequisite to cellular interaction leading to organogenesis.

The importance of gastrulation forming the central event of animal development is highlighted in the statement of Lewis Wolpert," It is not birth, marriage, or death, but gastrulation, which is truly the most important time of your life". The tissue produced by gastrulation forms the basis for all adult tissues except germ cells. The ectoderm gives rise to the epidermis and nervous system, the endoderm contributes mainly to the digestive system and its related organs, and the mesoderm develops into the variety of tissues that fill the space between the epidermis and digestive tract.

The animal half of the blastula consists of prospective ectoderm that will form the outer layer by virtue of its spreading ability. The cells of the vegetal half tend to move inside the ectoderm to form the internal structures such as the gut, coelom and skeleton.

MOLECULAR MECHANISMS DETERMINING GERM LAYERS FORMATION

In the African clawed frog, *Xenopus laevis*, there exist two molecular mechanisms responsible for the induction of germ layers: the maternally supplied intracellular factors that promote mesodermal and endodermal determination (such as the gene-regulatory factor VegT), and cellular interactions involving extracellular inductive signals (maternal and embryonic), such as members of the transforming growth factors (TGF-) protein family. Both mechanisms activate mesodermal and endodermal development. Nodal and bone morphogenetic protein 4 (BMP-4) are the two important members of the TGF- family in determining mesodermal differentiation. Recent discovery of a region-specific maternal factor, namely Ectodermin is responsible for the determination of ectoderm in the animal pole.

FACTORS AFFECTING GASTRULATION

The extent of cellular movements during gastrulation depends on the number of cells that the blastula possesses. For example, the annelid gastrula is composed of only 30 cells and hence, the gastrulation is necessarily a simple process. Conversely, the vertebrate blastula, such as that of the frog consists of numerous cells necessitating a very large-scale cellular rearrangement. The cellular movement during gastrulation is also affected by the amount of yolk contained in the egg. These blastomeres especially in the vegetal half of the telolecithal eggs (e.g. frog) tend to move very slowly compared to the animal half cells which are free from the yolk materials exhibiting rapid movement.

CELLULAR AND MOLECULAR MECHANICS OF GASTRULATION MOVEMENTS

The mechanisms utilised during the cellular reorganization are changes in cell adhesiveness, cell motility and cell shape. Cell adhesiveness between the cells and the extracellular matrix occurs through interaction involving cell surface proteins. Changes in these proteins can determine both strength of cell adhesion and its specificity. The most prominent among the cell adhesive proteins are the class of cadherins which bind to identical cadherins on another cell surface in a calcium- dependent manner. Cadherins also interact with the cell's cytoskeleton through the connection of their cytoplasmic tails with intracellular β-catenins, and thus can be involved in transmitting signals to the cell's interior. Cellular adhesion to extracellular matrix is accomplished by another protein, integrin which interacts with extracellular proteins such as collagen, fibronectin, laminin, and tenascin as well as proteoglycans. Integrins play another role of transmitting information about the extracellular environment by complexing with cytoplasmic proteins which are associated in the cytoskeletal filaments such as actins.

In addition to cell adhesiveness, changes in the cell shape and cell motility also play major role in gastrulation. Here again, the cytoskeletal protein polymers such as microfilaments (actin) and microtubules (tubulins) play a major role. Actin filaments are organised into bundles and three dimensional networks which lie beneath the plasma membrane. They interact with myosin to act like a miniature muscle. Localized concentration of the actin network in the cortex causes changes in the cell shape. Similarly, the cellular movements are brought by the formation of cytoplasmic processes called filopodia or lamellipodia. They contain actin/myosin bundles and their contractile and extensive properties make the cells to move over the extracellular surfaces by periodic anchorages over the surface.

TYPES OF GASTRULATION MOVEMENTS

Although the pattern of gastrulation varies considerably from one phylogenetic group to another, the basic mechanisms are almost similar. The mechanism of cellular movements to bring about the three germ layered structure of the gastrula can be categorized into the following types.

EPIBOLY

The epithelial cells of the ectoderm spread as a sheet to surround the inner layers of the embryo.

INVAGINATION

The infolding of a region of cells in the blastoderm, much like the pushing in of the one side of a soft rubber ball.

INVOLUTION

Inturning of an expanding outer layer so that it spreads along the inner surface of the superficial layer.

INGRESSION

Separation of a small group of individual cells from the blastoderm, followed by their migration into the embryonic interior.

TYPES OF GASTRULATION IN REPRESENTATIVE ANIMAL EMBRYOS

The mechanism of cellular reorganization with their new neighbouring cells can be best understood by a comparative study of gastrulation in different animal groups with varied developmental potentials.

SEA URCHIN

The sea urchin blastula is a hollow ball of cells forming into a single-layered blastoderm. The animal half of the blastula consists of prospective ectoderm which will be the outer layer of the pluteus larva; whereas, the cells of the vegetal half will form the internal structures of the pluteus. In general, invagination of the endodermal cells in the vegetal half to form the archenteron will be the first step in the gastrulation process. However, in sea urchin, the inward migration of the primary mesenchyme cells, which are the descendants of the micromeres at the vegetal pole of the blastula is the first step to begin morphogenesis. The primary mesenchyme cells form the skeleton of the pluteus. The second stage is gastrulation proper involving invagination of the whole vegetal half to form the gut of the pluteus from which the coelom arises by budding.

    The migration of the primary mesenchyme cells begins when these cells lose contact with each other and with the surrounding hyalin layer. Before gastrulation, all the ventral plate cells including the micromeres that give rise to primary mesenchyme cells bind tightly to one another and to the hyaline layer, but adhere only loosely to the inner basal lamina, secreted by blastoderm. At the beginning of gastrulation, whereas other ventral plate cells (prospective endoderm cells) retain their tight binding to the hyaline layer and to the neighbouring cells, the prospective primary mesenchyme cells lose their affinity for these structures, and increasing their affinity for basal lamina as well as extracellular matrix (fibronectin) significantly. As a result, the micromeres release their attachments to the external hyaline layer and their neighbouring cells, and are drawn in by the basal lamina to migrate into the blastocoel.

    Once inside, these mesenchyme cells become ameboid and move about by means of long pseudopodia, termed the filopodia. The filopodia move actively along the inner blastodermal surface as if exploring its contours. The movement of the cells is caused by attachment, retraction and reattachment of the filopodia to the blastoderm wall. The cells continue to move until they reach an area where adhesion is thought to be stronger and then, become stationary. By fixation to the inner surface of the blastoderm, they form a ring at the base of the invaginating archenteron. The filopodia of the primary mesenchyme cells in the ring fuse with one another to form a syncytial cable-like structure. Two branches then rise from the ring, one on each side of the archenteron. The cables formed by the fusion of the filopodia are the sites for the deposition of skeletal matrix. The contractile movement of the filopodia is brought about by the parallel arrangement of the microfilament in them. Different steps in the gastrulation movements leading to the formation of ciliated gastrula are illustrated in Figure 7.1.

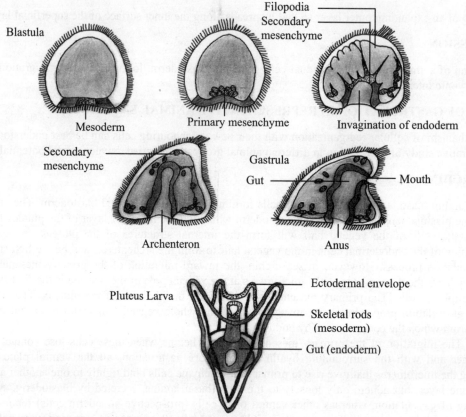

Fig. 7.1 Gastrulation movements in the sea urchin embryo with final steps in the pluteus larvae formation (Modified from Gilbert, 1997)

An autonomous property of the vegetal plate cells causes invagination into the blastocoel by a reduction of mutual contact between the columnar cells. Simultaneously, the inner borders of these cells become rounded up. Bundles of actin filaments with their contractile action help in this process. At the same time, their outer contact with the hyaline layer in the ventral plate is retained. The hyaline layer is made up of two laminar layers. The outer lamina is made primarily of hyaline protein (originating from the cortical granule exocytosis during fertilization) and an inner lamina composed of fibropellin proteins. Fibropellins are secretory granules released from the egg cytoplasm, following the cortical granule exocytosis. Thus, fibropellins have formed a mesh-like network over the embryo surface, but below the hyaline layer, as the inner lamina. At the time of invagination in the ventral plate, the ventral plate cells secrete a chondroitin sulphate proteoglycan into this inner layer of the hyaline layer. These hygroscopic proteoglycan molecules absorb water and swell the inner lamina, but not the outer lamina. This causes the vegetal region of the hyaline layer to buckle. Coincident with this, a second force arising from the movements of epithelial cells adjacent to the vegetal plate may also facilitate this invagination by drawing the buckled layer inward. By these processes, a cup-shaped archenteron is formed.

Further elongation of the archenteron into a tube is made possible by the pseudopodia that are formed by the cells at the tip of the archenteron. The continued contractile action of the pseudopodia at the tip of the archenteron, pull the archenteron tip to the gastrula wall near the animal pole where they eventually fuse with the mouth. Just like the primary mesenchyme cells, the secondary mesenchyme pseudopodia also possess longitudinally oriented microtubules. As the archenteron tip nears the gastrula wall, the secondary mesenchyme cells disassociate from it and migrate individually into the blastocoel, where they proliferate to form the mesoderm. Thus the archenteron is composed entirely of endodermal cells. Finally a mouth is formed at the point of contact of the archenteron with the gastrula, thus producing a continuous tube called the gut. This gives the basic plan of the pluteus larva.

AMPHIOXUS

The cleavage in *Amphioxus* is holoblastic and the blastula consists of a single layered blastomere surrounding a large blastocoel. Even before the commencement of the gastrulation three presumptive germ layer cell types could be discerned. The animal hemisphere cells have clear cytoplasm, being destined to become the ectoderm. These cells are tightly packed to form a columnar epithelium. The cells in the vegetal pole are larger and the cytoplasm granulated. These cells will later form the endoderm. The third type of cell, which are the smallest cells having basophilic cytoplasm, is found on one side of the egg roughly corresponding to the marginal zone. These cells are spherical in shape and are loosely packed. The external surfaces bulge out of the embryo. These cells originate from the crescent shaped cytoplasm found in the marginal zone.

Gastrulation is initiated when the blastoderm at the vegetal pole becomes flat and subsequently bends inwards so that the spherical embryo becomes cup shaped. Thus a large cavity that is formed in the vegetal pole of the embryo establishes open communication with the exterior. The cup shaped embryo now has a double wall with its internal layer lining the newly formed cavity which is called the primitive gut or archenteron. The external and internal epithelial layers are continuous over the rim of the cup shaped embryo. In this stage, there is still a space between the external and internal walls representing the remnants of the blastocoel. These outer and inner epithelial layers represent the ectoderm and endoderm respectively. The presumptive material of the notochord and the mesodermal crescent at first lie on the rim of the cup, but very soon they shift inward so as to occupy a position on the internal wall of the cup. In this way, the endoderm and the mesoderm disappear from the surface of the embryo to the interior of the embryo. The external surface of the embryo now consists only of the ectoderm.

The movements of infolding or inward bending of the endoderm and the mesoderm are known as invagination. The cavity arising through the invagination of the endoderm and mesoderm is called archenteron and the opening to the exterior is called the blastopore. Thus the blastopore represents the pathway by which the endoderm and mesoderm pass into the interior of the embryo. The blastopore has dorsal, lateral and ventral lips. The blastopore is very broad in the initial stages of gastrulation. But soon, its lips begin to contract so that the blastoporal opening becomes reduced. The contraction of the blastoporal lips starts after the disappearance of the mesodermal crescent and the presumptive notochord from the rim of the cup shaped embyro. After getting into the interior of the embryo the positions of the notochord and the mesoderm change.

At this stage, the embryo becomes elongated in the anterior-posterior direction. Simultaneously, the mesodermal crescent converges towards the dorsal side of the embryo and comes to lie on both sides of the presumptive notochord. The embryo is now completely enveloped by the ectodermal cells, thus bringing about the final positioning of the three germ layers in such a way that the endoderm occupies the inner most region, lining the archenteron, the ectoderm covering the embryo externally and the mesodermal cells with its presumptive derivatives occupying the intermediate positions.

68   Molecular Developmental Biology

AMPHIBIANS (*XENOPUS*)

In the unfertilized eggs of amphibians, the egg cytoplasm is homogeneous with the yolk forming a gradiance relative to the animal- vegetal axis. However, after the sperm entry into the egg during fertilization, the egg cortex rotates 30 degrees relative to the inner cytoplasm which remains unaltered. Sperm entry during fertilization probably triggers cortical rotation by generating a signal which is passed onto the cytoskeleton. Recent studies have shown that cortical rotation also specifies the Neuwkoop centre in the vegetal region on the side of gray crescent, just below the equator. Thus a bilateral symmetry results from the cortical rotation. UV irradiation blocks cortical rotation by disrupting the microtubule array responsible for the cortical movement. Embryos developing from such UV treatment are ventralized and are deficient in structures normally found on the dorsal side. This rotation is effected from vegetal cortex to the animal half on the side of the egg opposite the sperm entry point, bringing about the formation of the gray crescent (Fig.7.2).

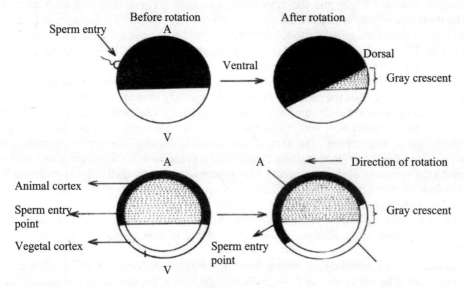

Fig. 7.2 Cortical rotation and Gray crescent formation in frog eggs. The egg cortex rotates by about 30 degrees relative to the cytoplasm.The cortical rotation forms gray crescent and specifies dorsoventral polarity
(Redrawn from Gilbert and Raunio,1997).

Gray crescent in the frog *Rana pipiens* is formed by the movement of pigmented animal cortex in the marginal zone and its subsequent replacement by nonpigmented vegetal cortex, which gives a less pigmented appearance in the marginal zone. Gray crescent marks the future dorsal side of the embryo. Characteristic of vertebrates, the number of cells that partake in the gastrulation process in the amphibian embryos is quite numerous and hence, the process is much more complicated than that of invertebrates such as sea urchin. Again, the characteristic behaviour of three germ layer cells is exhibited more vividly than in any other embryo.

The pregastrulation *Xenopus* embryo is a hollow ball of some 104 cells with the inner blastocoel slightly displaced towards the animal pole. The cells at the animal (dorsal) region are small, while those at the vegetal (ventral) region are larger due to considerable amount of yolk. The amphibian blastula is regionalised into three major zones. They are (1) the animal zone which constitutes several layers of cells in the blastocoel roof (BR), (2) the ventral zone consisting of large vegetal blastomeres below the blastocoel and (3) the marginal zone (MZ) which separates the animal and vegetal zone cells. The vegetal limit of the marginal zone corresponds with the gray crescent. The marginal zone cells that involute into the embryo during gastrulation constitute the involuting marginal zone (IMZ). During gastrulation, the blastoporal lip is formed at the lower boundary of the IMZ.

Gastrulation in amphibians begins in the marginal zone, where the animal and vegetal hemispheres meet. The first external sign of gastrulation is the formation of the dorsal lip of the blastopore. Gastrulation in frog embryo is initiated just below the equator in the region of the gray crescent, where the animal and vegetal hemispheres meet. The first external sign of gastrulation is the formation of the dorsal lip of the blastopore. Here, the local endodermal cells invaginate to form a slit-like blastopore. These cells change their shape dramatically. The main body of each cell is displaced towards the inside of the embryo while it maintains contact with the outside surface by way of a slender cytoplasmic strand. These cells are called bottle (flask) cells and they line the initial archenteron. Microtubules which arrange parallel and lengthwise along the apical-basal axis and the circumapical actin microfilaments, presumably functioning in contraction, play significant role in bottle cell formation. As gastrulation proceeds, the bottle cells continue to invaginate producing the lateral lips and finally the ventral lip of the blastopore. Invaginating bottle cells show two distinct phases of behaviour, namely, formation of their unique shape and respreading, in which they return to their original cuboidal shape. Bottle cells are moved inside as the archenteron deepens, and there they respread to form a large area on the periphery of the archenteron (Fig.7.3). Thus, respreading of bottle cells recapitulates the order of their formation. Apical contraction and subsequent relaxation of these apices and respreading of bottle cells are autonomous and intrinsic to the bottle cells at the late blastula stage. Among the amphibians studied so far, most bottle cells of urodeles and anurans other than *Xenopus* are of mesodermal rather than endodermal in fate, and they ingress to deeper layers where they form notochord and contribute to somitic mesoderm rather than remaining on surface. The endodermal cells encircled by the blastopore form the yolk plug (Fig.7.4).

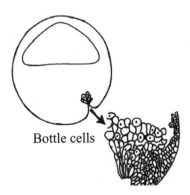

Fig. 7.3 Mechanism of gastrulation in amphibians. The bottle cells which migrate into the interior of the blastoderm through the blastoporal lip as well as their respreading to assume original shape as shown in the figure (arrow). (Modified from Gilbert and Raunio 1997)

The next phase of gastrulation involves the migration of the marginal zone cells towards the blastoporal lip. They then turn inward and travel along the inner surface of the outer ectodermal cell sheath by the process of involution. The cells constituting the lip of the blastopore are constantly changing. The first cells to compose the dorsal lip are the endodermal cells that invaginated to form the leading edge of the archenteron. These cells later become the pharyngeal cells of the fore gut. As these first cells pass into the interior of the embryo, the blastopore lip becomes composed of involuting cells that are the precursors of the head mesoderm. The next cells involuting over the dorsal lip of the blastoderm are called chordamesoderm cells. The rolling-in of the endoderm and mesoderm at the blastopore is by a process of convergent extension.

As the new cells enter the embryo, the blastocoel is displaced to the side, opposite the dorsal blastopore lip. With the movement of endodermal and mesodermal cells inside, the ectoderm expands by a process called epiboly. As a result the ectoderm becomes thinner and spreads downward to cover the entire embryo. The coordinated movement of the three germ layers results in the three-layered embryo with the inner archenteron completely lined by endoderm and the ectoderm enveloping the intervening mesodermal layer.

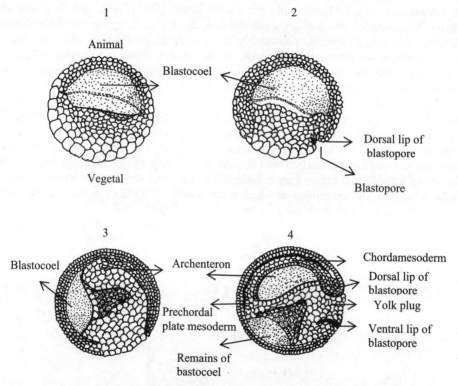

Fig. 7.4 Important stages in the gastrulation of frog embryo up to yolk plug formation
(Modified from Gilbert and Raunio, 1997)

Cellular adhesiveness has an important function in governing the arrangement of the germ layers in the amphibian gastrula. Experimental evidence indicates that the ectoderm cells have the lowest total cohesiveness and hence they take up the external position with respect to other two cell types in the embryo. However, the ectodermal cells have high cohesiveness on their inner surfaces. Mesodermal cells have the highest total cohesiveness because they sort out internally between the ectoderm and endoderm. If the ectoderm cells are experimentally removed from the gastrula, the mesodermal cells sink into the endoderm. The position of the mesoderm, in the gastrula is determined by its strong adhesion to the internal surfaces of the ecdodermal cells. The endoderm, which is less cohesive than the mesoderm takes up the internal position because it is more cohesive than the ectoderm, but it can not compete with the mesoderm for adhesion to the internal ectodermal surface. An interesting aspect of mesodermal adhesion to the inner surface of the ectoderm during morphogenic movements is the presence of fibronectin receptors on the inner surface of the ectodermal cells as well as the presence of fibronectin as a glycoproteinaceous cell surface coat, on the surface of the mesodermal cells, facilitating cell surface interaction between these two cell types. Since the endodermal cells do not produce fibronectin they cannot directly adhere to the inner surface of the ectoderm.

The strategic location of mesoderm in between the ectoderm and endoderm will have important developmental consequences as the ectoderm and endoderm depend on mesoderm for their differentiation.

NEUWKOOP CENTER

In the *Xenopus* embryos, the cortical rotation formed in consequence of sperm entry into the egg results in the formation of gray crescent, specifying the dorsal axis of the embryo. Soon after the 32 cell stage in *Xenopus laevis*, the Neuwkoop center is formed in the vegetal region on the side of the gray crescent. The Neuwkoop center is formed by endoderm cells, but induces more animal cells to form dorsal mesoderm. The signals involved in this mesoderm induction include growth factors such as activin and fibroblast growth factor. The formation of Spemann's organizer is centred at the dorsal lip of the blastopore and has powerful inductive properties. The cells constituting the dorsal lip undergo extreme convergence and extension movements during gastrulation, and they later form dorsal axial structures such as notochord, somitic mesoderm, and pharyngeal endoderm.

The experiments of Nieuwkoop demonstrated that the dorsal blastopore lip acquired the organizer properties from the vegetal cells underlying them. He showed that the animal cap cells (presumptive ectoderm) were induced by the vegetal cap cells (presumptive endoderm) to form the mesodermal structures such as notochord, muscles, kidney cells and blood cells. While the ventral and lateral vegetal cells induced ventral (forming mesenchyme and blood) and intermediate (forming muscle and kidney) mesoderm, the dorsalmost vegetal cells specified dorsal mesoderm components (somites and notochord), including the organizer. The dorsalmost vegetal cells of the blastula, specifying the organizer is called as Nieuwkoop center. Experimental studies further indicated that the cells of the Nieuwkoop center, when transplanted into the vegetal side of another blastula, gave rise to a new embryonic axis comprising gut, neural tube, notochord and somites, on the ventral side. This result is similar to the transplantation experiments of Hans Spemann and Mangold using dorsal lip as the organizer. Evidently, the dorsalmost vegetal cells of the frog blastula induced the overlying animal pole cells of the prospective dorsal blastopore lip to become dorsal mesoderm. Consequently, this cellular inductive event initiated gastrular movement in amphibians.

The prospective mesodermal cells, on induction from the vegetal endodermal cells, express the *Xenopus Brachyury* (*Xbra*) gene. Xbra protein is a transcription factor that activates the genes to produce mesoderm- specific proteins. The dorsalmost vegetal cells in the Nieuwkoop centre expresses yet another protein, β- catenin. β- catenin which is part of the *Wnt* signal transduction pathway, is necessary for this formation of dorsal axis. Experimental depletion of β- catenin transcripts with antisense oligonucleotides results in the lack of dorsal structures. Interestingly, injection of exogenous β-catenin molecules into the ventral side of the embryo produced a secondary axis. The organizer tissue responds to Nieuwkoop centre by activating the *goosecoid* gene. This gene encodes a DNA-binding protein that activates the migratory movements of dorsal blastopore lip cells into the interior. During gastrulation, the Nieuwkoop centre cells remain endodermal, whereas the cells of the organizer became the dorsal mesoderm and migrate underneath the dorsal ectoderm.

72 Molecular Developmental Biology

EXOGASTRULATION

The necessity that the mesoderm to induce development of neural tube should lie immediately beneath the dorsal ectoderm in the early gastrula was clearly explained in an embryonic abnormality, called exogastrulation. Exogastrula can be induced by treating the gastrulating embryos with hypertonic salt solution. In exogastrulation, the mesoderm, instead of moving inward, moves outward. This mesoderm gives rise to axial structures such as notochord and somites. However, the left out ectoderm does not form neural tissues, suggesting the importance of mesoderm to induce the ectoderm for neurectoderm transformation (Fig.7.10). Recent studies have however disclosed that neural cell adhesion molecules such as N-CAM are expressed in the ectoderm of the exogastrula. These neural genes are also expressed in an anterior- posterior order. This neural expression is suggested to be due to the planar signal passing transversely through the ectoderm from the dorsal lip.

Gastrulation in amphibians is initiated by the formation of the blastopore, which also sets in the organization of the body axis. The blastopore marks the future posterior end of the embryo. The archenteron, which is initiated by the invagination of the presumptive endoderm, extends forward from the blastopore to the future anterior end of the embryo. The dorsal lip of the blastopore, formed in the marginal zone, is the site of involution of the presumptive chordamesoderm. These cells spread underneath the overlying ectoderm and later form the somite mesoderm, while the dorsal ectoderm forms the neural plate.

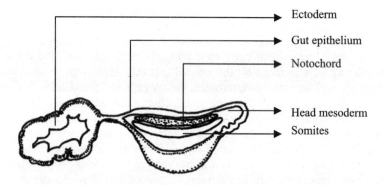

Fig. 7.5 Exogastrulation in amphibians The primary organizer and Spemann experiments (Modified from Gilbert, 2000)

Two important signalling paths from Spemann's organizer are involved in neural induction. They are called vertical and planar signalling. In the vertical signalling, the involuting mesoderm induces the overlying ectoderm to form neurectoderm. In planar signalling, signals from the dorsal lip pass through the ectoderm towards the animal pole. These signals, possibly in the form of morphogenetic gradients, lead to spacially restricted expression of regulatory genes that begin the process of regional specification.

The dorsal lip of the blastopore plays an important role in the organization of the body axis. The German scientists, Spemann and Mangold did several experiments to prove the ability of the dorsal lip or the involuting chordamesoderm in forming the anterior posterior axis of the embryo. They transplanted the dorsal lip of the urodele embryo to the ventral or lateral region of another embryo. This resulted in the formation of a secondary embryonic axis consisting of a gut, neural tube, notochord and somites. Obviously, the invagination of the transplanted dorsal lip produces a secondary archenteron and the dorsal lip mesoderm gave rise to the notochord and somites. On the contrary, the neural tube is not a derivative of the dorsal lip, but derived from the induction of the ventrolateral ectoderm by the dorsal lip material. Spemann and Mangold used for these experiments donor and host embryos that differed in the amount of pigments they contained facilitating easy distinction between donor and host cells in the secondary axis.

The above results of Spemann and Mangold indicate that during normal development, the dorsal lip (or its derivative, chordamesoderm) induces the dorsal ectoderm to develop into neural ectoderm. This fixation of developmental fate is called determination. By virtue of the ability of the dorsal lip to establish the anterior-posterior axis in the embryo, Spemann termed it, the primary organizer.

Another feature of interest is the transplantation experiments of Spemann on the competence of the dorsal ectoderm as well as the lateral and ventral ectoderm to undergo induction to form the neural tube by the chordamesoderm. This suggests that the ectoderm might consist of two different types of determined cells, neural and epithelial, which are intermingled throughout the ectoderm. Evocation could select the neural cells in the region overlying the chorda, which would respond by forming the neural plate. In the areas the chorda does not contact the ectoderm, the cells that are determined to become the skin would be selected, since the neurally selected cells would not be stimulated and hence would subsequently perish.

To be more specific, the chrodamesoderm, which is derived from the subsurface of the dorsal lip, induces the overlying dorsal ectoderm to develop as neural ectoderm instead of epidermis. This is further evidenced by the investigation of Spemann who implanted post-involution chordamesoderm under the ventral ectoderm. Furthermore, there are qualitative differences in the inductive influence exerted by different regions of chordamesoderm. These differences appear to be responsible for regional specialization along the anterior posterior axis. Thus, dorsal lip of the early gastrula contains the anterior chordamesoderm which induces the formation of the head with brain and sense organs at the anterior end. However, the dorsal lip of the late gastrula contains the posterior chordamesoderm which induces the formation of the trunk containing the spinal chord as well as the posterior end with the tail. It thus seems that the neural induction is an example of extrinsic influence over cell differentiation and, therefore, over differential gene expression.

The nature of the inducing substance from the dorsal lip has remained enigmatic, since the dorsal lip can function as an inductor even after it has been killed by heating. This again indicated that the inductor is a chemical which could still diffuse from the dead tissue. When a membrane was placed between the dorsal lip and the ectoderm, the inductive influence of the dorsal lip continued to diffuse through the membrane.

Although the dorsal lip is able to produce a secondary embryo with a clear anterior-posterior and dorsal-ventral axis, it is now known from experimental evidence that the interaction of chordamesoderm with ectoderm is not sufficient to organise the entire embryo, but it rather initiates a series of inductive events. The organizer ability of dorsal lip is not limited to urodeles only, but it occurs in the embryos of *Amphioxus*, Cyclostomes and a variety of other amphibians. The anterior portion of the primary streak, namely Henson's node, which is the homologue of dorsal lip of amphibians, is similarly effective in organizing secondary embryos in birds and mammals.

A clear indication from the Spemann's experiment on organizer in amphibians is that it initiates patterning and regionalization of the anterior- posterior axis. Positional identity of cells along the anterior- posterior axis is encoded by the combined expression of *HOX* genes. *HOX* genes are a particular family of homeobox - containing genes that are involved in patterning the anterior- posterior axis. *HOX* gene expression establishes positional identity for mesoderm, endoderm, and ectoderm.

GASTRULATION OF AVIAN EMBRYOS

The enormous quantity of yolk present in the avian eggs has not only influenced the cleavage but also affected the gastrulation process considerably. Since the yolk mass does not participate in the cleavage, which is confined only to the blastodisc, the resulting blastula appears very different from those of the holoblastic embryos. The main difference is the absence of a blastocoel in the avian embryos. However, the central cells (blastomeres) of the avian blastodisc are completely separated from the underlying yolk creating a clear subgerminal cavity in between the yolk mass and the blastodisc. This central area of blastodisc is called the area pellucida. The cells at the margin of the area pellucida appear opaque because of their contact with the yolk. This area is hence called area opaca. The cells in the area pellucida constitute the epiblast. Certain cells of the epiblast migrate individually by a process of ingression into the underlying subgerminal cavity to form a layer called the hypoblast (Fig.7.6 and 7.7). Thus, a two-layered blastoderm (epiblast and hypoblast) is formed. These two blastoderm layers are joined together at the margin of the area opaca and the space between the enclosed layers is called blastocoel. It may be noted here that the holoblastic embryos possess a blastocoel, as a result of cleavage. On the contrary, in the avian embryos, the blastocoel is formed only after the cell migration occurring during gastrulation. Another interesting aspect is, only epiblast contributes to the embryo proper. The hypoblast marks the vegetal extent of the blastoderm, thereby defining the dorsoventral axis of the future embryo.

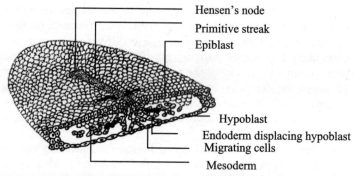

Fig. 7.6 Gastrulation movements found in the avian embryos. Note the completion of hypoblast layer and the ingression of endodermal and mesodermal layers (Redrawn from Gilbert, 2000)

Further cellular movements in the epiblast start after the formation of a primitive streak, which is formed as a thickening of the cell sheet at the central posterior end of the area pellucida. The primitive streak extends 60 -75% of the length of the area pellucida and marks the anterior-posterior axis of the embryo. As the cells converge to form the primitive streak, a depression forms within the streak. This is called primitive groove which serves as a blastopore for cell migration into the interior of the embryo. At the anterior end of the primitive streak is a regional thickening of cells called the primitive knot or Hensen's node. A funnel-shaped depression found in the centre of the Henson's node is again equivalent to the dorsal lip of the amphibian blastopore in allowing migration of more cells.

A comparison between the amphibian blastopore and the avian primitive streak is instructive at this juncture. Just like the amphibian blastopore, all the mesodermal and endodermal cells migrate through the primitive streak. The first cells to migrate are the endodermal cells which move to the base of the blastocoel and displace the hypoblast cells. Although a typical archenteron is not formed in the avian embryo, as in amphibians, the anteriorly migrated endodermal cells give rise to the gut later. Next to the endodermal cell migration, the mesodermal cells migrate through the Henson's node and the lateral portions of the primitive streak. An important difference to be seen in avian gastrulation is that the mesodermal cells do not migrate as sheets of cells into the blastocoel, as found in the amphibians. Instead, the ingressing cell population creates a loosely connected mesenchyme in the avian embryos.

As more cells move through the primitive streak, the latter elongates anteriorly to the future head region. Simultaneously, the underlying hypoblastic cells proliferate anteriorly parallel to the primitive streak.

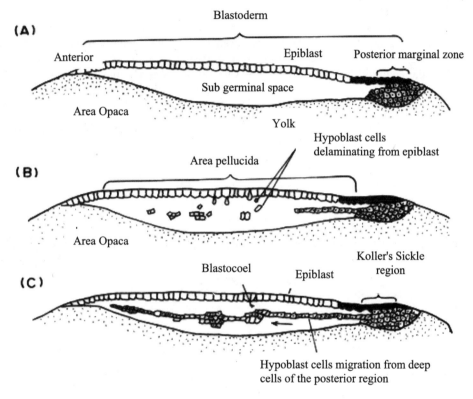

Fig. 7.7 Formation of the two-layered blastoderm of the chick embryo. (A, B) Primary hypoblast cells delaminate undividually to form islands of cells beneath the epiblast. (C) Secondary hypoblast cells from the posterior margin (Koller's sickle and the posterior marginal cells behind it) migrate beneath the epiblast and incorporate the polyinvagination islands. As the hypoblast moves anteriorly, epiblast cells collect at the region anterior to Koller's sickle to form the primitive streak. (Redrawn from Gilbert, 2000).

After most of the endodermal cells have been internalized, and the mesodermal cells continue to migrate, the primitive streak starts to regress moving the Henson's node to a more posterior region. As the primitive streak retracts to the posterior region, a notochord is formed in its position, thus forming a dorsal axis to the embryo. Ultimately, the primitive streak has completely regressed and has given rise to the anus in the most posterior position. By this time, the epiblast is composed entirely of presumptive ectodermal cells. While the presumptive mesodermal and endodermal cells were moving inward, the ectodermal precursors surround the yolk by emboly. As gastrulation process comes to the final stage, the ectoderm has surrounded the yolk, the endoderm has replaced the hypoblast and the mesoderm has positioned itself between these two regions.

At the end of the blastocyst stage, the blastodisc is attached to the interior of the trophoblast epithelial cells. However, by the commencement of gastrulation, the blastodisc gets detached from the trophoblast layer and gets thickened. The blastodisc now consists of the upper epiblast and the lower hypoblast, very similar to that of the avian embryo. The epiblast consists of a thick plate of columnar epithelial cells whereas the hypoblast cells become cuboidal or even columnar (Fig. 7.8).

Gastrulation process is evident with the appearance of a primitive streak with a Hensen's node at the anterior end. The primitive streak is however shorter than in birds, and does not extend beyond the midpoint of the blastodisc. The edge of the blastodisc from which the streak originated will be the future posterior end of the embryo and the advancing tip of the streak is the future anterior end.

As in birds, the primitive streak is a transient structure. The primitive streak having given rise to the mesodermal layer and notochordal rudiment recedes and shrinks with its Hensen's node reaching the posterior border of the blastodisc. With the disappearance of the primitive streak, the axial organization of the embryo becomes evident as the neural plates and somites are formed. Like the other vertebrates, the central nervous system is established first and initiates the development of bilateral organization. While such changes are occurring in the blastodisc region, the epiblast cells (ectoderm) give rise to the lining of the extraembryonic cavity, namely the amniotic cavity, whereas the endodermal cells of the hypoblast give rise to the yolk sac (Fig. 7.8).

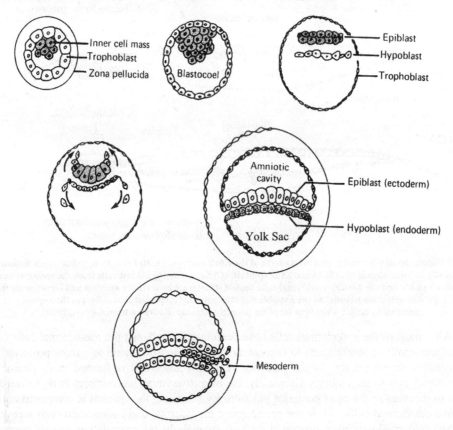

Fig. 7.8 Gastrulation in human embryo. Elaboration of hypoblast and epiblast and formation of amniotic cavity and yolk sac (Redrawn from Davenport 1979).

COMPARATIVE ASPECTS OF GASTRULATION

Just like the cleavage pattern changing almost in a species specific manner among the animal species, the process of gastrulation that brings about the morphogenetic movements in the epithelial cells of the blastula, is also species specific with considerable variation existing with different types of organisms. Incidentally, a comparative study of gastrulation in various animal species is also illustrative of different types of cellular migration and movements that characaterize the gastrulation. It is therefore instructive to explain the various types of morphogenetic movements of blastula epithelial cells with reference to different animal species (Fig.7.8).

Two other types of movements may occur in which cells act more independently. One of these is cavitation, in which one of the cells move to the periphery of a solid blastula (stereoblastic) creating small cavities in the interior that becomes confluent. The second case is almost the reverse of the first. Cells of the blastopore move singly into the interior, and this is termed as "ingression". The ingression may be widespread over the blastoderm or may be localised in one region.

The developmental importance of the morphogenetic movements lies in the fact that the cells of the epithelial layer of the blastula are brought into new spatial relationship. Such changes in the positioning of the blastodermal epithelial cells together with the concurrent changes in the shape of the blastula is an absolute requirement for the continuation of the ontogenetic processes since all subsequent patterns of morphogenesis depend on them. Morphogenetic movements like cytoplasmic rearrangements and cleavage occur along the guidelines established by the primary axis of the symmetry of the egg.

FATE MAPS

Due to the regularity of the morphogenetic movements and their constancy within a species, it is possible to trace specific regions of the blastoderm during gastrulation and therefore relate the three layered structure of the gastrula to the blastula. This relationship is accomplished using fate maps that depict the eventual deployment of various regions of the blastoderm. Since the blastula of the egg has been formed by subdivision of the original cytoplasmic mass without spatial rearrangements, the germ layers can actually be mapped in uncleaved eggs as specific cytoplasmic region in most eggs.

FATE MAP OF SEA URCHIN BLASTULA

The fate mapping of sea urchin blastula was done by directly observing the cell movement in the blastoderm followed by noting what its descendants became. However, recent fate mapping was made by injecting the fluorescent dyes into the 60-cell stage blastula when most of the embryonic cell fates were specified. At this time, although the cells consistently produced the same types of cells in each embryo, these cells remained pluripotent and could give rise to other cell types if experimentally grafted in different parts of the embryo.

Figure 7.9 depicts the fate map of the 60-cell sea urchin embryo. As a result of the holoblastic radial cleavage, the sea urchin embryo at this stage is consisted of different tiers of blastomeres between the animal and vegetal poles. By virtue of size difference, these layers of cells comprise three different types namely, mesomeres, being further divided into animal pole an_1 and an_2 sublayers; macromeres of vegetal hemisphere, again comprised of two layers such as veg_1 and veg_2. The lower most tier consists of the micromeres which are further divided into large and small micromeres.

The determination of these blastomeres during gastrulation into three germ layers (ectoderm, mesoderm and endoderm) started from the micromeres which secreted certain signalling molecules and, in effect, a cascade of induction propagated upwards into the macromeres and finally mesomeres. The molecules responsible for specifying the micromeres appear to be the transcription factor, β -

catenin, which is often activated by the *Wnt* pathway. It is now confirmed that β- catenin is essential for giving the macromeres their inductive ability by activating the genes necessary for producing the inductive signals such as a Notch ligand.

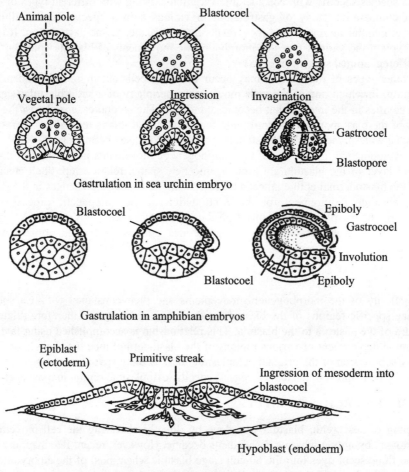

Fig. 7.9 A comparative representation of gastrulation movement in the representative animal species to describe various mechanisms of cellular movement. (Modified from Davenport, 1979)

As seen from the Fig. 7.10 the animal half of the embryo (an_1 and an_2), gave rise to the ectodermal derivatives such as the larval skin and its neurons. The lower layer veg_1 produces cells that can enter into either the ectodermal or endodermal organs. The descendants of veg_2 layer populated three different structures namely, endoderm, the coelom (body wall) and secondary mesenchyme (pigment cells, immunocytes and muscle cells). The first tier of micromeres (large micromeres) produces the primary mesenchyme cells that form the larval skeleton, while the second tier (small micromeres) contributed cells to the coelom. Among these cells, only cells whose fates are determined automatically are the skeletogenic micromeres. For example, if these micromeres are transplanted into the animal pole region of the blastula, their descendants will form skeletal spicules, in addition to altering the fates of nearby cells by inducing a secondary site for gastrulation by making the animal cap cells, generating endoderm, and producing a more or less normal secondary larva. Thus a homology could be established between the sea urchin embryo micromeres and the dorsal lip cells of amphibian embryos in respect of initiating gastrulation movements and cell specification. Evidently, the micromeres produce a signal that tells the cells adjacent to them to become endoderm and induces them to invaginate into the embryo.

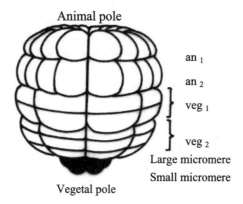

Fig. 7.10 Fate map and cell lineage of the sea urchin *Strongylocentrotus purpuratus* for abbreviations see text (Redrawn for Gilbert 2000)

FATE MAPPING IN AMPHIBIAN EGGS

In many amphibian species, the animal region is the most deeply pigmented, while the marginal zone, also pigmented, is typically grayish, in contrast with the commonly white vegetal region. The gray crescent present in the marginal zone is important in establishing bilateral symmetry in the fertilized eggs. The three regions roughly representing the future three primary germ layers are depicted in Fig.7.11

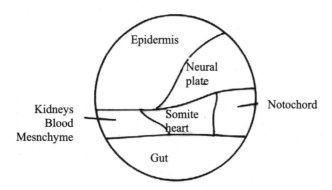

Fig. 7.11 Fate mapping of *Xenopus* late blastula (Lateral View)

The animal region constitutes the presumptive germ layer (ectoderm) and the marginal zone becomes the middle layer (or *chorda mesoderm*). The movement of the cells from these externally distinguishable regions could be better visualised by the technique of vital dye staining. Vogt saturated agar chips with neutral red or Nile blue sulphate which will stain the embryonic cells without having any harmful effects. These stained agar chips were pressed to the surface of the blastula, so that the dye was transferred to the contacted cells. The movements of each group of stained cells were followed throughout gastrulation and the results were summarised in fate maps. These maps have been confirmed by scanning electron microscopy and dye injection techniques. Vital dye studies of Lovtrup and his associates have shown that cells of *Xenopus* blastula have different fates depending on the superficial layers of the embryo. In *Xenopus* (but not in other amphibians) the mesodermal precursors exist in the deep layer of cells while the ectoderm and endoderm arise from the superficial layer on the surface of the embryo. In urodeles, the notochord and mesoderm precursors are found in both the surface cells and the deep marginal cells.

FATE MAP OF CHICK EMBRYO

The fate map of avian embryo is restricted to the epiblast, as it only contributes to the developing embryo. Again, since the blastoderm is organised into a flat disc, lying over the uncleaved yolk mass, invagination as a mechanism of gastrulation has been abandoned. Hence, unravelling the pattern of cell movements during gastrulation has proved to be difficult. Inaccessibility of the lower layer of cells in the intact embryo is also another difficulty. Since the vital dye and carbon particle fail to penetrate both epiblast and hypoblast layers, these classical methods to follow the gastrulation could not be employed for chick embryos.

More recently, transplantation of cells that can be identified in the host embryo is used to chart cell movement. Radioactive labelling of the nuclei with ^{3}H-thymidine in the cells transplanted from a portion of the embryo to a comparable region of an unlabelled embryo has also proved profitable in fate mapping studies. The transplantation experiments have established that the epiblast is the source of the ectoderm, mesoderm and endoderm. The fate mapping of the chick epiblast has shown that the presumptive notochordal cells are centrally located between the anterior presumptive ectoderm and the posterior presumptive endoderm.

8

EMBRYONIC INDUCTION AND NEURULATION

As seen in the previous chapter cellular movements that occurs during gasturlation constitute the key event in the development of the embryo. These cellular movements during gasturlation form the basis of the germ layer formation. The movements of the germ layers to occupy the final positions as well as the inductive action of the organizer region are crucial to the establishment of the vertebrate body plan. During gastrulation, the cells of the dorsal-most mesoderm in the region of the organizer or the dorsal lip of the blastopore are internalised. This internalized marginal mesoderm gives rise to a rigid rod-like notochord along the dorsal mid-line flanked on each side by blocks of somites. The somites are derived from the marginal zone mesodermal cells lying on either side of the organizer. The formation of the notochord leads to the induction of overlying dorso-medial ectoderm cells into neuro-ectoderm, which consequently gives rise to the neural tube. The neural tube finally develops into the brain and spinal cord. The somites are now positioned on either side of the neural tube. Thus the main axial structures characteristic of vertebrate embryo consists of the neural tube, notochord and the somites. These three structures exhibit regional organization along the anterior- posterior axis. Furthermore, the structures derived from both mesoderm and ectoderm also shows distinct anterior-posterior organization. For example, the brain develops in the anterior most part of the neural tube followed by the spinal cord. The development of axial structures is important in the furtherance of organogenesis involving successive inductive interactions between the germ layers. Several cell signalling events also take part during the axialization of the embryo (See chapter 12 for details).

INDUCTIVE INTERACTION AND NEURULATION

The concept of inductive interaction between cells originated from the experiments of Spemann and Mangold on amphibian dorsal lip of the blastopore. During embryogenesis, inductive interaction is the basic requirement for cellular differentiation and tissue patterning. Inductive interactions between the germ cell layers during post-gastrulation development could be categorised into two types. In the first type, the induced tissue is so committed to its future course of development that the inductive signal, which may be even non-specific, is only a cue that stimulates activity. This type of induction is called permissive and the example is neurulation. In the inductive interaction that leads to neurulation, the ectoderm is inducted to become neural tissue by the underlying mesenchyme cells. Experimental evidence indicates that the direct cellular contact between the two tissues is not required for induction to take place. When agar blocks soaked in some chemicals are placed below the ectoderm, the overlying ectodermal tube forms the neural tube. The specificity of the interaction here clearly resides in an ectoderm earlier committed to a particular development pathway and the inductive cue from the mesenchyme clearly triggers the process.

The second type is the secondary induction, which is also called directive, wherein there is requirement of close physical contact between cells of the two tissue rudiments. In this type, the tissue has a choice of developmental pathways and the inductive interaction both stimulated morphogenesis and determines that choice. The secondary induction normally results from interactions between mesenchyme and the epithelia derived from the ectoderm or endoderm.

NEURULATION IN AMPHIBIANS

In amphibians, the gastrula is a sphere that is surrounded by ectoderm and contains an inner archenteron lined with endoderm. Between the two layers, a sheet of mesoderm is formed. After gastrulation process is completed, the presumptive neuronal system begins to form from the dorsal ectoderm. This process of neurulation brings about characteristic change in shape and appearance of the embryo for the latter to be known as neurula.

There are two ways in which the neural tube could be formed. In primary neurulation, the chordamesoderm influences the overlying ectoderm to proliferate, invaginate and pinch off from the surface to form a hollow tube. In secondary neurulation, the neural tube arises from a solid cord of cells that sinks into the embryo and subsequently hollows out to form a neural tube. In *Xenopus*, most of the tadpole neural tube is made by primary neurulation excepting the tail neural tube, which is made by secondary neurulation. On the contrary, neurulation in fishes is exclusively secondary. In birds, the anterior portion is constructed by primary neurulation, whereas the caudal region is made of secondary neurulation.

The first indication of the amphibian neurulation is the flattening of the dorsal ectoderm to form the neural plate (Fig. 8.1). The edges of the neural plate thicken to form neural folds which flank a central depression called neural groove. The neural groove extends along the entire mid-dorsal line of the embryo. There is a remarkable change in the cells of the neural ectoderm during neurulation. Prior to neurulation, they are made of low columnar epithelium, like the non-neural ectoderm. However, during neurulation, the neural ectodermal cells elongate dramatically to become high columnar cells (Fig.8.1). The ability of the neural ectoderm cells to undergo shape change is an intrinsic property of these cells as the elongated neural plate cells retain their columnar shape and continue to elongate in culture. The lengthening of the neural plate is achieved by their cells undergoing convergent extension by intercalating several layers of cells into a few layers.

The neural folds slowly rise and eventually meet above the deepening neural groove, where they fuse to form the neural tube, which is the rudiment of the central neuronal system. Fusion of the neural folds separates the neural ectoderm from the presumptive epidermis, which now completely encircles the embryo. During this time, a group of neural ectoderm cells are excluded from the neural tube and lie in between it and the overlying epidermis along the mid-dorsal line. These cells are called 'neural crest cells' (Fig.8.1). They undergo an epithelial to mesenchymal transition and migrate laterally and ventrally to give rise to a variety of cell types, scattered throughout the body. Derivatives of the neural crest cells include (1) most of the craniofacial cartilage, skeletal and connective tissues; (2) neurons and supporting cells of the peripheral nervous system; (3) pigment cells in the dermis, and (4) the medullary cells of the adrenal glands. The migration of neural crest cells from the neural tube is effected by the loss of adhesion molecules such as N- cadherin and E- cadherin from the neural crest cells at the time of their migration. The epithelial to mesenchymal transition of neural crest cells also involves the expression of *slug* gene.

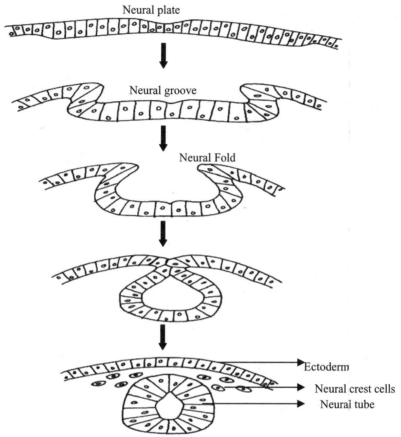

Fig. 8.1 Neural tube formation in *Xenopus* (Redrawn from Wolpert, 1978)

MECHANISM OF NEURAL TUBE FORMATION

Neural tube formation is ultimately linked to the changes in cell shape. Microtubules and microfilaments are both involved in these changes. As the ectodermal cells begin to elongate, the randomly arranged microtubules of these cells align themselves parallel to the long axis. The role of microtubules in cell lengthening can be experimentally proved by blocking the neural tube formation using colchicin, which is a potential inhibitor of microtubule polymerization. A second change in cell shape involved the apical constriction of elongated cylindrical cells to form wedges. These changes occur mainly in the lateral sides of the neural tube (Fig.8.2). These changes are caused by the contraction of the microfilaments, which consequently produces a purse string effect, constricting the apical end of each cell. The contraction of microfilaments is due to the presence of the contractile protein, actin, which is linked to the myosin molecules from which the actin derives energy for contraction. The effect of actin on cell constriction was experimentally demonstrated by inhibiting the contractile function with the drug, cytochalasin B. As a result, although the embryonic neural ectodermal cells elongated, they failed to undergo constriction at the apical end. The presence of a membrane protein, spectrin, is shown to bind with the microfilaments inside the cell to accomplish the constriction effect of the cell.

In addition to changes in the cell shape caused by the contractile proteins, other principles such as convergent extension and cell division in the neural plate can also contribute to the neural tube formation. Furthermore, external forces applied by the surrounding tissues might also be responsible for the correct folding of the neural tube.

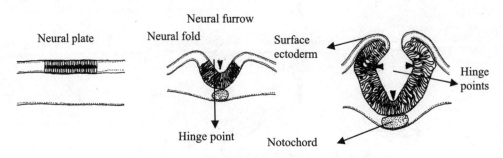

Fig. 8.2 Mechanism of neural groove formation in chick (Redrawn from Wolpert, 1978)

Yet another mechanism underlying the neurulation process is attributed to the cell adhesiveness in the dorsal ectoderm. The separation of neuroectodermal cells from the presumptive epidermis involves changes in cell adhesion due to differential expression of cell adhesion molecules. For example, the cells of the neural plate express N-cadherin and N-CAM, whereas, the adjacent ectoderm expresses E-cadherin. As a result, the neural tube cells not only separate out from the surrounding ectodermal cells, but also sink beneath the ectodermal epidermis to form the neural tube.

9

NEURAL CREST CELLS

The neural crest cells are a migratory cell population, found only in vertebrates. The neural crest arises at the neural plate border, the boundary between the neural plate and ectoderm. As the neural plate rolls up, these cells are found at the crests of the neural folds, hence their name. At around the time of neural plate closure, the neural crest cell population undergoes an epithelial to mesenchymal transition and migrates laterally and ventrally to give rise to a variety of cell types, scattered throughout the body. Derivatives of the neural crest cells include (1) most of the craniofacial cartilage, skeletal and connective tissues; (2) neurons and supporting cells of the peripheral nervous system; (3) pigment cells in the dermis, and (4) the medullary cells of the adrenal glands. Table.9.1 summarizes the different cell types and tissue contributions of the neural crest derivatives. By virtue of this multipotent activity, the neural crest is sometimes called the fourth germ layer. Neural crests are divided into four main functional domains, such as the cranial neural crest, trunk neural crest consisting of general and vagal/sacral trunk crest, and cardiac crest.

Table. 9.1 showing cell types and tissue contributions from neural crest cells in vertebrates

Cell type	Tissues
Sensory neurons	Spinal ganglia
Cholinergic neurons	Thyroid gland
Adrenergic neurons	Ultimobranchial body
Rohon Beard cells	Adrenal gland
Schwann cells	Teeth
Glial cells	Dentine
Chromaffin cells	Connective tissue
Parafollicular cells	Adipose tissue
Calcitonin producing cells	Smooth muscles
Melanocytes	Cardiac septa
Chondroblasts, chondrocytes	Dermis
Osteoblasts, osteocytes	Cornea
Odontoblasts	Endothelia
Fibroblasts	Adipocytes
Cardiac mesenchyme	Mesenchymal cells
Striated myoblasts	Smooth myoblasts

FORMATION AND MIGRATION OF NEURAL CREST CELLS

In vertebrates, the neural crest arises along the neural plate border at all positions caudal to the forebrain. At the level of the forebrain, the neural plate border gives rise predominantly to ectodermal placodes which form into sense organs. Specification of the neural plate border is accomplished by inductive signals from the adjacent presumptive epidermal ectoderm. The neural crest induction appears to occur through a tissue interaction between the neural plate and non-neural ectoderm. This was evidenced by the fact that when the presumptive epidermis and neural plate in urodeles are juxtaposed, neural crest cells were generated *de novo*.

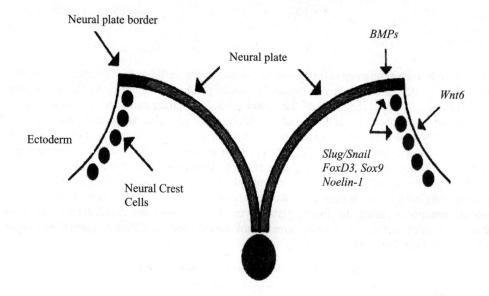

Fig. 9.1 Showing the formation of neural crest cells during neural tube formation. Note the expression of different signaling molecules in the neural plate and non-neural ectoderm. see text for details
(Redrawn from Lanza et al., 2002)

A variety of molecular markers are expressed at the border of the neural and non-neural ectoderm to generate neural crest cells. The zinc finger transcription factor, Slug, is one of the earliest known neural crest markers, and is expressed in the neural folds as they elevate and close to form the neural tube. Furthermore, as shown in Fig.9.1, Wnt6 in the non-neural surface ectoderm and BMPs in the neural plate are expressed to induce neural crest cell formation. Wnt and BMP signalling act either synergistically or independently during neural crest induction to generate individual lineages within the neural crest. Evidently, combinations of multiple genes, functioning cooperatively, are necessary to generate the neural crest forming ability. During their formation, neural crest cells express the transcriptional repressors Snail/ Slug, which regulate the epithelial to mesenchymal transition and delamination of neural crest cells from the neural tube. A variety of other transcriptional factors, including *SOX9*, Ford3, Noelin-1 are expressed in either premigratory or migratory neural crest cells. They are important for neural crest specification.

For example, the signal proteins BMPs are strongly inhibited to form the neural plate by Noggin, chordin, and Follistatin secreted by the organizer in *Xenopus*. Further away from the neural plate, the inhibition is weaker and moderate levels of BMP signalling may specify the cells as neural plate border. Further away still, there is no inhibition, resulting in the specification of epidermis.

In addition to BMPs, there are other signals (FGFs and Wnts) originating from the ectoderm as well as paraxial mesoderm playing a role in initial specification of the neural crest precursor cells. Both BMPs and Wnts continue to be expressed in the dorsal region of the closed neural tube, and this may be required for the maintenance and proliferation of the crest cells. Following this, the neural crest cells begin to express a series of transcription factors such as Snail, Slug, Pax3, Pax7, and members of the Zinc family such as Msx1 and Msx2. In the posterior part of neural tube, the neural cells are not determined as neural crest cells until the cells actually undergo the epithelial to mesenchymal transition and begin to migrate.

CONTROL OF NEURAL CREST MIGRATION

The ability of neural crest migration arises following the epithelial to mesenchymal transition, which appears to be controlled by changes in cell adhesion and cytoskeletal architecture. The transcription factor Slug plays a pivotal role in this transition, because when expression of the *Slug* gene is inhibited, neural crest cells are unable to migrate. Evidently, genes encoding cell adhesion molecules are regulated by *Slug* gene. For example, the genes for N-cadherin and several other genes are expressed generally by cells of the dorsal neural tube, whereas other cadherin genes are expressed only during the migratory phase. Cadherins associate with many regulatory proteins including β- catenin, which is activated by Wnt signalling. Thus, several Wnts are expressed in the dorsal neural tube at the time of neural crest cell migration and may play a role in modulating the adhesive properties of the cells. Migration is also facilitated by changes in the extracellular matrix, which guides the cells towards their destination. For example, the migration of cranial crest cells requires matrix components such as fibronectin, tenascin, and laminin in chicken embryos.

THE CRANIAL NEURAL CREST

It should be remembered that the neural crest cells emanate from all along the neural tube including the differentiating brain parts. Cranial neural crest cells contribute to a variety of different structures in the head and neck, depending on their position along the anterior- posterior axis of the central nervous system. The cranial crest cells have three major migration routes. The neural crest cells migrate in discrete, segregated streams from the neural tube into the adjacent pharyngeal arches. Cranial neural crest cells migration and patterning rely on a balance between the signals they acquire in the neural tube during their formation and their competency to respond to the signals they contact in the environment during their migration and differentiation. Some examples of extracellular signals to which the neural crest cells respond are (1) Neuegulin-1 which promotes Schwann cell differentiation, (2) BMP2/4 which promotes autonomic neuronal and smooth muscles, and (3) TGF-β that promotes smooth muscle differentiation.

The most anterior crest cells migrate to various positions in the developing head and face, giving rise to craniofacial mesenchyme that will form dermal bones, connective tissue surrounding the eye, the eye's ciliary and papillary muscles, and dermis of the face. The neural crest cells originating from the hindbrain consisting of the rhombomeres migrate predominantly to the pharyngeal arches. Cells from rhobomeres 1 and 2 migrate to the first pharyngeal (mandibular) arch, forming the jaw bones as well as the incus and malleus bones of the ear. The neural cells of the frontonasal process generate the bones of the face. The remainder gives rise to the hyoid cartilage of the neck, thymus, parathyroid, and thyroid glands. In mammalian embryos, cranial neural crest cells migrate before the neural tube is closed. Some hind brain crest cells also migrate laterally to form the cranial ganglia.

TOOTH DEVELOPMENT

Tooth development provides an important example to illustrate the epithelial-mesenchymal interactions that occur during morphogenesis. During such interactions, the mesenchyme influences the epithelium; the epithelial tissue, once changed by the mesenchyme, can secrete factors that change the mesenchyme. During tooth development, the neural crest-derived mesenchyme cells become the dentin-secreting odontoblasts, while the jaw epithelium differentiates into the enamel-secreting ameloblasts. The mandibular epithelium causes neural crest-derived ectomesenchyme to aggregate at specific sites to give rise to different types of teeth. The interaction between the signal molecules, BMPs and FGF8 determines whether the teeth will be of molar or incisor type. Again, the expression of the transcription factor, Pax9 in the ectomesenchyme is critical for the initiation of tooth morphogenesis. In Pax9-deficient mice, tooth development ceases early.

NEURAL CREST CELLS AND AXIAL PATTERNING

Neural crest cell derivatives generate a variety of cell types and structures along the anterior-posterior axis of the vertebrate embryo. Gene expression studies have revealed that *HOX* family of transcription factors display ordered expression in cranial and trunk neural crest cells, as they migrate from the neural tube. In animal embryo development, *HOX* genes have a highly conserved role in the regulation of axial patterning and the specification of regional identity. Accordingly, *HOX* expression provides a combinatorial effect that specifies the anterior-posterior character of the neural crest cells, thus patterning the neural crest cells.

It thus appears that the combinations of *HOX* genes expressed in the various regions of neural crest cells specify their fates. When *HOXa-2* is knocked out from mouse embryos, the neural crest cells of the second pharyngeal arch are transformed into those structures of the first pharyngeal arch. Similarly, when the *HOXa-3* gene is knocked out of the mice embryos, the resultant mutant mice had severely deficient or absent thymuses, thyroids, and parathyroid glands, shortened neck vertebrae, and malformed major heart vessels. Furthermore, *HOXa-3* genes are responsible for specifying the cranial neural crest cells that give rise to the neck cartilage and pharyngeal arch derivatives. On the other hand other *HOX* genes, such as *HOXa-1* and *HOXa-2* are required for the migration of rhomomere 4 neural crest cells into the second pharyngeal pouch.

Moreover, altering *HOX* gene expression patterns vitiates neural crest cells specification. In the mice, retinoic acid alters *HOXb-1* expression patterns and mediates transformation of hindbrain regions. In mutants of *HOXa-1* and *HOXb-1* that fail to express these genes in rhombomere 4 (r4), the mouse fail to develop ears.

CARDIAC NEURAL CREST

A specialized population of cardiac crest is located in between the posterior region of the hindbrain and the trunk neural crests. Cardiac neural crest cells have the unique ability to form endothelial tissue of the aortic arch as they arise from the heart and form the truncoconal septum between the aorta and the pulmonary artery. If the cardiac neural crest cells are removed and replaced by anterior cranial or trunk neural crest, cardiac abnormalities occur suggesting that these cells are already determined to generate cardiac cells, and that the other regions of the neural crest cannot substitute for it. Recently, it has been realized that the cardiac neural crest cells have also a role in heart formation. In mice, the cardiac neural crest cells express the transcription factor Pax3. Mutations of *Pax3* result in persistent truncus arteriosus (the failure of the aorta and pulmonary artery to separate), as well as defects in the thymus, thyroid, and parathyroid glands. In addition, the cardiac cells can develop into melanocytes, neurons, cartilage, and connective tissues of the third, fourth, and sixth pharyngeal arches.

TRUNK NEURAL CREST

The trunk crest cells disperse from the neural tube soon after its closure. There are two major pathways taken by the migrating trunk neural crest cells. Those cells migrating along the dorsolateral pathway becomes the melanin forming cells, the melanocytes. They travel through the dermis and enter the ectoderm, where they colonize the skin and hair follicles. The other pathway is called ventral pathway, wherein trunk neural crest cells become sensory and sympathetic neurons, adrenomedullary cells, and Schwann cells. An exciting feature of trunk neural crest cells is their pluripotency. A single neural crest cell can differentiate into different cell types, depending on its location within the embryo. For example, the parasympathetic neurons formed by the vagal (neck) neural crest cells produce acetylcholine as their neurotransmitter, and hence are cholinergic. The sympathetic neurons formed by the thoracic (chest) neural crest cells produce norepinephrin, and hence are adrenergic. But when chick vagal and thoracic neural crests are reciprocally transplanted, the former thoracic crest produces the cholinergic neurons of the parasympathetic ganglia, and the former vagal crest forms adrenergic neurons in the sympathetic ganglia.

MECHANISM OF TRUNK NEURAL CREST MIGRATION

The migratory pathway of the trunk neural crest cells is controlled by the extracellular matrices surrounding the neural tube. One set of proteins found in the extracellular matrices promotes migration. They are fibronectin, laminin, tenascin, collagen molecules, and proteoglycans. There is another set of proteins that impedes neural crest cell migration. The main proteins involved in this restriction of neural crest cell migration are the ephrin proteins. Ephrin proteins are expressed in the posterior section of each sclerotome and hence neural crest cells do not migrate through this region. The neural crest cells recognize the ephrin proteins through their cell surface Eph receptors. Binding to Eph ligand activates the tyrosine kinase domains of the Eph receptors in the neural crest cells, and these kinases phosphorylate proteins that interfere with the actin cytoskeleton, preventing cell migration. This patterning of neural crest cell migration generates the overall segmental character of the peripheral nervous system, as reflected in the positioning of the dorsal root ganglia and other neural crest-derived structures. In addition, chemotactic and maintenance factors are also important in neural crest cell migration. Thus, stem cell factor is critical in allowing the continued proliferation of those neural crest cells that enter the skin. It may also serve as an anti-apoptosis factor and chemotactic factor.

10

CELL DIFFERENTIATION

As the fertilized eggs begin to cleave, the biparental chromosomes duplicate themselves and all daughter cells usually receive identical sets of chromosomes. Yet, a complex organism with several organs is formed from these genetically identical cell mass. The key event, underlying such a development is cellular differentiation, a process by which the descendants of the single celled zygote come to differ from one another and to form tissues and organs performing specialised functions. The embryonic cells, during their differentiation into their final specialised conditions, are influenced by successive inducing stimuli and thus become more determined to differentiate along specific pathways. At the early stage of differentiation, cells become determined or committed with respect to their developmental potential. Thus the mesoderm of somites can give rise to muscle, cartilage, dermis, and vascular tissue, but not other cell types. The determined cells then pass on that determined state to all their progeny.

The central feature of cell differentiation is a change in gene expression. Early embryonic cells can contribute to a wide range of tissue types. But as development proceeds they become less and less pluripotent, as they are tailored towards specific functions. This restriction in developmental potential is associated with turning on of a small number of genes and the turning off or switching off of a large number of genes. A typical specialised cell thus expresses a minority of all the genes in the genome. In other words, gene silencing is an important aspect during cell differentiation. Once cells have differentiated into specific types, the silencing of unwanted gene expression is very tightly controlled. For instance, the modification of DNA with methyle groups (methylation) is commonly associated with gene silencing (see below). This eventually leads to the production of cell-specific proteins that characterize a fully differentiated cell. The examples are: haemoglobin in red blood cells, keratin in skin epidermal cells, and muscle-specific actins and myosins. Thus, a differentiated cell is characterized by the protein it contains.

The earliest step for cells to undergo differentiation is to stop proliferation. Differentiation and proliferation are inversely related, i.e., differentiation programmes are initiated as cell proliferation decreases. In molecular terms, the initiation of differentiation programme that results from transcriptional activation of defined genes is closely linked to the molecular mechanism that lead to the cell cycle arrest.

MECHANISM OF GENE ACTION DURING CELL DIFFERENTIATION

In order to understand cellular differentiation, an overview of gene expression and its regulation in eukaryotes is given here. Transcription is the first stage in gene expression and the principal step at which it is controlled. It involves the synthesis of a RNA chain identical in sequence with one strand of duplex DNA. RNA polymerase is the enzyme responsible for this DNA directed synthesis of RNA. In eukaryotes, there are three types of nuclear RNA polymerases. Of these three, RNA polymerase I, a nuclear enzyme synthesizes precursor of rRNA; RNA polymerase II, which occurs in the nucleoplasm, synthesizes mRNA precursors, and RNA polymerase III, also present in the nucleoplasm, synthesizes precursors of 5S RNA, tRNA and many other small nuclear and cytosolic RNAs. Transcription is initiated at promoter sites which have consensus sequences. None of the 3 RNA polymerases recognises its promoters directly. Transcriptional initiation is achieved by cell specific DNA binding proteins.

Differentiated eukaryotic cells possess the capacity for the selective expression of specific genes. The selective binding of cell specific proteins called transcription factors to promoters and enhancers in the genes modulates the rate of initiation of transcription, and thereby gene expression.

It is now well established that differential gene expression results in the cellular differentiation during embryogenesis. The environment in which the genes lie appears to be responsible for activating or repressing certain genes. The cytoplasmic factor to induce gene expression need not always be present in the respective cell cytoplasm, but can be derived by cellular induction from the neighbouring tissues to the responding tissue. The best example is the diffusible morphogen in the frog *Xenopus*, in which the mesoderm inducing factor, activin ,appears to diffuse from a localised source to activate the expression of the genes *'gooscoid'* and *'xbra'*. Thus, to discuss differentiation is in effect to discuss the regulation of gene expression of multicellular organisms.

A process of DNA methylation also plays a selective role in gene activation. In 1943, R.D.Hotchkiss discovered a 'fifth base' in DNA, 5-methyl cytosine. This base is made enzymatically after the DNA is replicated and about 5% of the cytosine in the mammalian DNA is converted to 5-methyl cytosine. Recent studies have shown that the degree to which the cytosine of the gene is methylated may control the gene's transcription. In other words, DNA methylation might change the structure of the gene and in doing so regulates its activity.

The first evidence that DNA methylation helps regulate gene activity came from studies showing a correlation between the gene activity and nonmethylation especially in the promoter region of the gene. In developing human and chick red blood cells, the DNA involved in globin synthesis is completely nonmethylated, whereas the same genes are highly methylated in cells that do not produce globin. Foetal liver cells that produce haemoglobin during early development have nonmethylated genes for foetal haemoglobin. These genes become methylated in the adult tissues. Organ specific methylation patterns are seen in the chick ovalbumin gene. The gene is nonmethylated in the oviduct cells but is methylated in the other tissues.

EPIGENETICS

As defined by C. H. Waddington, epigenetics is the branch of biology which studies the causal interactions between genes and their products, which bring the phenotype into being. The genome does not have just the sequential information stored in the arrangement of the four bases A, C, G, and T in its DNA chain. But there are additional layers of information stored in the mammalian genome, and this is referred to as the epigenome. Epigenetics is the study of this part of information carried in a cell, which is heritable during cell division but does not constitute part of the DNA sequence.

Epigenetic modifications involve 1) chemically attaching a methyl group (-CH3) to the DNA base cytosine, particularly cytosines that come before the base guanine (the so called CpG sequence) ; 2) chemicals attaching an acetyl group (CH3CO-) to the amino acid lysine at the end of two histone proteins that wrap around the DNA (causing the otherwise tightly folded DNA chain to open up a little) and 3) jumping or transposition of certain stretches of the DNA sequence itself within the long chain, leading to subtle changes in the way the genetic message is read or processed in the cell.

In addition, the binding of specific protein complexes to DNA also occurs to form stable and heritable chromatin structures that ensure efficient silencing of genes that are no longer required for determination of cell fate, allowing expression of only those genes that define properties of specific, differentiated cell types. Taken together, all these epigenetic modifications found in the chromosome make alterations in gene expression patterns, without changes in DNA sequences.

The additional layer of the above discussed epigenetic information is also referred to as genomic imprints, as they carry a molecular memory of their parental origin that is acquired in the germ line. These imprints are erased and reinitiated normally in the germ line, and passed on to the offspring in which they survive into adulthood.

FACTORS INFLUENCING CELLULAR DIFFERENTIATION

There are two major ways by which cell differentiation during embryogenesis takes place. The first involves the cytoplasmic segregation of determinative molecules (morphogens) during embryonic cleavage, wherein the cleavage planes separate qualitatively different regions of the zygote cytoplasm into different daughter cells. The second mode of differentiation is embryonic induction, which involves interaction of cells or tissues to restrict the fates of one or both of the participants.

The cytoplasmic segregation of specific morphogenetic determinants into specific blastomeres was demonstrated by Wilson in the mollusc, Dentalium sp. The mollusc has a spiral cleavage pattern. During cleavage, it extrudes a bulb of cytoplasm immediately before the first cleavage (Fig.10.1). This protrusion is called polar lobe. As shown in the figure, the first cleavage splits the zygote asymmetrically, so that the polar lobe is connected only to the CD blastomere. The CD blastomere then absorbs the polar lobe material but extends it again prior to second cleavage. As a result the polar lobe is now attached to D blastomere, which ultimately absorbs the polar lobe material into the cytoplasm. The removal of polar lobe cytoplasm either in the 3 cells stage (connected to CD) or 4 cells stage (connected to D) embryo resulted in the formation of the normal trochophore larva, which totally lacked all mesodermal organs, such as muscles, mouth and foot. Even the shell gland, which is a derivative of ectoderm, but formed by the induction of the mesoderm, is not formed. It may be concluded from this experiment that the polar lobe cytoplasm contained the mesodermal determinant and that these determinants give the blastomeres the mesoderm forming capacity. Wilson showed that the localization of the mesodermal determinants was established shortly after fertilization, thereby suggesting that a cytoplasmic factor also determined the cleavage pattern to package these mesodermal determinants only into the D cells. Experimental results further indicate that these mesodermal determinants, apart from initiating the mesodermal and intestinal differentiation, is responsible for permitting the inductive interaction leading to the formation of the shell gland and the eye. These determinants are also necessary for specifying the dorsal-ventral axis of the embryo. This example typifies the influence of cytoplasmic factors on the genomic differentiation leading to cell differentiation, early in the embryonic development.

In general, the morphogenetic determinants influence cell differentiation in a gradient principle. In other words, the cells differentiate according to the concentration of a diffusible morphogen which is believed to be secreted from a localised source.

Several studies have been made on the morphogenetic determination of cell differentiation in the fruit fly *Drosophila melanogaster* . Christiane-Vollard and Wiesehans discovered a morphogen specifying the anterior portion of the *Drosophila* embryo. This is called bicoid protein. The mRNA encoding this protein is localised in the anterior portion of the oocyte during oogenesis. After fertilization, this message is translated into bicoid protein, which forms a gradient with the highest concentration at the anterior end of the embryo. Under its influence the pattern of the front part is determined. Mutations in the bicoid genes cause loss of the bicoid gradient and the embryo lacks head and the thorax. Its concentration gradually decreases towards the posterior end. Bicoid protein is a transcription factor, and binds to the promoter of the target gene. One of the target genes of the bicoid protein is the 'hunchback' that is transcribed in the front part of the embryo. In fact, a large number of target genes respond to single morphogen like bicoid depending on its concentration gradient producing many different zones of activation.

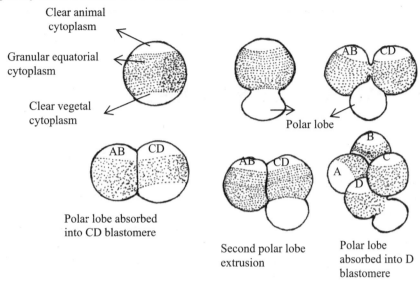

Fig. 10.1 **Polar lobe formation and cytoplasmic segregation in Dentalium development (Redrawn from Davenport, 1979)**

The progressive way of the restrictive gene activation during cell differentiation is exemplified by the development of skeletal muscle. It arises from myoblast cells located at a specific region of the embryo. These cells express several genes including the contractile protein genes, to synthesize myosin, actin, tropomyosin, troponin, but they are not contractile in the early stages. The myoblast cells gradually fuse with each other to produce myotubules, which are multinucleated. They produce more contractile proteins, as their genes get expressed preferentially. At the same time, the expression of the genes for noncontractile proteins decreases considerably. The skeletal muscles now begin to perform contractile functions and are dedicated to the movements of the various parts of the body. Harold Weintrand identified a single gene myoD, which can elicit the entire programme of muscle differentiation when introduced into a non-muscle cell like fibroblasts. myoD encodes a transcription activator protein that promotes myogenesis by directly binding to a control region shared by muscle specific genes. Thus myoD acts like a regulatory gene that plays a central role in the flow of myogenic information from the early embryo to the mature myofibril, a concept of general importance in cellular differentiation.

It may be concluded from the above account that the differential gene expression is controlled at transcriptional level by certain transcription factors which originate in the cytoplasm of the differentiating cells. Cell differentiation, thus, establishes a pattern of gene activity which passes on to daughter cells by various mechanisms described above.

11

DEVELOPMENTAL GRADIENTS

As the multicellular organism develops, its genome progressively expresses the phenotype resulting in the formation of the various organs with different physiological functions. Various embryonic regulatory factors influencing such differential gene action were discussed in the preceding section on cell differentiation. A major question in the embryonic cell differentiation is how the cells recognise their position in the embryo and then differentiate according to their positional information. In other words, positional information is a developmental signal that a cell receives by virtue of the position in the embryo it has arrived at and which directs its further development. The concept of positional information is most commonly conceived as utilising some type of quantitative variation along an axis, which is known as a gradient. The animalizing and the vegetalizing gradients of the sea urchin embryo are the best examples of the gradients that are known to operate in the early development and to control the patterns in the spatial organization and differentiation.

After it was possible that the blastomeres of the sea urchin could be isolated in the early stage and can be developed into whole larvae, experiments on animalization and vegetalization started. Separation of the animal and vegetal cell layers following the third or initial equatorial cleavage showed that segregation of morphogens have already occurred by this time. The four blastomeres of the animal half develop to form a hollow blastula but develops no further; whereas, the four vegetal blastomeres together develop into a blastula which gastrulates and may become a whole larva.

The 16 cell stage consisted of a ring of eight blastomeres or macromeres, and four small micromeres at the vegetal pole. At the next division, the animal half consists of two rings of eight cells each called as an1 and an2 for identification. At the 64 cell stage, the macromeres have also divided into 2 rings of 8 cells called veg1 and veg2. For convenience, therefore, the developing eggs may be said to consist of five layers: an1, an2, veg1 and veg2 and the micromeres. The development of these several layers representing strata along the animal-vegetal axis, has been studied separately, and in various combinations.

ISOLATION RESULTS ARE ESSENTIALLY AS FOLLOWS:

an1   develops to form a blastula entirely covered with a long stiff cilia, but it develops no further; i.e. it is unable to invaginate to form a gastrula.
an2   develops like an1, except that the long cilia do not cover the surface.
veg1   may become a ciliated blastula only or may develop a small archenteron by invagination.

veg2 gives rise to an ovoid larva with an archenteron and one or two spicules, *i.e.* veg2, which normally gives rise only to the endoderm and mesoderm, now produces ectoderm as well Micromeres continue to divide to form a specific number of small isolated cells, representing the primary mesenchyme cells.

From the above experimental results, the existence of a double gradient system is assumed to control morphogenesis and differentiation in the early development of sea urchins. At the animal pole, the power to form structures characteristic of that region is most intense; at the vegetal pole, the power to invaginate and form structures normally arising there (archenteron, mesenchyme) is at a maximum. These two gradients (animal and vegetal) are considered to extend to the opposite pole, so that the gradients overlap through the whole egg. They are qualitatively different in nature, as related to differences in metabolism. The existence of this double gradient system running in the opposite direction was proposed by Horstadius and Runnstrom independently. The theory of double gradients is indeed a modification of the originally proposed metabolic axial gradient of C.M.Child. Using Janus green as an indicator of oxidative metabolism, Child found the metabolic rate higher in the animal pole, thus facilitating the higher frequency of cleavage division. This metabolic gradient gradually diminished towards the vegetal pole, thus confirming the formation of metabolic axial gradient.

In gradient models, a soluble substance called morphogen diffuses from a 'source' (where it is produced) to a 'sink' (where it is degraded), establishing a continuous range of concentrations within that region. Morphogens are thus analogous to chemotactic substances which also operate by establishing a concentration gradient from a source. A characteristic feature of morphogens is that it is readily diffusible and degraded or inactivated. The sea urchin pattern formation in the isolated blastomeres of the animal and vegetal halves thus, provides an example for such a gradient model to exist in the early part of the embryogenesis.

In such a gradient model, the concentration of the morphogen changes over distance, the highest concentration being near the source. The cells with their 'sensors' would respond differently to different concentrations of the gradient. This is achieved by having enhancer elements that can bind the morphogen at different levels. For example, if a morphogen is made at the anterior part of the body, the genes responsible for the organization of head development may have an enhancer that binds the morphogen poorly. Only if there is a large concentration of the morphogen the genes will be active. The genes responsible for thorax formation could have an enhancer that binds the morphogen rather well. This would enable it to respond to relatively low levels of that morphogen. The cells of the head would express both of these genes, while the cells of the thorax would express only that gene whose enhancer could bind low levels of the morphogen. The genes in the posterior portion of the body would not be activated, as there will not be any morphogen in this region.

DEVELOPMENTAL GRADIENTS IN HYDRA

The ability of *Hydra* to form a head or a foot is due to developmental gradients that exist in the tissue. Head gradients originate from the head and control head formation. Similarly, the foot gradients originate from the foot and control foot formation. In the head end of *Hydra* a gradient of head-forming potential, called head activation, is maximal, resulting in the formation of an anterior head. This head gradient decreases down the body column towards the foot. A second gradient, gradient of head inhibition, also originates from the head; this gradient is also maximum in their head and decreases down the body column. Head inhibition prevents body column tissue from forming additional heads. Similar gradients of foot activation and foot inhibition arise from the foot end of *Hydra*. They are maximal in the foot and decrease up the body column. Foot activation promotes foot development at the basal end of the animal, while foot inhibition prevents body column tissue from forming multiple feet. The head and foot gradients work together to maintain the overall morphology of *Hydra* and to prevent the formation of spontaneous structures along the body column. It is the relative levels of activation and inhibition along the body axis that determine whether a structure will form or not. The developmental gradients of activation and inhibition are established by factors called morphogens. Recent studies have identified the two activators to be small peptides with a molecular weight of 1000. The inhibitor morphogens are non-peptide molecules with molecular weight less than 500. These morphogens are produced in nerve cells and released into the intercellular spaces of *Hydra*, where they act on target cells.

12

AXIS SPECIFICATION

DROSOPHILA

The axis determination in the insect *Drosophila melanogaster* is evident already in the egg stage by the identification of anterior to posterior axis and the dorsal to ventral axis. Such a patterning of the body plan is accomplished by preformed mRNAs and proteins that are synthesized and laid down in the egg by the mother fly. The genes responsible for the synthesis of these mRNAs are called, maternal effect genes. These genes are present in the ovarian cell types such as the nurse cells and follicle cells and their expression products are channeled into the oocytes in the form of mRNA enclosed in ribonucleoprotein particles, and translated into protein after the egg is fertilized and laid. In *D. melanogaster*, cleavage commences after the fusion of male and female pronuclei. This involves a series of rapid nuclear divisions without cytokinesis, generating a syncytial embryo. Hence, early events of pattern formation occur in the context of this syncytium. Therefore, when the cellular blastoderm is formed, the future segmental body plan is already established. After gastrulation, the embryo undergoes germ band extension by which the ventral blastoderm (germ band) extends towards the posterior and curves over the back of the embryo. At this time, the metameric organization of the embryo becomes apparent with the appearance of a series of ectodermal grooves. These grooves demarcate the developmental compartments known as parasegments. The parasegments do not correspond to the future segments but each of which represents the anterior two thirds of one segment and the posterior third of the next.

ANTERIOR – POSTERIOR POLARITY

The anterior posterior polarity in the *Drosophila* eggs is specified by the maternal effect genes expressed in the mother's ovary and their mRNAs placed in different regions of the egg. The proteins translated from these mRNAs act as the regulatory proteins and diffuse through the syncytial blastoderm to activate or repress the expression of certain zygotic genes. Two such proteins, Bicoid and Hunchback regulate the production of anterior structures, while another pair of maternally specified proteins, Nanos and Caudal, regulates the formation of the posterior parts of the embryos (Fig.12.1).

BICOID PROTEIN

The bicoid mRNA is sequestered into the egg by means of its untranslated 3'- region and is anchored at the egg's anterior pole. After fertilization, the bicoid mRNA is translated into bicoid protein. The bicoid protein has restricted capacity to diffuse through the egg cytoplasm and thus a gradient is established having its highest concentration at the anterior pole and extending over half the length of the egg. The bicoid protein is a transcription protein that acts as a morphogen all along the length of the egg. The gradient of the bicoid protein provides positional information by switching on certain zygotic genes at different threshold concentrations, thus initiating a new pattern of gene expression along the longitudinal axis. Thus, a large concentration of bicoid protein in the anterior end switches on head specific combinations of genes, whereas, a reduced bicoid concentration activates the thorax-specific gene combinations. In the bicoid mutant, where the mother lacked the bicoid gene, the larva lacked a head and thorax and the acron (anterior most head structure) is replaced by an inverted telson.

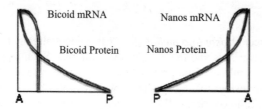

Fig. 12.1 Distribution of maternal effect gene products bicoid and nanos in the fertilized eggs of *Drosophila*. The two gradients of bicoid mRNA and protein and nanos mRNA and protein run in opposite direction. A= Anterior, P=Posterior axis (adapted from Wolpert 1998)

Among the zygotic genes switched on by the bicoid transcriptional factor, hunchback is one of the first to be expressed. The hunchback protein, also being a product of maternal effect genes, is widely distributed in the syncytial egg blastoderm initially, but its spatial expression domain in the embryo becomes restricted to the anterior two thirds of the eggs by the suppressing influence of the Nanos protein.

NANOS AND CAUDAL PROTEINS

Just as the anterior portion of the embryo is specified by the bicoid proein, the posterior portion of the embryo is organized by the maternal effect gene product, nanos protein, which diffuses away from the posterior tip of the egg, forming a concentration gradient running in an opposite direction to the bicoid gradient. Unlike bicoid, nanos is not a transcription factor and does not bind to the DNA. Another maternal posterior group gene, *oskar*, helps to localize the nanos mRNA at the posterior pole of the egg, by its product. Furthermore, nanos protein does not act directly as a morphogen to specify the abdominal pattern. Instead, it suppresses the translation of the maternal hunchback RNA in the posterior region of the egg. Under the influence of high concentration of bicoid proteins, the zygotic hunchback is also activated, resulting in an antero- posterior gradient of hunchback protein, which acts as a morphogen for the next stage of patterning. The inhibition of maternal hunchback RNA in the posterior region by the nanos protein enables the establishment of such an anterior-posterior gradient of zygotic hunchback protein. Similarly, another maternal product, *caudal* mRNA is also uniformly distributed throughout the embryo initially. Nevertheless, a posterior to anterior gradient of caudal protein is established by the inhibition of caudal protein synthesis by the bicoid protein in the anterior end.

FORMATION OF TERMINAL BODY STRUCTURES

The Torso receptor, a product of the torso gene mediates the formation of the terminal body structures, such as the acron (anterior) and the telson (posterior). Torso is a tyrosine kinase transmembrane protein serving as a receptor for an extracellular signal molecule. The signal molecule is synthesized by the follicle cells of the ovary and deposited in the perivitelline space at each end of the egg during oogenesis. After fertilization, this ligand is released and diffuses across the perivitellin space to activate the dorso receptor protein at the end of the embryo only. The binding of this ligand with the torso receptor initiates the differentiation of the terminal body structures such as acron and telson. The homeobox gene *caudal* is also involved in the specification of the telson.

DORSAL-VENTRAL POLARITY

Like the anterior- posterior axis, the dorsal- ventral axis is specified by the maternal genes at about the same time. Just as the bicoid protein establishes anterior-posterior axis, another maternal protein, dorsal, controls the patterning of dorsal- ventral axis. Dorsal is a transcription factor forming a gradient to establish the dorsal- ventral polarity in *Drosophila* embryo. Initially, the dorsal protein is present in the cytoplasm of the peripheral blastoderm throughout the embryo. However, during the 14th cleavage division, there is a translocation of cytoplasmic dorsal protein into the nucleus only on the ventral region. This specificity of dorsal protein entering into the ventral cells nuclei is mainly due to the expression of several other maternal effect genes only in the ventral region. In the unfertilized eggs, dorsal protein is bound to another maternal protein, called cactus, which prevents the entry of dorsal into the nucleus. However, after fertilization and blastoderm formation, a membrane bound toll receptor is activated only in the ventral region of the embryo. This receptor activation is accomplished by an extracellular ventral signal called spatzle fragment, originating from the ventral perivitelline space. The signals emanating from the activated toll receptor is then transmitted along an intracellular signalling pathway involving two other maternal gene products, tube and pelle, resulting in the degradation of the cactus protein. Now the dorsal protein is set free to enter the nucleus only in the ventral region. Since the toll signal is very weak in the dorsal blastoderm cells, there is little or no dorsal protein in the nuclei in the dorsal regions of the embryo. In embryos lacking cactus protein, the dorsal protein is found in all nuclei, resulting in the absence of a gradient for dorsal protein. Consequently, the embryo is ventralized all over, without any development of dorsal structures.

Once inside the nucleus, the dorsal protein binds to certain genes to activate or repress their transcription. If dorsal does not enter the nucleus, the genes responsible for specifying ventral cell types (*snail* and *twist*) are not transcribed. Similarly, in the dorsal region of the embryo, two genes, *decapentaplegic* and *zerknullt* are responsible for specifying dorsal cells.

GENETIC CONTROL OF BODY SEGMENTATION

Along its main axis, the body of insects is composed of repetitive modules which eventually become visible as segments (see above). The zygotic genes, switched on by the transcription factors encoded by maternal effect genes, such as *bicoid*, are involved in pattern formation, eventually resulting in the establishment of the segments. Actually, there is a hierarchy in the sequential activation of the zygotic genes in determining the pattern formation leading to segment specification.

GAP GENES

As seen in the fig. 12.1, a gradient of hunchback protein is formed as a morphogenetic field all along the main axis of the embryo, under the influence of the anteriorly accumulated bicoid proteins and the posteriorly placed nanos protein. This gradient of morphogenetic determinants differentially activates the gap genes, which define broad territories of the embryo. The gap genes are expressed in spatially restricted zones called parasegments. Each parasegment is a developmental compartment i.e., cells in one parasegment remain restricted therein and give rise to clones that never cross the boundary into the next parasegment (Fig.12.2). Gap genes, when mutated, cause large contiguous gaps in the segment pattern of the embryo. Examples of gap genes are *hunchback*, *Kruppel*, and *knirps*. Mutations of *hunchback* lead to loss of head and thoracic structures whereas, mutations in *Kruppel* lead to a loss of thoracic and abdominal structures. Mutation in *knirps* causes loss of most of the abdominal structures.

PAIR-RULE GENES

The expression of the gap genes is followed by the expression of the pair-rule genes. In fact, many of these zygotic genes are expressed as proteins that function as gene regulatory factors. The pair-rule genes are expressed in patterns of seven stripes in alternating segments: one vertical stripe of nuclei expresses a gene, the next strip does not express it, the next adjoinig segment expresses it, and so forth (Fig. 12.3). For an example, there is an alternating expression of two genes, *eve* (*even-skipped*) and *ftz* (*fushi tarazu*) in the parasegments. Mutations in *pair-rule* genes affect alternating segments. For example, in embryos mutant for *even-skipped*, the denticle belts in the first and third thoracic, as well as in all even-numbered abdominal segments are deleted.

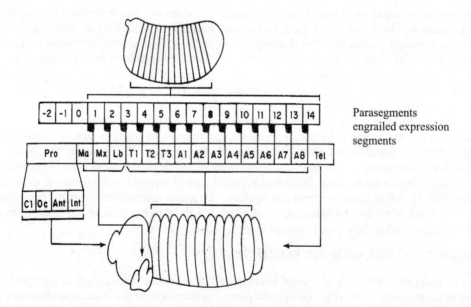

Fig. 12.2 Relationship between segments and parasegments in the syncytial embryo (top panel) and cellular blastoderm (bottom panel), with reference to early engrailed expression in the trunk region of *Drosophila melanogaster*. Pro=procephalon (occipital (Oc), antennal (Ant) and intercalary (Int) segments). Ma=mandibles, Mx=maxillae, Lb=labium. T and A represent thoracic and abdominal segments respectively. Tel=telson.
(Redrawn from R.M.Twyman 2001).

SEGMENT POLARITY GENES

The *pair-rule* gene expression is followed by the activation of segment polarity genes which divide the embryo into segment–sized units along the anterior- posterior axis. Some of the important segment polarity genes involved in the demarcation of final visible segment boundaries in the middle of the parasegments is *engrailed* (*en*), *wingless* (*wn*) and *hedgehog* (*hh*). In the normal embryos, the stripes expressing *engrailed* are adjacent to stripes expressing the wingless protein. The border between these adjacent stripes marks the future boundary between two visible segments. In *Drosophila*, mutual interactions between *engrailed* and *wingless* expression is implicated in the segregation of mesoderm from the ectoderm.

THE HOMEOTIC GENES

The homeotic genes finally define the individual identity of each segment. For example, these genes determine the wingless prothorax, a winged mesothorax, and a metathorax with halters. The genes responsible for this function are called homeotic selector genes or homeotic master genes. The proteins encoded by these genes contain a DNA-binding domain namely helix-loop-helix domain in their respective homeobox sequences. In *Drosophila*, most of the homeotic genes are on the third chromosome, arranged in two clusters. One cluster is called the *Antennapedia complex* (*Antp-C*) and the other is called *bithorax complex* (*BX*). Flies with the bithorax mutation have part of the haltere (the balancing organ on the 3rd thoracic segment) transformed into part of a wing, whereas flies with the dominant Antennapedia mutation have their antennae transformed into legs. Genes identified by such mutations are called homeotic genes, because, when mutated, they result in homeotic transformation, a process in which a whole segment or structure is transformed into another related one, as in the transformation of antenna to leg. Interestingly, the sequential positions of the genes along the chromosome correspond generally to the sequence along the body of locations where the genes are expressed (see Fig.12.4).

After their discovery in *Drosophila melanogaster* , the *homeobox* genes were also found in vertebrates for pattern determination along the anterior-posterior axis. These genes are collectively called as *HOX* genes and occur in all vertebrates, including prochordates. In the vertebrate, the *HOX* genes occur as four clusters and in the mouse they are designated as *HOX* complexes such as *HOX a*, *HOX b*, *HOX c* and *HOX d*. They are located on chromosomes 6, 11, 15 and 2 respectively. A feature of *HOX* gene expression in both insects and vertebrates is that the genes in each cluster are expressed in a temporal and spatial order that reflects their order of occurrence in the chromosome.

AXIAL GRADIENTS IN VERTEBRATES (*XENOPUS*)

Unlike *Drosophila*, vertebrate axes are not formed from preexisting localized determinants, but occurs through a sequence of interactions between neighbouring cells. These inductive interactions have been first demonstrated by Hans Spemann by his Nobel Prize winning work in 1935. The experiments of Spemann and Mangold showed that the dorsal lip of the blastopore, and the notochord that forms from it, constituted an organizer that could instruct the formation of new embryonic axes. This was further evidenced by the fact that when the dorsal lip tissue was transplanted into the ventral region of the embryo, the blastoporal lip initiated gastrulation and embryogenesis leading to the formation of another embryo. Spemann concluded that the transplanted dorsal lip induced the host's ventral tissue into a secondary embryo with clear anterior-posterior body axis. During normal development, these cells organize the dorsal ectoderm into a neural tube and transform the flanking mesoderm into the anterior- posterior body axis. Recent evidence indicate that the inductive properties of dorsal blastopore lip to give rise to presumptive mesoderm is derived from the vegetal cells (presumptive

102 Molecular Developmental Biology

endoderm) underlying them. This inducing factor with organizer effect originating from the dorsal most vegetal cells of the blastula is called the Nieuwkoop centre. The factor emanating from the Nieuwkoop centre is β-catenin, which is part of the *Wnt* signal transduction pathway. Injection of exogenous β-catenin into the ventral side of the embryo produces a secondary axis, suggesting its axialising potentiality.

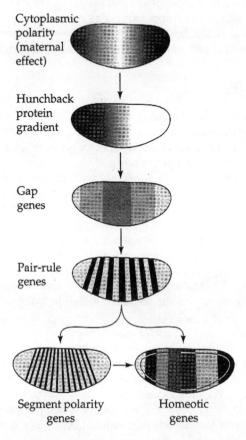

Fig. 12.3 Temporal sequence of expression of master genes controlling embryo pattern formation in *Drosophila*. The stripes show the distribution of proteins encoded by such genes. The pattern is established by maternal effect genes that determine the main body axes and induce the expression of zygotic gap genes; these genes define broad territories and turn on pair-rule genes that are expressed in alternating stripes and forecast the location of future segments. The segment polarity genes initiate the actual segmentation and their subdivision into smaller units. The homeotic genes ultimately define the individual identities of the segments. (Redrawn from Gilbert and Raunio 1997)

In the frog embryos, the patterning of anterior-posterior axis involves cell intercalation activity, thereby aligning trunk extension with anterior-posterior polarity. Choradamesodermal cells (prospective notochord in vertebrates) in the early amphibian embryos undergo dramatic narrowing (convergence) and lengthening (extention) movements parallel to the anterior-posterior axis. Establishment of such an anterior-posterior polarity along with the position dependent chordamesoderm cell activities are due to countergradients of two expressed zygotic genes namely, *Xenopus Brachyury (Xbra)* and *Chordin (chd)*. In the midgastrulae, *Xbra* expression is strongest at the blastoporal lips and fades off towards the anterior chordamesoderm. The converse is true for *chd* expression. The activities of these two genes are under the control of a signalling molecule, activin. Activin boosts the expression of chordin, but reduces *Xbra* expression in the anterior end of the embryo. Interestingly, the anterior-posterior patterning of genes in the frog is similar to those of *D. melanogaster*, namely the pair-rule genes.

AXIS FORMATION IN MAMMALIAN EMBRYOS

Like the amphibians, mammalian embryos (before gastrulation) have also two signalling centres, one in the Henson's node and another in the anterior visceral endoderm. The node appears to be responsible for the creation of all the body, but the two signalling centres work together to form the forebrain. These two signalling centres express several genes during the axis specifications. The node produces two important transcription factors namely Chordin and Noggin, whereas the anterior visceral endoderm expresses several genes necessary for head formation. To name a few are the genes producing the transcription factors Hesx-1, Lim-1 and *Otx*-2 together with the gene for the paracrine factor Cerberus.

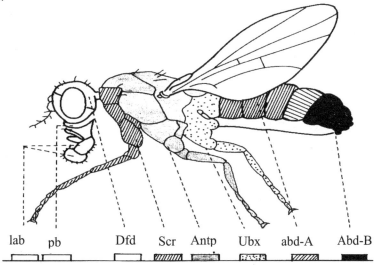

Fig. 12.4 The *HOX* genes of a fruit fly. Eight genes (abbreviated lab, etc.) are located on one chromosomes in the fruit fly, and each gene shapes the development of body regions at different positions along the body axis (indicated by shading and stippling) (Redrawn from Gilbert 2000).

HOX GENES

Vertebrate *HOX* genes are arranged and expressed in the same order as the body parts they help to produce. These genes are grouped into clusters on chromosomes, and are lined up in these clusters in the same order as they are temporally and spatially expressed during development. The *HOX* gene clusters encode proteins with a DNA- binding region known as the homeodomain, which controls the transcription of genes that regulate specific embryonic patterning processes. In vertebrate embryos, all four *HOX* gene clusters are expressed in a temporally and spatially collinear manner along the anterior-posterior axis. In general, the gene at the 3'end of each cluster is expressed first and most anteriorly to dictate the identity of anterior tissues. The gene at the 5' end is expressed finally and posteriorly to pattern the posterior structures.

After commencement of gastrulation, the next set of developmental genes namely *HOX* genes start expressing specification of the anterior-posterior polarity, as in all other vertebrates. These genes are homologus to the homeotic gene complex of the fruit fly *Drosophila*. The homeotic gene complex of *Drosophila* is resident on chromosome 3 as antennapedia and bithorax clusters, serving as a single functional unit. On the other hand, mouse and human genomes contain four copies of the *HOX* complex, located on four different chromosomes. However, the order of these genes on their respective chromosomes is remarkably similar. In addition, the expression of these genes follows the same pattern: those mammalian genes homologous to the *Drosophila labial, proboscipedia,* and *deformed* genes are expressed anteriorly, while those genes that are homologous to the *Drosophila Abdominal-B* gene are expressed posteriorly. Another set of genes that controls the formation of the fly head (orthodenticle and empty spiracles) has homologues in the mouse that show expression in the midbrain and forebrain. *HOX* gene expression can be seen along the dorsal axis from the anterior boundary of the hindbrain through the tail. In the mammalian embryos, different sets of *HOX* genes are necessary for the specification of any region of the anterior-posterior axis. Furthermore, the members of a paralogous group of *HOX* gene may be responsible for different subsets of organs within these regions. Similar to the *HOX* genes, retinoic acid also plays a role in axis specification during normal development. However, teratogenic doses of retinoic acid could cause homeotic changes in mouse embryos. Thus exogenous retinoic acid given to mouse embryos in utero can cause certain *HOX* genes to become expressed in groups of cells that usually do not express them. Thus, the homology of gene structure and similarity of expression pattern between *Drosophila* and mammalian *HOX* genes suggest that this patterning mechanism is highly conserved.

HOX GENE REGULATION OF AXIAL MORPHOGENESIS IN MOUSE

The regulation of the patterning of the trunk and tail as they develop is a function of the *HOX* gene family, which has been evolutionarily conserved among the metazoans. The *HOX* genes confer positional information to the axial and paraxial tissues as they emerge gradually from the posterior aspect of the mouse embryo. During the formation of the vertebral column, pairs of mesodermal blocks are established sequentially on either side of the neural tube as the vertebrate embryo develops. These blocks, called somites, will differentiate into distinct mesodermal tissues, depending on their axial level. The identity of these blocks is specified by their unique combinatorial expression of the *HOX* genes. For example, the first trunk somites to form will give rise to the most anterior prevertebrae. Their anterior (or rostral) identity is achieved through the exclusive expression of the 3' *HOX* genes. The next somites to form acquire a more posterior or caudal identity through the expression of these 3' *HOX* genes, together with the following more 5' *HOX* genes. All axial and paraxial tissues between the middle of the hindbrain and the tip of the tail acquire differential and combinatorial *HOX* expression patterns. Lateral plate mesoderm and spinal cord cells also express a differential combination of *HOX* genes depending on their ultimate axial level.

The combination of *HOX* genes expressed in a specific anterior-posterior region has been called its "*HOX* code". The correspondence between the order of the *HOX* genes on their chromosome and the anterior- to- posterior sequence of the structures that express them has been called 'spacial colinearity'. Similarly, the first expression of 3' *HOX* genes followed by sequential expression of 5' *HOX* genes in the chromosome in a time related manner during embryogenesis is called 'temporal co- linearity'. The expression of *HOX* genes themselves is regulated by certain signal molecules such as Fgf and Wnt. In addition, the transcriptional activation of the *HOX* genes during mesoderm and neurectoderm is accomplished by *Cdx* genes.

13

DEVELOPMENTAL CYCLE AND MORPHOGENESIS IN THE SLIME MOULD *DICTYOSTELIUM DISCOIDEUM*

In the metazoans, multicellularity is achieved by repeated cell division that follows the formation of zygote by sexual reproduction. The multicellular embryo then differentiates into tissues and organs by way of morphogenesis and cell differentiation. However, in certain unicellular organisms, multicellularity is achieved not by cell division but by aggregation. One example is the cellular slime mold, a primitive eukaryotic organism, belonging to the group of " social amebae'(myxamoebae, to distinguish from amoebae species that always remain solitary).These cellular aggregates, nevertheless, exhibit complex features of morphogenesis, cell differentiation and pattern formation, as found in phylogenetically more advanced multicellular organisms. These features include cell movement and chemotaxis, specific intercellular adhesion, cell-cell and cell- extracellular matrix associations, cell type- specific gene expression, morphogens, polarity and pattern formation. A study on the developmental cycle and morphogenesis in the slime mould *Dictyostelium discoideum* illustrates this developmental phenomenon.

LIFE CYCLE

The well studied *Dictyostelium discoideum* is a social soil amoeba whose life cycle consists of two distinct phases- growth and development (Fig.13.1). The haploid amoebae live in films of water on decaying logs, feeding on bacteria and multiplying by binary fission (vegetative phase). These amoeboid cells continue to divide, remaining solitary and undifferentiated as long as the nutrient sources are abundant and adequate. However, following the depletion of nutrients or removal of the food source, the cells initiate their developmental program. The starving cells aggregate and turn into a migrating slug containing many thousands of cells that not only move in a directed fashion, but are amazingly sensitive to their environment, so that they can produce their spores in a spot for optimal dispersal. The 'slug' or grex is surrounded by a slimy acellular sheet which enables them to move much like a slug. It migrates to a bright location where it transforms into a fruiting body. The smaller group of anterior cells of the slug is clearly different from the posterior cells; the former becoming the stalk cells, and the latter, the spores. They fruit by having the slug stop its forward motion with anterior amoebae becoming vacuolated to form the cellulose enclosed stalk cells. The posterior cells are pulled up as though they were in a bag, and each one becomes encapsulated to form a spore, which rests atop the long stalk. The basal plate and the stalk consist of somatic cells, which eventually die. But the spore cells survive; they constitute the generative cells whose formation and release perform the function of asexual reproduction. In the presence of nutrients, the spores germinate, giving rise to feeding amoebae, and thereby restarting the cycle (see Fig.13.1). The whole cycle takes about 2 to 3 days in the laboratory and the slug and the fruiting bodies are on the order of magnitude of a millimeter. However, the size is dependent on how many amoebae entered an aggregate.

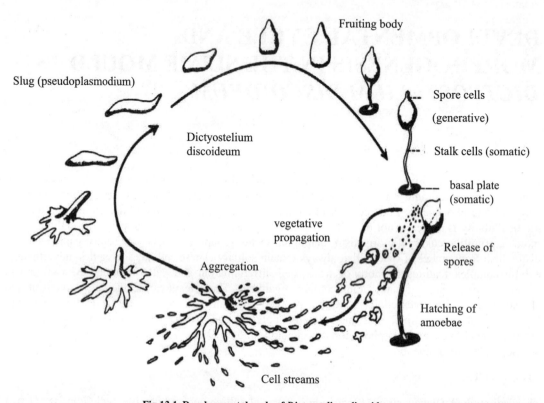

Fig.13.1. Developmental cycle of *Dictyostelium discoideum*.
Haploid cells hatch from the spore envelope, live as soil amoebae, and reproduce asexually by mitotic cell division. As the food supply is exhausted, the amoebae present within a certain area assemble at a point into an aggregate. The size of the area is determined by the range of chemotactic signals emitted by the cells in the center of the aggregate. The aggregate takes the form of a slug, migrates to an appropriate place, and forms a fruiting body that releases new spore cells. The spore cells are born by a stalk and fixed to the substratum by a basal plate. The spores are released, hatching into the `amoebae which migrate and form the aggregation through cell streaming. (Redrawn from Gilbert, 2000)

POLARITY, PATTERN FORMATION AND MORPHOGENESIS

The cellular slime mold is an excellent model system for studying the construction and patterning of multicellular structures. As said before, the morphogenesis and the cytodifferentiation start when the food source is removed from the exponentially growing *Dictyostelium* amoebae. They start a catabolic process that includes the metabolism of intracellular macromolecules, such as glycogen, protein and RNA. Following this, some cells begin to secrete 3', 5' cyclic adenosine monophosphate (cAMP), which now acts as a chemoattractant. Amoebae adjacent to the secreting cells respond by moving centrally to the cAMP source, and start secreting cAMP themselves, thus relaying the chemoattractive signal to other moving amoebae. This leads to the establishment of aggregation territories, containing approximately 105 cells. A family of four proteins, with galactose binding activity, called the discoidin lectins, accumulates during aggregation. The ability of this lectin to act as a carbohydrate binding protein suggested an obvious cell adhesive function whereby, the lectin on one cell binds to a carbohydrate ligand on another cell. Thus, the discoidin molecule may act as extracellular matrix materials to increase the cell cohesion and also involve themselves in the directed migration of aggregating cells. However, the cells retain individuality and hence cell fusion does not occur. An apical tip now develops on the newly formed aggregate that acts like an embryonic organizer and gives the structure- specific polarity to the moving slug. Interestingly, transplantation of a tip of one slug to the body of another slug induces a second morphogenetic axis, much like the transplantation of the dorsal lip of the blastopore in amphibian embryos; the recipient slug dividing into two smaller slugs, eventually.

The multicellular aggregate surrounds itself with an extracellular matrix or slime sheath composed of cellulose and specific proteins. The slime mold is now ready for proceeding along one of two alternative developmental pathways: either forming a fruiting body directly at the site of aggregation, or becoming a different structure called a migrating slug (Fig.13.1). The pathway selected is based on the environmental conditions, such as pH, light, humidity, salt concentration, and the accumulation of ammonia, in the extracellular fluid. The slugs are photo- and thermotactic. If the environmental conditions favour fruiting body construction, the slug stops migrating and starts constructing a fruiting body. During this morphogenetic process, the originally undifferentiated growth- phase cells differentiate into two discrete regions of anterior pre-stalk and posterior pre-spore cells. The anterior pre-stalk cells move down through the aggregate to form the cellular stalk while the posterior nascent spore mass rises on the developing stalk, differentiating into fruiting bodies.

The slime mold formation and the fruiting body construction involve morphogens to give polarity to the aggregating amoebae and morphogenetic events coupled with specific gene expression leading to pattern formation and ultimate cell differentiation.

ROLE OF MITOCHONDRIA IN GROWTH/DIFFERENTIATION TRANSITION

Growth and differentiation are mutually exclusive, but they are cooperatively regulated during the course of development in *Dictyostelium*. A variety of intercellular and intracellular signals are involved positively or negatively in the initiation of differentiation. In the *Dictyostelium*, mitochondria play a major role in such activities. In the eukaryotic cells, mitochondria are self- reproducing organelles with their own DNA and they play a central role in ATP synthesis by respiration. In *D.discoidium*, mitochondria are the major regulatory machinery of growth/differentiation transition (GDT), cell- type determination, cell movement and pattern formation. Firstly, the expression of the mitochondrial ribosomal protein S4 (rps4) gene is required for differentiation of *Dictyostelium* cells from the GDT point. Other functions of mitochondria or their regulatory products are listed in the Table. 13.1 which summarises the mitochondrial molecules along with their purpoted functions.

GENE EXPRESSION DURING CELL GROWTH

During growth, *Dictyostelium* cells continuously synthesize and secrete autocrine factors that accumulate in proportion to cell density. This includes a glycoprotein called pre-starvation factor (PSF) whose effect is counteracted by food bacteria. The "pre-starvation response" in the cells occurs during the increasing levels of PSF with a corresponding decrease of food source. The discoidin I gene family is among the first genes to be activated in the pre-starvation response. When the food is depleted and cells stop growing, the PSF production declines and the starving cells secrete another glycoprotein, called condition medium factor (CMF). CMF is essential for establishing of cAMP(Cyclic adenosine monophosphate) signalling and the initiation of aggregation.

Tab.13.1 showing the multiple functions of mitochondria in *Dictyostelium* development
(Modified from Yasuo Maeda 2005).

Mitochondrial Molecule / Structure	Function
Mitochondrial ribosomal protein subunit S4 (RPS4)	The expression is essential for the initiation of cell differentiation from the GDT point
Mitochondria-localized molecular chaperone (TRAP-1)	Its translocation into mitochondria induces the prestarvation response and subsequent cell differentiation
Tortoise (TorA)	This novel mitochondrial protein is required for the efficient chemotaxis toward Camp
Mitochondrial large ribosomal RNA (mtlrRNA)	The inactivation of mtlrRNA impairs photo taxis and thermotaxis of the migrating pseudoplasmodium (slug)
Prespore-specific vacuole (PSV)	This cell-type-specific organelle is formed from a mitochondrion as the structural basis in alliance with the Golgi apparatus
Cyanide-resistant respiration	Specific inhibition of CN-resistant respiration induces prestalk differentiation, suppressing completely differentiation of prespore cells

CHEMOTAXIS AND cAMP SIGNALLING

It is now well established that *Dictyostelium discoideum* uses a chemotactic mechanism for aggregation and that cAMP is the chemotactic molecule. In *D. discoideum*, cAMP is secreted as pulses from cells and travels as a wave from cell to cell. Cells detect the increasing concentration gradient of cAMP via cell surface receptors. The cAMP receptor, being a transmembrane protein, interacts with the intracellular guanine nucleotide binding proteins or G- proteins. Thus, a complex signal transduction system is initiated to propagate the cAMP signal and to direct migration and aggregation of the cells. Evidently, cAMP acts both as a chemoattractant and a signalling molecule in *D. discoideum*.

Cyclic AMP also activates specific genes during morphogenesis. Several genes have been found to regulate the growth- differentiation transition in *D.discoideum*. Protein kinase A (PKA) is a major regulator of early development. Over- expression of PKA leads to premature aggregation. Similarly, another protein kinase, YakA is also required for the turning off of growth stage genes and induction of developmental genes. Thus, there is a change in the pattern of gene expression during the transition from growth to differentiation in *Dictyostelium*. While new genes are activated for aggregation and morphogenesis, other growth-related genes are turned off.

MORPHOGENETIC GRADIENTS AND PATTERN FORMATION AND GENE EXPRESSION

As described earlier, the migrating slug is divided into two sections: the anterior pre-stalk and the posterior pre-spore region, the latter accounting for approximately 80% of all the cells. This proportion of the two cell types are maintained relatively constant throughout the long period of slug migration. The mechanism for generating the linear pattern of pre-stalk and pre-spore cells involves the use of morphogenetic gradients that specify cell fate in a concentration- dependent manner. There are four such morphogens identified in *Dictyostelium*, and they are: 1) the cell differentiation- inducing factor (DIF), 2) cAMP, 3) ammonia and 4) adenosine. Each molecule has both specific and interacting effect on cell type determination and cell type gene expression. DIF promotes stalk cell differentiation, by inducing specific stalk cell gene expression and at the same time suppressing pre-spore specific gene expression. Interestingly, DIF can induce pre-spore cells to re-differentiate into stalk cells. The presence of cAMP is also necessary for stalk cell differentiation, but its requirement precedes that of DIF. Additionally, cAMP also plays a pivotal role in cell proportioning and morphogenesis. Many *In vitro* studies have shown that cAMP is essential for pre-spore cell gene expression and spore differentiation. The third morphogen, ammonia is involved in maintaining the migrating slug developmental pathway. Ammonia may act as an antagonist of DIF. The localized exposure of pre-stalk cells in migrating slugs to DIF or the depletion of ammonia causes these cells to prematurely differentiate into stalk cells. The fourth morphogen, adenosine, appears to have an effect on cell type patterning by acting as an antagonist of cAMP. Adenosine may act by competing with cAMP for binding to the cAMP receptor. Adenosine is also responsible for the overall size of the fruiting body.

Apparently, there exists a complex system of effector molecules that interact to control gene expression and cell fate bringing about cell patterning and morphogenesis in *Dictyostelium*. The construction of fruiting body in *Dictyostelium* exhibits similarity with organ formation in multicellular organisms. The process of signal transduction pathways are similar to those used during the embryogenesis of other multicellular organisms. The interplay of the morphogens and gene expression ultimately determining a cell to become a stalk or spore is a characteristic regulative type of development in *Dictyostelium*. The expression of genes during *Dictyostelium* morphogenesis is as complex as in other organisms. However, homeobox containing genes that specify the anterior-posterior polarity of the metazoan embryos have not been identified in the slime mould, *Dictyostelium discoideum*.

Evidently, these unicellular organimsms have achieved multicellularity by aggregation and not by cell division, as it happens in the metazoan embryos. However many of the features of morphogenesis and pattern formation found in the true mutlicellualr organisms are exhibited in these unicellular amoebae during their aggregation and cell differentiation into the fruiting body and stalk formation.

14

ORGANOGENESIS

Gastrulation not only brings in the internalization of the endodermal and mesodermal cells from the peripheral blastoderm, but also establishes the conditions favourable for the primary induction which sets in motion a series of inductive interactions between the germ layer cell types, giving rise to the primary organ rudiments. These interactions essentially occur between the mesodermal derivatives, the mesenchyme and, the epithelial cells of ectodermal and endodermal origin. Two major events are involved in the formation of organ rudiments: (a) morphogenesis, which involves the co-ordinated shape changes of individual cells, and, (b) differentiation of the component cells, by attaining the ability to perform specialised functions.

Functional specialization of the cell assemblages leading to organogenesis may result from the differential utilization of the genome. It is to be pointed out here that there is an identical distribution of the genomic materials into each and every cell of the embryo by the mitotic multiplication during cleavage. The differential gene action during organogenesis, therefore, should depend upon the extra chromosomal factors residing in the cytoplasm of the cleaved cells. Necessarily, there are specialised regions of the egg cytoplasm that determine the fate of the cells. The pattern of cell division during cleavage establishes the distribution of the cytoplasm and hence the location of presumptive tissues and organs. In addition to these cytoplasmic determinants, cellular interactions also occur during morphogenesis influencing differential gene action. The fates of the embryonic cells in different organ rudiments are thus getting restricted progressively. Evidently, synthesis of cell specific proteins is the means by which such restrictive cell differentiation is accomplished.

FORMATION OF VERTEBRATE BRAIN

Neurulation and the accompanying processes of body plan organization is a consequence of primary induction by the chordamesoderm. Once the neural tube is formed, it sinks beneath the surface to occupy a position between the chordamesoderm and the overlying nonneural ectoderm. The neural tube continues to differentiate linearly along the anterior-posterior axis, forming the characteristic structures of the central nervous system namely, the brain and the spinal cord. The chordamesoderm differentiates into the notochord, which lies immediately beneath the neural tube.

The axial differentiation of the neural tube is influenced by the underlying chordamesoderm, as revealed by the following experiment done in amphibian embryo. When the neural ectoderm or the underlying chordamesoderm was excised and rotated 180° with respect to the anterior posterior axis, the rotation of the ectoderm had no effect on the subsequent development of the neural tube. However, the rotation of the underlying chordamesoderm produced the brain in the posterior end of the larvae.

Thus, the regional specificity of the neural tube is conferred on the ectoderm by the underlying mesoderm The formation of the neural tube does not occur simultaneously throughout the ectoderm. While the neurulation in the cephalic region is well advanced, the caudal region of the embryo is still undergoing gastrulation. Regionalization of the neural tube occurs as a result of changes in the shape of the tube. In the cephalic end, where the brain will form, the wall of the tube is broad and thick. Here, a series of swellings and constrictions define the various brain compartments. Neural tube in the caudal region remains as a simple tube. The two open ends of the neural tube are called the anterior neuropore and the posterior neuropore.

FORMATION OF BRAIN REGIONS

In the anterior, the neural tube balloons into three primary vesicles, viz., the fore brain or prosencephalon, mid-brain or mesencephalon and the hindbrain or rhombencephalon. The expansion of the neural tube wall is thought to be caused by fluid pressure pressing against the walls of the neural tube by the fluid within it. As the neural tube closes, secondary bulges such as the optic vesicle are formed laterally from each side of the developing fore brain. The brain now bends at both the regions, in accordance with the overall shape of the embryo especially at the cephalic region. These bends are called (i) cephalic flexure, and (ii) cervical flexure, demarcating the boundaries of the brain ventrioles.

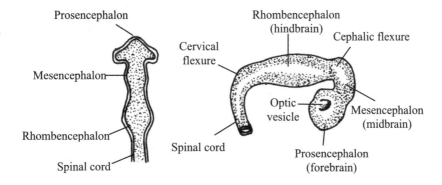

Fig. 14.1 Brain development in human embryo (4 weeks) to show the brain parts and their flexures (Redrawn from Gilbert, 1997)

As further development proceeds, the fore brain becomes subdivided into the anterior telencephalon and the posterior diencephalon. The telencephalon will eventually give rise to the cerebral hemispheres, whereas, the diencephalon will form the thalamic and hypothalamic brain regions as well as the region that receives neural input from the retina. The midbrain is not subdivided but its lumen becomes the cerebral aqueduct. The rhombencephalon becomes subdivided into the anterior metencephalon and the posterior myelencephalon. The myelencephalon eventually becomes the medulla oblongata. The neurons originating from this region will regulate the gastrointestinal, respiratory and the cardiovascular movements. The metencephalon gives rise to cerebellum which co-ordinates movements, postures and balance. The hindbrain develops a segmental pattern that specifies the places where certain nerves originate. Periodic swellings called rhombomeres divide the rhombencephalon into smaller compartments. The regional specialization of the human brain is depicted in the figure 14.1.

PLACODES

The major sensory organs of the head develop from the interactions of the neural tube with a series of epidermal thickenings called the cranial ectodermal placode. Lens placode is the most conspicuous in the early development of the embryo. Besides the lens, the olfactory (nasal) placode gives rise to the sensory olfactory epithelium. The auditory (otic) placode gives rise to the labyrinth of the inner ear and connects to the rhombencephalon via. the auditory nerve.

DEVELOPMENT OF THE EYE

Neurulation is a typical example of primary induction in which the ectodermal cells are inducted by the underlying mesoderm. The development of the vertebrate eye, especially the lens provides an excellent example for the secondary induction, in which, the ectodermally derived neural tube induces lens formations in the ectodermal epithelial cells. Just like the eye development, the majority of the sense organs in the head develop from the interactions of the neural tube with a series of epidermal thickenings called the cranial ectodermal placodes.

Optic development begins even at gastrulation when the involuting endoderm and mesoderm interact with the adjacent prospective head ectoderm. However, the first sign of eye formation appears with the development of two optic vesicles which bulge from the lateral walls of the embryonic diencephalon. These bulges continue to grow and change their shape to produce the cup shaped optic vesicles which are connected to the diencephalon by the optic stalks. Subsequently, when these vesicles contact the head ectoderm, the ectoderm thickens to form the lens placodes (Fig.14.3). The importance of the close contacts between the optic vesicles and the surface ectoderm to form the lens is revealed by the fact that mouse mutant "eye-less" fails to form the eye because of the absence of contact between the optic vesicles and the overlying ectoderm. In the normally developing embryos, the lens forms from the surface that overlies the optic vesicle. The lens primordium first appears as a thickening of the surface ectoderm which results from the ectoderm of these cells. This is achieved by the longitudinal alignment of microtubules in these cells. The lens placode in turn induces the invagination of the optic vesicle which becomes cup-shaped and develops into the pigmented and neural retina of the eye. The neural retina sends back information to the brain via the optic stalk. Concurrently, the lens placode transforms into a cup-shaped pocket called the lens cup. The lens cup eventually constricts from the surface ectoderm and forms a spherical lens vesicle. The lens vesicle ultimately differentiates into the definitive lens. The co-ordinated development of the lens and the retina is due to the induction of the head ectoderm by the optic cup (Fig.14.3). In the amphibians, the formation of lens vesicle is different. Here, the inner layer of the epidermis thickens to form a solid mass of cells. These cells subsequently rearrange themselves into a vesicle which then undergoes lens differentiation. The differentiation of the vertebrate lens starts with the elongation of the cells lining the inner face of the lens vesicle. The cells first become columnar and are later transformed into the long "fibre cells", which occupy the centre of the lens. The cells lining the outer face of the lens vesicle remain epithelial and form a sheath covering the outer surface of the fibre cells. The fibre cells produce the lens specific protein crystallin. In birds and reptiles, in addition to crystallin, and crystallins are also produced. These lens proteins are important for the optical properties of the lens. Certain gene activity controls the lens formation. *Pax-6* gene is expressed in the lens and it regulates crystalline gene expression by encoding a transcription factor. Mutation in the *Pax-6* gene is responsible for the small eye (Sey) phenotypes in the mouse. In the Sey heterozygotes, the lens vesicle and the optic cup are formed but do not undergo further growth and differentiation. *Pax-6* proteins thus appear to be important in the development of the retina and serve as a common denominator for photoreceptive cells in all the phyla.

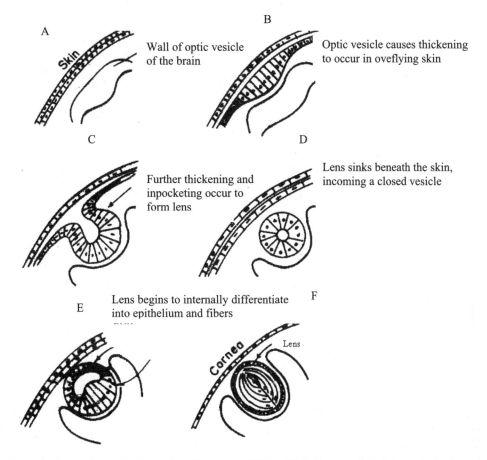

Fig. 14.2. Lens formation in vertebrates (Redrawn from Davenport, 1979) A: Wall of optic vesicle of the brain, B: Optic vesicle causes thickening to occur in overlying skin, C : Further thickening and inpocketing occurs to form lens, D : Lens sinks beneath the skin, forming a closed vesicle, E : Lens begins to internally differentiate into epithelium and fibres, F : Fully formed eye.

HEART FORMATION IN FROG AND CHICK

Cellular induction seems to be the primary source of cell differentiation in organogenesis. Thus, in neurulation, the ectodermal cell is inductively influenced by the chordamesodermal cells to differentiate into the brain and other neural tissues. In the case of the limb formation, the inner mesodermal cells are induced to differentiate into multifarious tissues, such as cartilage, muscles, blood vessels and connective tissues. Yet another instance, where the mesodermal cells are induced by the endodermal cells is the heart formation. Unlike the other organ forming systems, in the heart formation, only the loose mesodermal cells and not sheets of mesodermal cells take part. Though induced by endoderm the heart and the associated structures are formed only by mesodermal cells.

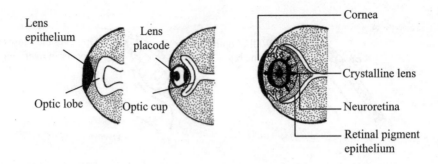

Fig. 14.3 Inductive interactions (signalling) between eye tissues during the eye lens formation
(Redrawn from Jeffery and Martasian, 1998)

FROG

In the frog, at the completion of the gastrulation, the lateral mesoderm in the region of the pharynx consists of loosely arranged cells which are involved in the formation of the gill arches. In the floor of the pharynx, the mesoderm forming the ventral edges of the lateral plates participate in the formation of the heart. The pharyngeal floor occupying the space between the endoderm lining the pharynx and the ectoderm is initially free from any mesodermal cells. Slowly, the mesodermal cell derived from the lower edge of the two lateral mesodermal layers occupies the centre of the pharyngeal floor space. These scattered cells are destined to give rise to the endothelial lining of the heart or endocardium. Evidence indicates that this overlying endoderm has an organising or inducing effect on their development into the heart. By experimentally removing the endoderm of the pharynx, Mangold showed in newt embryos that the heart failed to form from the intact mesoderm and the ectodermal cells probably due to the lack of inductive influence emanating from the endoderm. Another line of evidence from the hanging drop culture of mesodermal cells shows that there was no heart differentiation in the mesodermal materials taken from gastrula, but materials taken from neurula stage embryos showed spontaneous contraction, a process so characteristic of heart forming cells. The mesodermal cells along with ectoderm, taken from a neurula stage, when cultured in saline solution, developed cardiac muscle tissue, which is capable of autonomous rhythmic contraction. They even produce pulsating tube with resemblance to normal hearts. The extracts of endodermal tissues also have the same influence in promoting heart differentiation in these tissues, thus suggesting that the endodermal induction is mediated by soluble substances. The mesodermal cells derived from two lateral parts will give rise to the two separate hearts, if they are separated medially by a piece of foreign mesoderm. Conversely, if a second layer of heart forming endoderm is superimposed by transplantation of the heart forming mesoderm from one embryo upon the heart-forming mesoderm in another mesoderm in another embryo, the two layers will fuse and a single normal heart will be formed.

FORMATION OF THE PRIMITIVE CARDIAC TUBE

The scattered endothelial cells in the floor of the pharynx organise into the two tubes first. Simultaneously the lateral mesoderm descends down to meet in the ventral medial region. In this process, the lateral mesoderm is split up into two layers. First is the somatic mesoderm, which forms the upper layer above the primitive heart and separates the later from the endodermal floor, after fusion with its counterpart from the opposite side. The second layer is splanchnic mesoderm which forms the outer border of the pharyngeal region, after meeting its counterpart ventromedially. As a result, a cavity lined above by somatic mesoderm and below by splanchnic mesoderm is formed in the floor of the pharynx. This cavity is called as the pericardial cavity, in the centre of which is formed the primitive heart, which is now formed as single tube, resulting from the fusion of the two endothelial cell-walled heart rudiments. The endothelial lining of the heart is called endocardium. The splanchnic mesoderm surrounds the endocardium and gives rise to the inner muscular myocardium and the outer visceral pericardial wall. At this time, the somatic mesoderm makes the outer lining of the pericardial cavity. The centrally located endocardial tube is suspended both dorsally and ventrally to the pericardial walls by mesodermal tissues called respectively as dorsal and ventral mesocardium. They, however, disappear later on as the heart formation progresses (Fig.14.4).

FORMATION OF THE HEART PROPER

In all vertebrates, the heart develops essentially in the same manner from a straight tube. This straight tube does not show subdivisions into its various chambers. In the frog, the tube becomes inflected in a very characteristic way, in order to become coiled in the shape of letter 'S'. The degree of the twisting in higher vertebrates is greater than in lower ones. The tubular heart rudiment becomes constricted in some places and becomes dilated in others, and is thus subdivided into its four major parts. Before this subdivision occurs, the functioning of the heart starts; it begins to pulsate at a regular rhythm. This pulsation starts even before the peripheral blood vessels is ready to receive the blood stream.

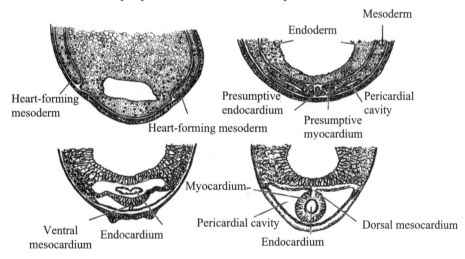

Fig. 14.4 Heart formation in the amphibians. A. Prospective heart on both sides of the embryo. B,C. Migration of the heart (endocardium) to the mid ventral region. D. Formation of myocardium
(Modified from Gilbert and Raunio, 1997)

DEVELOPMENT OF BLOOD VESSELS AND CORPUSCLES

In frog, the blood vessels develop out of the mesenchyme and the splanchnic mesoderm by a rearrangement and differentiation of the cells to form a flat endothelium which constitutes the inner lining of the vessels. It is entirely similar to, and continues with the endothelial lining of the heart. The muscular and the connective tissue coat are also differentiated from the mesoderm and are added later, the muscles being more abundant in the artery and the connective tissue in the veins. The endothelial tissues do not form as a continuation of the heart, but they are disconnected vesicles, which grow together, until they are united. They, however, appear first in the vicinity of the heart. The method of blood vessel formation is same in all the vertebrates. The corpuscles are formed chiefly from patches of splanchnic mesoderm on the ventral side of the yolk mass. These patches are called blood islands; the corpuscles migrate from here to the developing blood vessels.

CHICK

The periphery of the reptilian and avian blastodisc, lying on top of the yolk mass, is differentiated as the area opaca, whose external edge spreads over the surface of the yolk and completely covers it. The cells of the area opaca consist of the same germ layers (ecto-, meso- and endoderm) of the area pellucida. The endoderm specifically adheres to the yolk mass from the growing embryo. The nutrient supply from it to the embryo is facilitated by the blood supply established early in the area opaca. To start with, a network of blood vessels is developed in the inner part of the area opaca which becomes the area vasculosa, and the outer part of it is called area vitellina. Both the blood vessels and the blood corpuscles are formed from a group of cells called blood islands. The cells of the blood island differentiate into the thin epithelial layer which forms into the future endothelium of the blood vessels. The central cells of the blood islands become differentiated into blood corpuscles which lie inside the blood vessels. The blood vessels thus formed join together into a network which establishes connections to area pellucida and the embryo proper. This establishment is achieved by two lateral veins called the left and the right vitelline veins, joining in the unpaired sinus venosus of the heart. They also have connections to the dorsal aorta by means of right and left vitelline arteries.

During the first day of chick development, two masses of mesoderm may be identified on each side of the blastoderm at the anterior end of the primitive streak. This heart forming mesoderm arises from the epiblast cells that have moved through the primitive streak and assembled on either side of the node in the mesoderm layer. During the first 24 hours, this presumptive myocardium acquires the capacity to undergo self differentiation. That the mesodermal areas represent the typical embryonic fields for heart differentiation is shown by the fact that, if these cells when transplanted to an indifferent location, they differentiate into contractile heart muscles.

The folding movement of ectoderm and the endoderm in the formation of the foregut brings the two lateral heart rudiments together in the midline to form a single median tube with double walls. As in the amphibians, the endocardium composed of typical endothelium forms the lining of the heart. Similarly, the epimyocardium differentiates from the splanchnic mesoderm overlying the mesoderm. A thick myocardium and a nonmuscular epicardium or visceral peritonium of the heart is also formed from the splanchnic mesoderm (Fig. 14.5).

An interesting aspect of the heart formation in the chick is the forward migration of heart mesoderm. The migration of these cells is an intrinsic and indispensable facet of heart formation. This movement also occurs under the guidance of the endoderm. Thus, if the endoderm is separated from the heart forming mesoderm by treatment with sodium citrate, at the primitive streak stage, the heart mesoderm develops the vesicles of contractile tissue, but it fails to migrate forward. The role of the endoderm in guiding the migration of the mesoderm is also indicated by the absence of any tubulation movements in endoderm-less chick embryos. After obtaining its final position underneath the pharynx as in amphibians, they are also suspended from above by the dorsal mesocardium. After the fusion of the heart rudiments, the tubular heart undergoes the characteristic twisting, which is very similar to that of the amphibians.

URINOGENITAL ORGANOGENESIS

The urinogenital system consisting of kidney and gonad is derived from the intermediate mesoderm or mesomere. The development of these two systems is so closely interrelated that they are treated as a single system. The development of vertebrate kidney progresses through three major stages namely pronephros, mesonephros and metanephros. These three stages in the kidney development also reflect on the evolution of excretory systems among vertebrates. In addition, some of the structures used in the excretion also take part in the reproductive functions, especially in the transport of gametes at a later stage. The progressive evolution of urinogenital structures among the different vertebrate animals is described below.

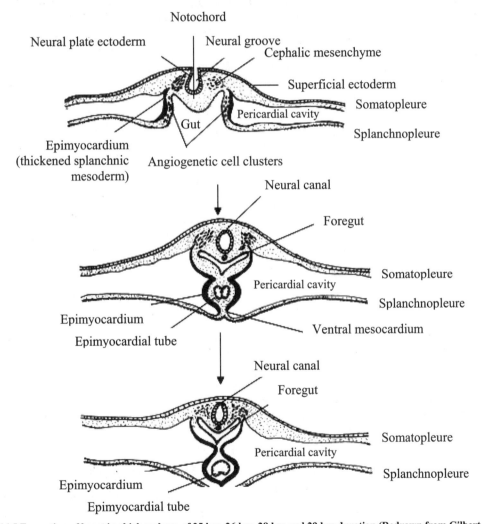

Fig. 14.5 Formation of heart in chick embryo of 25 hrs, 26 hrs, 28 hrs and 29 hrs duration (Redrawn from Gilbert, 1997)

PRONEPHROS

The most primitive type of functional kidney is known as pronephros, which is found in fish and amphibian embryos. The pronephros originates from the mesomere. The specific area of the mesomere that gives rise to the pronephros is called as nephrotome. The cells in the nephrotome give rise to the pronephric tubules which contain internal lumen called nephrocoels. The pronephric tubules are continuous with the coelom by ciliated funnels called nephrostomes. A network of fine blood vessels (glomus) is associated with the ciliated funnels of the pronephric tubules. Waste materials are filtered through the lining of the blood vessels and then enter into the opening of the nephrostomes. The pronephric tubes fuse at the outer end to form the pronephric duct. This duct elongates in a posterior direction and fuses with the cloaca. Although the pronephros develops in the initial stages of the kidney formation, they are never functional in the reptiles, birds and mammals.

MESONEPHROS

As larval fish and amphibians become adults, the pronephric tubules disintegrate, while the pronephric duct remains intact. A second set of tubules develops from the nephrostomes in the intermediate mesoderm by aggregation of the mesoderm cells posterior to the disintegrating pronephric tubules. The new tubules are the mesonephric tubules and the old pronephric duct is now called mesonephric duct or Wolffian duct which functions as the excretory duct since the mesonephric tubules are attached to it. It appears that the formation of the mesonephric tubules is under the induction of pronephric duct, since the prevention of the pronephric duct from the area that forms the mesonephric tubules, results in the failure of the mesonephric tubule formation. The contact area of the glomeruli expands and invaginates to form the Bowman's capsule. Each Bowman's capsule plus its glomerulus is the functional renal unit called Malphigian body. Wastes are filtered in the blood through the Bowman's capsule of the mesonephric tubules and then passed on to the mesonephric duct proper.

The mesonephros is the functional kidney of adult fish and amphibians. It is also the functional kidney of embryonic reptiles and birds and in some mammals like the pig, where the placenta is inefficient in filtration of the waste materials.

METANEPHROS

Metanephros is the functional kidney of adult reptiles, birds and mammals. The origin of the metanephric kidney in a mammal is illustrated in the figure (Fig. 14.6). A new duct, the ureter arises as an outgrowth at the point at which the mesonephric duct joins the cloaca. The outgrowth begins as an evagination, called the ureteric bud. The ureteric bud grows and branches into the metanephrogenic mesoderm which is also called metanephric blastema. This is borne out by the fact that metanephric tubules fail to develop if ureteric bud does not enter the metanephrogenic mesoderm. The metanephros includes the Bowman's capsule and the glomeruli as in the case of mesonephros. In all the above types of kidneys, the tubules are connected to the main excretory duct, which empties the excretory products into the cloaca and through any other means.

Inductive Interactions between the Ureteric Bud and the Mesenchymal Cells

The important event in the development of mammalian kidney is the interaction between two mesodermal tissues, the ureteric bud and the metanephrogenic mesenchyme (metanephric blastema). These two tissues not only interact with each other, but also reciprocally induce each other. First, the metanephrogenic mesenchyme causes the epithelial ureteric bud to elongate and branch (Fig.14.6). At the tip of these branches, the ureteric bud in turn induces the loose mesenchyme cells to form an epithelial aggregate. Each aggregate of about 20 cells will divide and differentiate into an intricate structure of renal nephron. The extent of ureteric bud growth and branching is proportional to the number of nephrons that eventually form. In the rodent metanephric kidneys, fibroblast growth factor (FGF-2) genes modulate the extent of ureteric bud growth as well as the number of nephrons that eventually form.

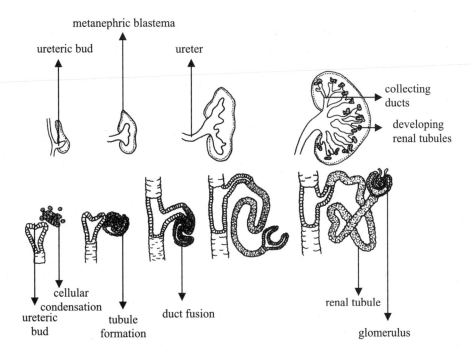

Fig. 14.6 Inductive interaction between ureteric bud and metanephric blastema leading to tubule formation. Each tubule develops a glomerulus (After Wolpert, 1998)

The kind of reciprocal induction between the ureteric bud and the mesenchyme has been demonstrated by *in vitro* studies. These studies show that in the absence of mesenchyme, the ureteric bud does not branch. Similarly, in the absence of ureteric bud, the mesenchyme does not condense to form tubules. When cultured together, the ureteric bud grows and branches and the tubules form throughout the mesenchyme. Furthermore, only the metanephrogenic mesenchyme has the ability to respond to ureteric bud to form kidney tubules. This is evidenced by the fact that the metanephrogenic mesenchyme will form into kidney tubules, even in the presence of the other epithelia-like embryonic salivary gland or the neural tube. This also suggests that the mesenchyme has already been determined, probably as the result of its interaction with the endoderm earlier in the development. Clearly, an innate capacity exists in the mesenchyme for self-organisation into tubules that simply require a relatively non-specific permissive signal for the development to proceed. All this suggests that an interaction between the epithelial ureteric bud and the metanephrogenic mesenchyme is necessary for the kidney tubulogenesis. In order to understand the nature of this induction, the epithelial ureteric bud was separated from the mesenchyme by a porous filter. It was seen that the stimulus was transmitted through the filter, if the pore size was large enough (0.1 µm). If the barrier was nonporous like a cellophane paper, there was no interaction and no tubulogenesis. These experiments clearly demonstrated that this epithelial mesenchymal interaction did not require cell to cell contact, but suggested the involvement of a chemical factor, which could pass through a porous filter.

The conversion of mesenchymal cells into epithelium is an uncommon event in organogenesis. Such changes would involve substantial changes in the extracellular matrix of the metanephrogenic mesenchyme cells by the ureteric bud. Before induction, the extracellular matrix contains fibronectin and collagen type I and III. Upon induction, these proteins disappear and are replaced by a basal lamina, characteristic of the epithelium containing laminin and type IV collagen. The transformation of the mesenchymal cells into the epithelial cells is also accompanied by a change in the expression of their adhesion molecules from N-CAM to E-cadherin. A corresponding change in the cytoskeleton also occurs, when the loose mesenchyme is converted into epithelium. This finally gives rise to the polarised epithelium on a basal lamina from the loose mesenchymal cells. The aggregates of the mesenchymal cells then form S-shaped tube, which elongate and differentiate into the functional unit of a renal tubule and a glomerulus. Thus, the epitheliomesenchymal interactions bring out urinogenital organogenesis in the vertebrates. Such interactions are common in the organogenesis of glandular cells, where the interaction is between the epithelial cells derived from either the ectoderm or endoderm or the mesenchyme derived from the mesoderm.

SEXUAL DIFFERENTIATION

In the mammals, the determination of sex is chromosomal. The female is XX and the male is XY. During mammalian embryogenesis, genetically XX and XY foetuses both develop two pairs of genital ducts associated with the mesonephros and undifferentiated gonad (Fig.14.7). The Mullerian ducts (paramesonephric ducts) have the potentials to differentiate into the oviducts, uterus and upper part of vagina of the female reproductive tract. The Wolffian ducts (mesonephric ducts) can differentiate into the vas deferens, epididymis and seminal vesicles of the male reproductive tract. Thus, each individual has the potential to develop both male and female reproductive systems, regardless of sex chromosome genotype. Thus, for normal male and female development to occur, one of the two genital duct systems in the fetus must differentiate while the other must regress.

In the XY individuals, expression of the Y chromosome-linked *SRY* gene determines the fate of the indifferent gonad to testis. The Sertoli cells of the fetal testis produce a peptide hormone, namely Mullerian inhibiting substance (MIS), also known as anti-Mullerian hormone. MIS activity induces the regression of the Mullerian ducts, thereby preventing the development of female reproductive organs. Mullerian duct cells possess receptors for MIS to facilitate apoptotic action. The degeneration of the Mullerian duct is characterised by the appearance of macrophages in the tissue, and lysosomes and condensed nuclei within the duct cells. Dissolution of the basement membrane surrounding the duct results in the loss of duct integrity followed by the transformation of surviving epithelial cells into mesenchyme. All these changes characterise the programmed cell death during embryogenesis.

Leydig cells of the testis also start producing another hormone, testosterone which induces the differentiation of Wolffian ducts into the vas deferens and associated male accessory sex glands. During female development, the fetal ovaries do not produce MIS, creating a permissive environment for the differentiation of the Mullerian ducts into oviduct and uterus. In addition, the absence of testosterone in female leads to the passive regression of the Wolffian duct system. XY mice that are MIS deficient differentiate both the Mullerian and Wolffian duct systems, creating pseudohermaphrodites.

FREEMARTIN AND MIS

The Mullerian inhibiting substance (MIS) – a 50 amino acid glycoprotein – made in the Sertoli cells is responsible for the disintegration of the Mullerian duct in the undifferentiated gonad, paving the way for male differentiation. The Mullerian duct in the absence of MIS will develop into the female reproductive tract. The antifeminine activity of MIS is well exemplified in certain species, such as bovine foetuses. Freemartin are produced in heterosexual foetuses with fused chorions and vascular anastomoses. When a female foetus is exposed to a male twin's blood by such chorio-allantoic anastomoses, regression of the Mullerian ducts results. This condition is known as Freemartinism. In addition, Freemartin ovaries cease to grow and become depleted of germ cells. They may also develop seminiferous tubules containing Sertoli cells. Recent experiments reveal that some aspects of Freemartin effects including the inhibition of the germ cells proliferation and the development of the seminiferous chord-like substances can be reproduced *In vitro* when foetal rat ovaries are exposed to purified MIS. The Freemartin is an example to illustrate the endocrine mediated sex differentiation in mammals where the sex is determined genetically.

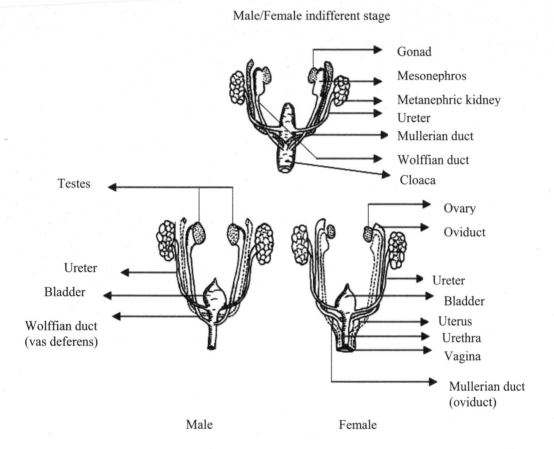

Fig. 14.7 Differentiation of testis and ovary from indifferent gonad (Redrawn from Gilbert, 1997)

GENES INVOLVED IN GONADAL DIFFERENTIATION

Embryonic development of gonad is under genetic control. Several genes are known to be involved in the process of testicular and ovarian development from the undifferentiated gonad (Fig.14.8). Obviously, the hormones that are responsible for the sex differentiation are also under such gene action. The mesoderm giving rise to bipotential undifferentiated gonad is under the control of the Wilm's tumour 1 (*WT 1*) - gene. Another gene involved in the development of bipotential gonad and kidney is *LIM 1* - gene. Further progression from the bipotential gonad towards testicular differentiation is mediated through gonosomal and autosomal genes. Y chromosome is male specific and bears several genes important for male function. A specific testis-determining factor essential for testicular development is encoded by a gene located in Y chromosome. This gene, termed sex-determining region of the Y chromosome (*SRY*) encodes a protein, which acts as a transcription factor regulating the expression of other genes, such as *LMH* gene. In addition, *SRY* controls the expression of steroidogenic enzymes necessary for testicular function. The *SRY* triggers the differentiation of the Sertoli cells in a cell autonomous manner from one of the somatic cell lineages in the genital ridge. In humans and mice, the onset of *SRY* mRNA expression defines testis determination. Expression of *SRY* gene in humans is not switched off and unlike mice, continues until adulthood.

Two other genes involved in the testicular differentiation are *SOX 9* and *DMRT 1* and *2*. Another gene *DAX 1* located in X chromosome is expressed during ovarian development, but is suspended during testicular formation, implying a critical role for this gene in ovarian formation. *DAX 1* expression during testicular development is repressed by *SRY* gene. In particular, *SOX 9* influences the differentiation of the precursor somatic support cells to differentiate into Sertoli cells. The upregulation of *SOX 9* is also controlled by a set of gonadal genes, *SRY*, *DAX1* and *TDA 1*.

ACTION OF SEX CHROMOSOMES IN MALE AND FEMALES

The mammalian X and Y chromosomes arose from one pair of autosomes some 3000 million years ago. Female mammals possess two Xs, whereas the male possesses a Y. After their recruitment into a chromosomal system for sex determination, mutations in genes on the Y made it the male- determining chromosome. Then, the sex chromosomal pair began to diverge. Still the X and Y share a few genes, mainly in the pseudoautosomal region on the tip of the X; here the two chromosomes still exchange DNA to maintain proper segregation in cell division. Over time, the Y disintegrated significantly, losing a major part of the chromosome. However, as long as the genes for maleness are preserved, most of the formerly autosomal genes seem to be largely extraneous because men already have one copy on the X. Interestingly, the X chromosome developed a way to inactivate most of the genes on one of the two Xs in females, so that males and females would have the same dosage of gene products. Early in female development, cells randomly choose either the maternal or paternal X to be the active X chromosome. Recent studies have shown that the inactive human X remains silenced even after being transferred into mouse cells, allowing for the formation of hybrid cells containing the inactive but not the active human chromosome. From this experiment, it has been found that 75% of genes are permanently silent, and about 15% permanently escape inactivation, meaning that they are expressed at twice the level in females as males. The remaining 10% are expressed in some inactive Xs but not others.

ORIGIN AND EVOLUTION OF SEX CHROMOSOME

Like other mammals, human females have two X chromosomes (XX) and males have a single X and a single Y chromosome (XY).The X is large and bears a proportional number of genes (3000 or 4000), which have a variety of functions much like those of genes located on other chromosomes. On the other hand, Y chromosome is much smaller than the X chromosome and contains only a few genes. The chromosomal pattern in snakes is completely opposite from the pattern in mammals: snakes with a ZW karyotype are female and snakes with a ZZ karyotype are male. Like the mammalian X chromosome, the Z chromosome is large, containing about 6% of the genome in all snake families. The W chromosome is much more variable. Some snake families have a tiny, heterochromatic W chromosome, whereas others have a W chromosome that is much the same as the Z chromosome. Birds also have a ZZ male and ZY female system. As in snakes, the W chromosome in birds has different sizes in different families; it is large in ratite birds and small in carinate birds. A process of W chromosome degeneration has therefore taken place both in bird and reptile lineages, independently.

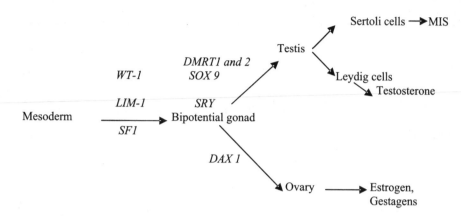

Fig. 14.8 Genes involved in gonadal differentiation

The human X and Y chromosomes originated from a pair of identical chromosomes some 3000 million years ago from the reptiles, long before mammals arose. The genes on these chromosomes were subjected to environmental stimuli for sex determination, which is reflected even today in certain reptiles like turtles. These genes acquired mutations over a period of time resulting in loss of responsiveness to environmental cues. In mammals, sex chromosomes probably arose with the differentiation of *SRY* genes from SOX3, which is a structural homologue on the mammalian X chromosome. Comparative sequence analysis and expression studies indicated that *SRY* and *SOX3* descended from a progenitor gene, with the more evolved *SRY* having gained and retained the male-determining function. Thus the dominant sex-determining allele of the proto-*SOX3*/ *SRY* gene effectively rendered an autosome pair into sex chromosomes. During the divergence of mammalian X and Y chromosomes, minor changes took place in the gross structure of the X chromosome but rapid degeneration occurred in the Y chromosome.

15

PATTERN FORMATION AND DEVELOPMENT OF LIMB

Cell differentiation due to differential gene action brings about cellular diversity during early development of the embryo. Thus, each cell type with its characteristic protein synthesizing ability will give rise to organ systems. Cellular differentiation is however the only one aspect of development. The various forms that characterise the adult animals arise rather from a different process namely pattern formation. It is the process by which the embryonic cells form ordered spatial arrangements of differentiated tissues. Consider the development of human arm. It is always attached to the shoulder and not to the belly, and it has always four parts, the upper arm, the fore arm, the wrist and the hand, which are always arranged in this order. It also has different cell types, such as muscles, cartilage, connective tissue, skin etc. It has the skin on the outside, muscles, nerves, blood vessels and the bone in the middle. It never has bone on the outside and the skin in the middle. Such a precise arrangement and pattern of the many parts and tissues in the adult organism is called the pattern formation. It is achieved through the spatial organisation of differentiated cell types. Though the man and the chimpanzee have exactly the same cell types, the difference between the man and the chimp is the spatial organisation of the cells. In short, the differentiating cells become organised in space during development, resulting in different forms of adult organisms.

DEVELOPMENT OF VERTEBRATE LIMB BUD

As the vertebrate limb is a self-contained system, manipulative experiments could be carried out on limb bud without affecting other regions of the embryo, and thus not affecting the viability of the whole embryo. Furthermore, limb development expounds many principles and processes of development such as axis specification and patterning, region-specific morphogenetic behaviour of cell types, as well as cell migration. As many human birth defects are characterized by limb abnormalities, the study of its development could unravel the genetic and developmental bases of such diseases. More importantly, the developmental anatomy of the limb could be easily visualized due to the presence of cartilaginous elements which can be seen by staining so that changes caused by mutation or experimental manipulation can be easily characterized. In recent years, molecular data generated by transgene misexpression and gene knockouts have also yielded a good understanding on the molecular basis of limb development.

ORIGIN OF LIMB BUD

The formation of vertebrate limb is a fascinating system to study morphogenesis and pattern formation during embryogenesis. Each vertebrate limb originates on the body wall from a circular area of the lateral plate mesoderm called the limb field. Interiorly, the first trace of limb bud development is found in the lateral plate mesoderm. The somatic layer of the lateral plate becomes thickened just underneath its upper edge. The cells of this thickening soon lose their epithelial connections and are transformed into a mass of mesenchyme cells. These loosened mesenchyme cells migrate outwardly and accumulate beneath skin epithelium in the region of the embryo corresponding to formation of fore- and hind-limbs. The epidermis over the mesenchyme mass becomes slightly thickened and bulges outwards and forms the limb-bud.

DEVELOPMENT OF WING BUD

The wing of the chick develops from a paddle-like bud very similar to the arm bud of a human embryo. It starts off as a small bulge that appears about two days after egg laying and at a time when the main axial structures such as somites are laid down. The pattern of cartilage, muscle and tendons take shape within a loose network of mesenchyme cells, encased in a sheet of ectodermal cells that will become skin. After 10 days of egg incubation, the basic pattern of the limb bones namely, humerus, radius and ulna, wrist and digits is well established in the form of a cartilage. Most of these cartilages will be subsequently turned into the real bones by a process of mesenchymal condensation. Cell adhesion molecules such as N-Cadherin and N-CAM bring about such cellular condensation.

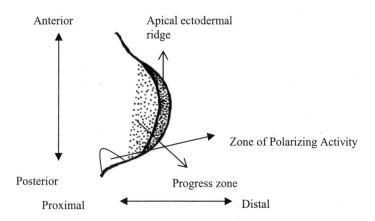

Fig. 15.1 Limb bud formation. The developmental axis are indicated by arrows (Redrawn from Wolpert, 1978)

The developing wing bud has two major components, namely, a core of loose mesenchymal cells enveloped by an epithelial ectodermal layer. Most of the limb structures develop from the mesenchymal core, although, the muscle cell in the limb has a separate lineage from the somite (Fig.15.1). Muscle cells after arriving in the limb bud, interact with the limb bud mesenchymal cells to give rise to the future muscle mass of the limb. At the tip of the limb bud is the progress zone consisting of rapidly dividing and proliferating undifferentiated cells. The progress zone is covered by a thickening of ectodermal cells, viz., apical ectodermal ridge (AER). In the posterior region of the progress zone lie a group of mesodermal cells which constitute the polarising region or zone of polarising activity (ZPA). Both AER and ZPA possess crucial organising functions as evidenced by the fact that their removal or transplantation has profound effects on the normal limb formation.

The limb (wing) bud has three developmental axes:

1) The proximal-distal axis which runs from the base of the limb to the tip. The cell determination along the proximal-distal axis of the limb bud is specified by the progress zone. The proliferatingmesenchymal cells acquire the positional information when they are within the progress zone. The cell proliferation within the progress zone is directly controlled by the AER with the release of fibroblast growth factor.

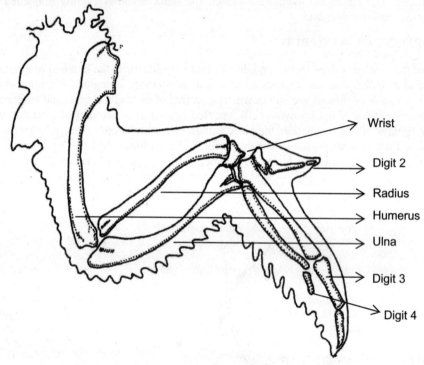

Fig. 15. 2 Wing structure to indicate the position of digits (Redrawn from Wolpert, 1978)

2) Anterior-posterior axis that runs parallel with their main body axis. In the chick limb, it starts from digit 2 to 4 (See Fig. 15.2). Anterior-posterior axis is specified by a small block of mesodermal tissue near the posterior junction of the limb bud and the body wall. This is called zone of polarizing activity. The zone of polarising activity possess crucial organising functions as evident by the fact that its removal or transplantation has profound effects on normal limb formation.
3) The dorsal-ventral axis running from the back of the hand to the palm is determined by the ectoderm encasing it. One signal molecule important in specifying the dorsal-ventral polarity is Wnt7a. The *Wnt7a* gene is expressed in the dorsal (but not ventral) ectoderm of the chick and mouse limb buds.

The positional information needed to construct the limb depends upon the coordinated functioning of the above three axes.

The tetrapod limb skeleton is laid down in a progressive manner that establishes a proximodistal axis that runs from body to extremity. The limb comprises four proximodistal elements: a zonoskeleton, stylopodium, zeugopodium and autopodium, each corresponding to different bone or set of bones in the arm or leg (Fig.15.3). In tetrapod limbs, the elements of the limbs, viz., humerus, radius and ulna and digits are specified in a proximal to distal order. They are formed respectively during three different phases of development, namely, phase I stylopod, phase II zeugopod, and phase III autopod. (Fig.15.3.)

ROLE OF *HOX* GENES IN LIMB PATTERNING

HOX genes play a pivotal role in the development of the three phases of limb formation as described above. The non-expression of any of these *HOX* genes during the limb formation would result in congenital limb deformality in the humans. The evolutionary role of *HOX* family genes in the origin of tertrapod limb has been explained in detail in the chapter No. 28 on Evolutionary Developmental Biology. It is now well established that the *HOX* genes in vertebrate embryos express in a temporally and spatially collinear manner along the anterior- posterior axis in organizing structures along the trunk axis. During the tetrapod limb evolution, the *HOX* gene clusters have however acquired a new, specialized function in the patterning of limb skeletal formation. For example, genes from the *HOX A* and *HOX D* clusters are essential for limb development and their collinear regulation is similar to that observed in the trunk. Evidently, the skeletal organization of the limb is prefigured in the genomic topology of these genes.

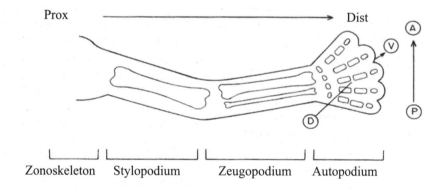

Fig. 15.3 The anatomy of vertebrate limb showing the three principal axis. The figure also shows the four proximodistal elements A-P anterior posterior axis, D-V dorsoventral axis: Prox-Dist proximal distal axis. The figure also indicates the different skeletal elements zonewise during development (redrawn from Twyman 2001).

The type of structure formed along the proximal-distal axis of the limb bud is accordingly specified by the *HOX* genes. In particular, they specify whether a particular mesenchymal cell will become stylopod, zeugopod, or autopod (see fig.15.3). When the stylopod is forming in the limb bud, *HOXd-9* and *HOXd-10* are expressed in the progress zone mesenchyme. When the zeugopod bones are formed, the posterior region expresses all the *HOXd* genes from *HOXd-9* to *HOXd-13*, while only *HOXd -9* is expressed anteriorly. When the autopod is formed, *HOXd-9* is no longer expressed in the autopod, whereas *HOXa-13* is expressed in the anterior tip of the limb. Simultaneously, *HOXd-13* products join those of *HOXa-13* in the anterior region of the autopod region of the limb bud whereas *HOXa-12*, *HOXa-11*, and *HOXa-10* are expressed throughout the posterior two thirds of the limb bud.

Sculpting of digits during autopod development is accomplished mainly by the process of programmed cell death. For example, the difference between a chicken's foot and that of a duck is the presence or absence of cell death between the digits. Furthermore, the signals for apoptosis in the autopod are provided by the BMP proteins. BMP2, BMP4, and BMP7 are each expressed in the interdigital mesenchyme, and blocking BMP signalling prevents interdigital apoptosis. BMP signalling is also important for the formation of joints.

CELLULAR INTERACTION WITHIN THE DEVELOPING WING BUD

The work of American scientist, Saunders pointed out the importance of apical ectodermal ridge (AER) in the cellular interactions that lead to the orderly arrangement of the parts in the wing. The AER is a thickened region of the ectoderm running like a ridge across the tip of the limb bud (Fig.15.1). If AER is removed as the cartilaginous parts are being laid down, parts of the limb will fail to develop. The later the time the ridge is removed, more the structures are developed. If the ridge is excised at an earlier stage only the humerus may be formed; if it is removed later, only the digits may fail to develop. This shows that the structures are specified in a proximal-distal sequence, *i.e.*, beginning near the point of attachment of the limb to the body and proceeding outward towards the tip of the limb. The AER appears to control such a proximal-distal sequence of wing structures. On the other hand, the formation and subsequent maintenance of the AER is promoted by the bud mesoderm present in the progress zone. It may also be construed from the above experiments that the AER emits qualitative information at different stages of limb development. Thus, a young AER should be able to promote the differentiation of both proximal and distal elements whereas the older AERs should be able to promote only the distal elements. However, the experiments in which the mesoderm is capped with AERs of various ages show that the mesodermal derivatives develop in proper proximal-distal sequence regardless of the age of AER. This clearly indicated that the information for the proper sequence of differentiation resides in the mesoderm itself. Nevertheless, the AER is necessary for realising this information by the mesoderm.

The polarising region of the vertebrate limb bud has signalling properties similar to that of Spemann Organizer in the amphibians. If the polarising region from the early wing bud is grafted to anterior region of another wing bud, a wing with a mirror image pattern develops showing a digital pattern of 4,3,2,2,3,4 (Fig.15.4). The pattern of muscles and tendons in the limb shows similar mirror image changes. Evidently, a diffusable morphogen originating in the ZPA and running along the posterior-anterior axis is responsible for specifying the position of cells which could interpret their positional values by developing specific structures.

Just like the proximal-distal axis, the wing bud has an anterior-posterior axis, especially to decide the placement of the digits in their correct positions. This is controlled by the mesodermal region at the posterior edge of the limb bud, designated as the zone of polarising activity (ZPA) (Fig. 15.4). The ZPA regulates the anterior-posterior axis of the limb. It is presumed to be a source of a morphogen that diffuses into the other part of the limb and influences the pattern of differentiation. This region secretes a signalling molecule *Sonic hedgehog (Shh)* which mimics the function of ZPA by inducing tissue growth and formation of posterior skeletal elements, when ectopically expressed. The regions closest to the ZPA differentiate into posterior parts and as it moves forward the anterior part, its concentration declines, thereby differentiating the anterior parts. The existence of positional information and their deliverance along the gradient is also thought to be responsible for the anterior-posterior axis formation. Recent studies have shown that retinoic acid also acts like a gradient along the anterior-posterior axis. Retinoic acid activates the homeotic genes thereby specifying the limb field.

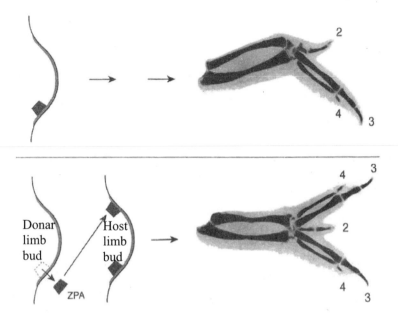

Fig.15.4 Showing induction of polydactyly in a chicken Transplantation of the zone of polarizing activity (ZPA) in the developing wing bud from a posterior site to a new anterior position induces extra digits pattern. (Redrawn from S.B. Carroll, 2005).

The development of wing development in chick, described above, is a typical example to illustrate pattern formation in vertebrates. Evidently, there are important interactions between the mesoderm and ectoderm. Furthermore, certain specialised regions such as the AER and ZPA appear to be important in the axial organisation and patterning of the pattern of limb structures.

16

DEVELOPMENT OF IMMUNE SYSTEM

Principal objective of immune system is to protect the species from foreign agents that are potentially harmful to the body. Protection not only involves the defense against invading microorganisms but also from spontaneous and dangerous changes such as mutation and malignant degeneration. Such an ability of an organism to protect or defend itself from invading agents or non-self substances is termed as immunity. In the words of David Hutson, "survival requires both the ability to mount a destructive immune response against non-self and the inability to mount a destructive response against self".

Anything that causes or elicits an immune response is called an antigen. An antigen may be harmless (inert carbon particles) or harmful, such as the flu virus. Disease-causing antigens are called pathogens. The immune system is designed to protect the body from all kinds of foreign materials. Though many animal species have developed immune system, among vertebrates, birds and mammals have been the subject of detailed investigation.

In humans, the immune system begins to develop in the embryo. The immune system starts with hematopoietic (from Greek, "blood-making") stem cells (HSCs). In the embryo, HSCs develop in the primitive yolk sac. These pluripotent HSCs, wherever present, will give rise to myeloid and lymphoid progenitors. These progenitors in turn differentiate into major cells of the immune system (granulocytes, monocytes, and lymphocytes). This process occurs throughout the life time, thus the developmental process associated with immune system is distinctly different from other systems of the body.

Conventional classification of immune system leads to the following two major types of immunity which are natural immunity and adaptive immunity. These two types are distinct from each other, yet possess overlapping networks in their role of defense. Natural immunity (also known as innate immunity), as the term implies, depends upon the defenses with which individuals or young ones are born. This type of immunity refers to essentially non-specific defense mechanisms that come into play soon after an antigen's appearance in the body. Such mechanisms include effective barriers (like skin, mucosal layers) against invasion, secreted products in the blood (such as serum proteins, cytokines and chemokines) and cellular elements (granulocytes, monocytes and natural killer cells). Each of these components, although non-specific, cooperates to provide rapid defense against any number of foreign or toxic agents.

Adaptive immunity (also known as acquired immunity) refers to antigen-specific immune response. Adaptive immune response has evolved more recently in evolutionary times and is far more complex than the innate immunity, and has been identified only in vertebrates. Salient features of adaptive immunity are; the ability to distinguish self from non-self; the ability to generate specific response against specific foreign intruder; the ability to provide more vigorous and rapid reaction on second or subsequent exposure to a foreign element. The latter property of aggressive response on re-exposure to a non-self object is attained through the phenomenon known as "immunologic memory" and this type of response is called as "anamnestic response". The cells that are responsible for adaptive immunity are T and B lymphocytes, which are distinguishable not by their morphology but by their genetic program and the expression of specific receptors on their surface.

The adaptive immune system can further be classified into two groups viz., humoral immunity and cell-mediated (cellular) immunity. Humoral immunity is primarily mediated by B-cells. The functional differentiation of B cells takes place in the bursa of Fabricius in birds and bone marrow in mammals. B cells secrete certain immunoglobulin or antibody molecules that bind to foreign molecules, coating their surfaces for convenient targeting and consequent destruction. Since the antibodies are secreted into the plasma, humoral immunity can be transferred to another member of the species through serum. Cell-mediated immunity comprises of T cells, which express T cell specific receptors. T cells develop in the thymus, where they acquire their unique specificity for antigen. In the periphery, the T cells further differentiate into T-cell subpopulations with different functions, including the secretion of soluble factors termed cytokines. Interestingly, in the absence of cytokine production, B cells fail to differentiate and secrete antibody. Thus, T cells' help (therefore the cellular immunity) is required for humoral responses, validating how significantly interconnected they are.

There is another type of classification within adaptive/ acquired immunity, which is in use though not widely propagated. Based on this there are two kinds of immunity: (1) active immunity, when the body is stimulated to produce its own antibodies, and (2) passive immunity, where the antibodies come from outside the person's body. Active immunity is usually permanent, and can be induced due to actual illness or vaccination. Passive immunity is not permanent because the antibodies are introduced from outside the body, thus the B-cells of the recipient never "learn" how to make them. Some examples of passive immunity include antibodies passed across the placenta and in milk from a mother to her baby. Because antibodies are only protein, they don't last very long and must be replaced if the immunity is to continue.

ORIGIN OF THE IMMUNE SYSTEM

Immune cells originate from the mesenchymal cells of hematopoietic system. It is the same HSCs that give rise to various blood cells, including immune cells by lineage specific differentiation. Until recently it was thought that the yolk sac produced the first cells of hematopoietic system. Yolk sac hematopoietic cells were thought to migrate to the liver, which acted as the predominant site of hematopoiesis during early fetal stages of development. The liver then organizes the hematopoitic tissue (HT) using HSCs, derived from yolk sac. The HT commences hematopoisis for the first time in the fetal liver. The liver continues hematopoiesis until the development of the spleen. Thereafter, spleen takes up the function of hematopoiesis, although liver continues to perform hematopoiesis to some extent. By the advent of bone marrow formation in the embryo, the HT is again relocated from spleen to red bone marrow establishing it as the chief and permanent site of hematopoiesis from embryonic stage to adult. Nevertheless, spleen and liver continue to perform hematopoiesis with the help of remnant HT retained by them.

In amphibians, the yolk sac analogue, namely the ventral blood islands produce the first hematopoietic cells in the embryo. Before the circulation is completed, the intraembryonic area containing the dorsal aorta and pro/mesonephros also produces hematopoietic activity. Both the ventral blood islands and the intraembryonic region contribute to different elements of the adult hematopoietic system. However, the intrabody region provides most of the cells of the blood.

In avian embryos, the yolk sac generates only transient hematopoietic cell populations, whereas the embryo body is the exclusive contributing source of long-lived adult hematopoietic cells.

Within the ventral wall of the dorsal aorta, large hematopoietic foci were identified and contain high frequency of CFU-C (colony-forming unit-culture assay) numbers supporting the idea that the area comprising dorsal aorta gives rise to the definitive hematopoietic system.

Among the mammalian vertebrates, mouse has been investigated in detail. Although two sites of hematopoiesis, namely, the yolk sac and AGM (dorsal aorta, gonads and mesonephros) region, function in mouse ontogeny prior to the onset of liver hematopoiesis, cellular interchange between these two tissues obscures the specific site of hematopoietic stem cell generation. However, various studies employing diversified methodologies demonstrated that, as in non-mammalian vertebrates, the intraembryonic AGM region generates the potent adult hematopoietic system, populating secondary hematopoietic tissues (liver and yolk sac), whereas the yolk sac itself, most likely contributes to low potency/transient hematopoiesis.

ORGANIZATION OF THE IMMUNE SYSTEM

PRIMARY LYMPHOID ORGANS

In mammals, bone marrow and thymus are the primary lymphoid organs. In birds, the Bursa of Fabricius is analogous to the mammalian bone marrow. In general, the process of hematopoiesis originates in the extraembryonic yolk sac, moves to the fetal liver, and continues throughout postnatal life in the bone marrow. The bone marrow is the source of stem cells which in the appropriate microenvironment and in the presence of colony stimulating factors (CSFs) proliferate and differentiate into erythroid, myeloid, and lymphoid cells (Fig.16.1). The bone marrow is also the site of differentiation of lymphoid cells into B lymphocytes. B cell differentiation is accompanied by Ig gene rearrangements, expression of surface IgM and IgD, Ig class switching, and secretion of IgG, IgM, IgA, or IgE antibodies.

The thymus, which originates in part from epithelial cells of the gut endoderm of the third embryonic pharyngeal pouch is encapsulated, divided into cortex and medulla, and located in the mediastinum. Thymus provides the microenvironment and secretes local hormones necessary for the differentiation of thymic (T) lymphocytes from lymphoid cells derived from the bone marrow. Thereafter, the mature T cells leave the thymus, circulate in the blood and lymphatic vessels, and colonize peripheral lymphoid organs. T cell differentiation in the thymus is accompanied by changes in surface markers (assayed by monoclonal antibodies), by DNA rearrangements which generate the diversity of T cell receptors, by development of the capacity to utilize self-MHC molecules as a means of antigen recognition, and by the induction of self-tolerance. In the neonate, a large portion of afferent lymphoid cells in the cell-packed thymic cortex become pycnotic, undergo programmed cell death (apoptosis), and are phagocytosed in the process of acquiring self tolerance by clonal selection and elimination of T cells which react with self antigens accessible in the thymic environment.

SECONDARY LYMPHOID ORGANS

The lymph nodes and spleen are encapsulated organs whose connective tissue framework contains lymphocytes (functionally separated into T cells and B cells), macrophages, and dendritic cells that localize and process antigens from the lymph and blood stream. The B cell areas of lymph nodes are located in the outer cortex and include lymphoid follicles and germinal centers. The germinal centers enlarge greatly in secondary antibody responses. They have a central role in the proliferation, mutation, and maturation of B cells into plasma cells and memory cells.

T CELLS

T cells differentiate in the thymus from lymphoid stem cells derived from bone marrow and, when mature, leave the thymus, circulate in blood and lymph, and spread to peripheral lymphoid organs. Resting lymphocytes are small round cells, and resting T and B cells are indistinguishable by ordinary light microscopy. Following mitogenic or antigenic stimulation, resting T cells are transformed into blast cells capable of division. In normal subjects, about 80% of peripheral blood lymphocytes are T cells and 10-20% is B cells.

T cells are activated by specific binding of the TCR to the immunogenic complex of antigenic peptide-class I or II MHC molecules, as previously mentioned. Stimulated by IL-2, activated T cells proliferate, undergo clonal expansion, and differentiate into functional classes: cytotoxic/suppressor T cells and helper/inducer T cells. Helper and suppressor T cells are regulator cells which, mainly through the secretion of cytokines, modulate the function of B cells and other T cells. Cytotoxic T cells are effector cells which kill virus-infected host cells and histoincompatible transplanted cells.

B CELLS

B cells differentiate from lymphoid stem cells in the bone marrow, circulate in the blood, and localize in peripheral lymphoid organs. The B cell receptor for antigen is surface membrane Ig. Surface membrane Ig is one marker for B cells which are identified by fluorescence labeling techniques using fluorescence microscopy or flow cytometry. Normally, about 10-20% of peripheral blood lymphocytes are B cells.

The initial step in the activation of B cells is the specific binding of antigen to the surface membrane Ig. In the presence of a second signal (T and B cell contact) and growth factors (IL-4, -5, -6) released by helper T cells, B cells proliferate, undergo clonal expansion, form germinal centers, and at full maturity differentiate into antibody-secreting plasma cells. Mature plasma cells are characterized by cytoplasmic and secretory Ig and morphologically by basophilic cytoplasm and an eccentric single, or sometimes double, nucleus often with a "cart wheel" pattern of dense chromatin.

NATURAL KILLER (NK) CELLS

NK cells, which are lymphoid cells found in the blood and peripheral lymphoid organs, are capable of killing virus-infected cells or tumor cells in the absence of prior immunization and without MHC restriction. NK cells are also described as "large granular lymphocytes" or as "null" cells because of the absence of surface markers characteristic of T or B cells. NK cells are able to lyse target cells by direct contact with them (in the absence of antibody) or by antibody dependent cellular cytotoxicity (ADCC). NK cells have surface receptors for IgG Fc and kill target cells which are coated with specific antibody and become bound to the Fc receptor, resulting in ADCC. NK cells and cytotoxic T cells produce pore-forming molecules called "perforin" or "cytolysin" which has structural and functional similarity to C9 of the complement system, binds to cell surface membranes, and forms transmembrane channels, leading to osmotic death of target cells.

MACROPHAGES

Tissue macrophages are derived from blood monocytes and belong to the widely distributed mononuclear phagocytic (reticulo- endothelial) system. Macrophages mediate critical functions both in innate immunity (non-specific phagocytosis and destruction of pathogens) and in the initiation of specific immunity (antigen processing and presentation). They secrete many biologically active products; among them, IL-1 promotes the differentiation of T cells and B cells and mobilizes other host defenses. Macrophages in turn are activated by interferon-gamma secreted by activated T cells.

Macrophages secrete tumor necrosis factor (TNF), kill some tumor cells, and by means of Fc receptors for IgG mediate ADCC.

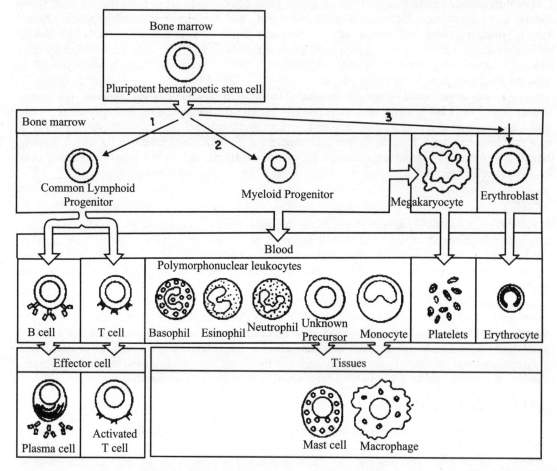

Fig. 16.1. The lineages of the cells that arise from hematopoietic stem cells. A model of differentiation pathways of blood cells and immune cells from heamatopoietic stem cells shown in relation to hematopoietic tissues.
(Modified from Morrison et. al., 1995)

DENDRITIC AND LANGERHANS' CELLS

Dendritic cells (DCs) are 'professional' APCs whose activation is an important step in innate immunity and in initiating acquired, lymphocyte-mediated, immune responses. DCs have elongated cytoplasmic processes; bind and localize antigen; process and present to T-cells and B-cells immunogenic peptide fragments complexed with MHC molecules; and express ligands and co-stimulatory signals required for T-cell and B-cell activation. DCs can migrate from non-lymphoid to lymphoid locations. They are present in skin (Langerhans' cells) and other locations and in T-cell areas of lymphoid tissue (interdigitating cells).

OTHER ACCESSORY CELLS

These include polymorphonuclear leukocytes and eosinophils, phagocytic cells with Fc receptors for IgG, and basophils and mast cells, which bear Fc receptors for IgE.

IMMUNOGLOBULINS (IG)

Secreted Ig molecules are composed of four-polypeptide chain units with constant and variable structural regions and consisting of two identical heavy (H) chains, whose constant regions define five classes of Igs, some with subclasses, and two identical light (L) chains, either kappa or lambda. Hypervariable regions of paired L and H chains form antibody combining sites, two for each four-chain unit. IgG, IgE, and IgD molecules contain one four-chain unit, IgA one or two (secretory IgA), and IgM five. Some Igs contain additional structural components, such as the secretory piece of IgA which is a major Ig in mucosal secretions.

Igs can be enzymatically cleaved into Fab fragments containing antibody combining sites and Fc fragments which consist of constant regions of H chains and convey the different biological activities of Igs, such as ease of placental transfer, complement fixation, and binding to Fc receptors of macrophages, polymorphs, platelets, or mast cells.

MAJOR HISTOCOMPATIBILITY COMPLEX (HUMAN LEUKOCYTE ANTIGEN COMPLEX)

The major histocompatibility complex (MHC) was originally discovered as transplantation antigens that predominantly determine the compatibility of tissues between different individuals. In all vertebrates there is a genetic region that has a major influence on graft survival. This region is referred to as the Major Histocompatibility Complex (MHC). Individuals identical for this region can exchange grafts more successfully than MHC non-identical combinations. There also exist many "minor" histocompatibility loci which also influence graft survival which are individually weak. Unlike minor histocompatibility antigens, the MHC products play an important role in antigen recognition by T cells. The MHC genes and their products are grouped into 2 classes on the basis of their chemical structure and biological properties. The two MHC proteins have a similar secondary and tertiary structure with subtle functional differences. Class I molecules are made up of one heavy chain (45 kD) and a light chain called ß2-microglobulin (12 kD) that contributes to the overall structure of the protein. Class II molecules do not contain ß2-microglobulin and consist of two (alpha and ß) chains of similar size (34 and 30 kD).

EXPRESSION OF MHC MOLECULES

MHC class I molecules are widely expressed, though the level varies between different cell types. MHC class II molecules are constitutively expressed only by certain cells involved in immune responses, though they can be induced on a wider variety of cells.

MHC LOCI

In man and mouse, as in most species, each class of MHC is represented by more than one locus; in man these are called HLA for Human Leucocyte Antigen. The class I loci are HLA-A,-B and -C and the class II loci HLA-DR, -DQ and -DP. All the MHC genes map within a single region of the chromosome (hence the term Complex).

All MHC loci are expressed co-dominantly, that is to say both the set of alleles inherited from one's father and the set inherited from one's mother are expressed on each cell. In about 97% of cases the entire linked MHC complex is inherited intact, i.e. without recombination; the set of linked MHC alleles found on the same chromosome is called a haplotype.

Human and murine class I molecules are heterodimers, consisting of a heavy alpha chain (45kD) and a light chain, beta-2-globulin (12kD). Alpha chain can be divided into three extracellular domains, alpha1, alpha2 and alpha3, in addition to the transmembranous and cytoplasmic domains. The alpha3 domain is highly conserved, as is beta-2-microglobulin. Both alpha3 domain and beta-2-microglobulin are homologous to the CH3 domain of human immunoglobulin.

Class II molecules are heterodimeric glycoproteins, alpha chain (34kD) and beta chain (29kD). Each chain has 2 extracellular domains, together with the transmembranous and cytoplasmic domains. The membrane-proximal alpha2 and beta2 domains are homologous to immunoglobulin CH domain.

There are 3 class I loci (B, C, A) in the short arm of human chromosome 6, and 4 loci (K, D (L), Qa, Tla) in murine chromosome 17. These loci are highly polymorphic. The variable residues are clustered in 7 subsequences, 3 in alpha1 domain and 4 in alpha2 domain. There are 3 major human class II loci (HLA-DR, HLA-DO, HLA-DP) and 2 murine loci (H-2I-A, H-2I-E). All class II beta chains are polymorphic. Human HLA-DQ alpha chain is also polymorphic.

Class I molecules present peptides to CD8-CD4+ cytotoxic T lymphocytes, whilst class II molecules present peptides to CD4+CD8- T helper lymphocytes. The composite peptide-MHC macromolecule is recognised by T-cell receptor (TCR) on the surface of T lymphocytes. The problem for the MHC molecules is to present a vast variety of peptides for specific recognition by TCR, with only a limited isotypes of each class in an individual. In terms of molecular structure, many peptides are to be pasted onto a single MHC dimeric protein. The consequence of these peptides pasting is the creation of different conformations of composite peptide-MHC macromolecules. In purely structural terms, one can view the peptide-MHC composite as a trimeric protein. Indeed, in the case of class I molecule, bare dimers are unstable. Class I molecules are stabilised by peptide binding and only the trimeric forms are expressed on cell surface. Classes I and II molecules are versatile in making a large variety of specifically distinct conformations by pasting different peptides onto themselves.

Class I or II molecules (glycoproteins) contain Ig-like domains (Ig homology units) as do TCRs. In short, the gene families of all molecules which recognize antigens - antibodies, TCRs, class I or II MHC - apparently belong to an "Ig gene superfamily" which presumably has evolved from a common ancestor for the function of self/non-self recognition.

COMPLEMENT SYSTEM

A chemical defense system that kills microorganisms directly, supplements the inflammatory response, and works with, or complements, the immune system. Protective proteins that are produced in the liver include the complement system of proteins. The complement system proteins bind to a bacterium and open pores in its membrane through which fluids and salt move, leading to swelling and bursting of the bacterial cell.

Complement proteins are made in the liver and become active in a sequence (C1 activates C2, etc.). The final five proteins form a membrane-attack complex (MAC) that embeds itself into the plasma membrane of the attacker. Salts enter the invader, facilitating water to cross the membrane, swelling and bursting the microbe. Complement also functions in the immune response by tagging the outer surface of invaders for attack by phagocytes.

The complement system is a major humoral component of innate immunity (and mediator of inflammation) and comprises about 20 proteins (including proteases) which are normally present in plasma in inactive form and become sequentially activated by classic or alternative pathways to mediate effector or amplifying functions.

CYTOKINES

This is a generic term for messenger molecules (polypeptides) which are secreted by lymphoid and non-lymphoid cells and form a mediator network regulating the growth, differentiation, and function of cells involved in immunity, hematopoiesis, and inflammation. Cytokines secreted by lymphocytes are also called lymphokines, and those secreted by monocytes/macrophages are known as monokines. An interleukin (IL) is a cytokine, which carries a message between leukocytes. Cytokines are involved in the regulation of T cells, B cells, and macrophages.

The bird's immune system mainly consists of lymphatic vessels and lymphoid tissue. Primary tissues are the thymus, located in the neck along the jugular vein, and the bursa of Fabricius, located adjacent to the cloaca. Secondary lymphatic organs and tissues would be the spleen, bone marrow, mural lymph nodules and lymph nodes. There is also a lymphatic circulatory system of vessels and capillaries that transport lymph fluid through the bird's body and communicate with the blood supply.

NEONATAL IMMUNITY

Birds differ from mammals in the way their immune system develops and becomes diverse enough to meet lifelong challenges, with a unique organ called the bursa and a process called gene conversion. Baby birds are born with an incomplete immune system.

MATERNAL IMMUNITY

Maternal immunity is passed on in the amniotic fluid and yolk of the egg. It is transferred to the embryo when it swallows amniotic fluid during hatch and in the absorption of the egg yolk after hatch. These antibodies give the newly hatched chick's immunity a start while their own system is developing.

The bird's immune system begins developing before hatch and is complete by sexual maturity. One of the most important stages of this development happens in the first six weeks of the chick's life, when gene conversion is taking place in the bursa.

GENE CONVERSION

Gene Conversion is the education of the B lymphocytes, cells that respond to a disease antigen by producing antibodies that then bind the nasty antigen for removal from the blood by the spleen and liver. Unlike mammals, when birds are born they do not have a "library" of genetic information for the B cells to use in their production of antibodies. Birds have only one variable function gene encoded in the germline DNA. If left this way, the B lymphocytes will not be able to produce the different antibodies needed to resist specific diseases. So, for the first six weeks of the bird's life, these B cells move to the bursa, where they are rearranged in order to provide the diversity needed to protect against the great variety of potential pathogens. These educated B cells leave the bursa to seed other organs of the immune system.

Although the B cell processing phenomena takes place for six weeks in the bursa it is the first three weeks that are considered the most critical. Whatever the diversity that has been achieved during this period is what the bird will have for its lifetime. The bursa continues to produce B cells until it involutes at sexual maturity and then the bone marrow takes over the task of producing the B cells. The bursa does continue to play a role in the immune system through the life of the bird, though not as critical a one as it does in the beginning.

DISORDERS IN IMMUNE SYSTEM

The immune system can overreact, causing allergies or autoimmune diseases. Likewise, a suppressed, absent, or destroyed immune system can also result in diseases.

ALLERGY

Allergy is an inappropriate and harmful response of the immune system to normally harmless substances. In other words, allergies result from immune system's hypersensitivity to weak antigens that do not cause an immune response in most people. Allergens are substances that cause allergies; include dust, molds, pollen, cat dander, certain foods, and some medicines (such as penicillin).

AUTO IMMUNITY

Autoimmunity is the response of an immune system against the "self", developing antibodies against its own antigens and destroying its own cells. For example, in myasthenia gravis, the person's immune system destroys the acetylcholine needed to transfer nerve impulses across the synapses. Multiple sclerosis (MS) is caused by antibodies attacking the myelin of nerve cells. Systemic lupus erythematosis (SLE) has the person forming a series of antibodies to their own tissues, such as kidneys (the leading cause of death in SLE patients) and the DNA in their own cellular nuclei. In systemic lupus erythematosis (SLE), the immune system attacks connective tissues and major organs of the body.

ACQUIRED IMMUNO DEFICIENCY SYNDROME (AIDS)

AIDS occurs when the virus causing this (Human Immunodeficiency virus- HIV) infects and kills helper T-cells. With fewer helper T-cells, the immune system can't form any new antibodies against any new invaders, thus people with AIDS usually die from some secondary infections or unusual forms of cancer.

17

EMBRYONIC NUTRITION

Depending on the type of embryonic nutrition, the sexually reproducing animals are categorised into two groups, namely oviparous and viviparous. In oviparity, the fertilization occurs internally, but the young hatch from the eggs deposited externally by the female. In this type, the growing embryo depends entirely on the stored yolk inside the egg to derive all its nutriments. On the contrary, in viviparity, the embryo develops inside the uterus or the equivalent part in the female reproductive tract, and there is total nutritional dependence of the embryo on the mother. In the highly evolved viviparous animals like mammals, the growing embryo within the uterus establishes special nutritive chords such as the placenta to derive directly the nutrients from the maternal blood. However, in the primitive form of viviparity such as found in many invertebrates like insects, the embryos in the genital tracts obtain nutrients via a placenta- like structure developed either by mother or embryo or both; this condition is referred to as pseudo placental viviparity.

OVIPARITY AND YOLK UTILIZATION

Embryonic development of oviparous animals occurs entirely within the confines of the laid eggs. As a consequence, the macro- and micronutrients provided by the egg contents must guarantee the production of viable offspring.

All oviparous animals accumulate enormous amount of nutritive materials within the egg cell during oogenesis. They provide energy to sustain the embryonic development until the embryo hatches out and obtains its own food. The egg nutrients, both organic and inorganic are referred to as the yolk. Most invertebrates and all submammalian vertebrates employ such a strategy of delivering nutrients to the developing embryo.

Vitellins, the principal component of egg yolk, are multi-subunit phosphoglycolipoproteins serving as the primary nutritive source for development of embryos in most egg laying animals. In this respect, the yolk proteins are comparable with the storage proteins of plant seeds. In addition to supplying growing embryos with amino acids, vitellins also carry covalently linked carbohydrates, phosphates and sulphates as well as non- covalently bound lipids, hormones, vitamins and metals.

Yolk proteins are endocytosed by the vitellogenic oocytes and stored in membranous organelles called yolk granules or yolk platelets, until used during embryogenesis. Vitellins are generally derived from a precursor molecule, vitellogenin, synthesized by extra ovarian organ such as liver in vertebrates and fat body in insects. Vitellogenesis, the process of yolk formation is under the control of female hormones such as estrogens in vertebrates. Besides vitellins, several non-vitellogenin yolk proteins are also stored with in the egg (see chapter on oogenesis for details).

BIOCHEMICAL COMPOSITION OF YOLK

The chemical composition of yolk varies among different animals. Basically, yolk is comprised by a lipoprotein component and hence it is commonly called as lipovitellin. Lipids are used during embryogenesis and postembryonic life as structural components and metabolic fuel. Hence, evolution of yolk proteins among oviparous animals involved mainly the transportation of lipids to the oocytes along with importation of the yolk precursor molecule, vitellogenin. The complexity of yolk and the extent of lecithality (density of yolk) are related to the time duration of embryogenesis. In the invertebrates with brief embryogenesis, such as the sea urchin, the yolk is simple and less in amount, whereas the yolk is complex and enormous in those animals such as reptiles and birds with extended embryogenesis and the development is direct (without involving any larval forms).

In general, the vertebrate egg yolk contains more lipid than that of the invertebrate eggs. The avian egg yolk contains two fraction of high density lipovitellins (α and β) as well as a low density lipoprotein. The β- lipovitellin contains around 21% of lipid (12% phospholipids and 9% triacylglycerol) with 78% of proteins by weight. On the other hand, the low density lipoprotein contains high amount of lipids (80%) with only 18% protein by weight. Both the yolk proteins contain cholesterol, in addition to phospholipids and neutral lipids.

The vertebrate vitellin also differs from that of invertebrates, in that it undergoes extensive phosphorylation especially of serine residues, catalysed by casein kinase II. Thus in chicken, *X. laevis* and the fish *Oncorhynchus kistuch*, the vitellin consists of a highly phosphorylated unit called phosvitin and lipovitellin moieties. Phosvitin is a highly phosphorylated phosphoprotein with a long polyserine domine. Phosvitin has high affinity for Ca^{2+} and hence it carries calcium phosphate in the yolk protein. During vertebrate embryogenesis these calcium salts support embryonic bone formation. On the other hand, the invertebrate vitellins generally lack a phosvitin domain; understandably invertebrates do not form endoskeleton. Differences in the amino acid composition, phosphorus, lipid, and carbohydrate levels of yolk proteins reflect differences in the nutritional demands of the diverse embryos. Another feature of vertebrate vitellins is that it carries hormones such as thyroid hormones to the egg. These maternal thyroid hormones have been shown to perform different morphogenetic roles in the developing embryos of quail. In mammals, during embryogenesis, estradiol, stored within the egg induces the synthesis of actin and collagen.

The invertebrate yolk is a high-density lipoprotein often conjugated to a variety of prosthetic groups such as glycogen. However, the oocytes accumulate large quantities of lipid droplets as an additional source of metabolic energy. In the molluscan eggs, the yolk is comprised of vitellogenin-derived lipoprotein and ferritin. Yolk ferritin is an iron transporter of molluscan blood, accumulated as the principal component of egg yolk. In the blood-sucking leeches, the formation of yolk is linked to the uptake of blood. The yolk, in addition to vitellogenin-derived lipoproteins, also contains ovo-hemerythrin. Hemerythrin is again an iron containing protein probably derived from the digestive products of the vertebrate blood, sucked by the leeches. Haem-binding yolk proteins have also been reported in the blood-sucking hemipteran insects like, *Rhodnius prolixus*. However, insects, like most other oviparous animals, accumulate yolk proteins in the form of lipoproteins often conjugated with carbohydrate prosthetic groups. In addition, phosphate or sulphate moieties may also be added to the yolk protein. Another important yolk component of insects is the lipid-transporting lipoprotein, lipophorin. Lipophorin is responsible for the transport of large quantity of lipid reserves accumulated within the eggs.

Crustaceans also accumulate large quantities of glycolipoproteins, conjugated with a variety of carotenoid pigments. Crustacean egg yolk is a high density lipoprotein with high percentage of lipids including cholesterol. An interesting aspect of insect and crustacean yolk proteins is that they carry significant quantities of the molting hormone, ecdysone. The ecdysteroids are conjugated with the hemolymph vitellogenins and are transported to ovary for deposition in a conjugated state. They are supposed to serve as morphogenetic hormones during the embryonic development, by inducing the embryonic cuticle synthesis and molting, before hatching of the larvae.

YOLK UTILIZATION

Yolk plays a nutritive role during the development of the embryo. Since the yolk proteins are stored inside the yolk bodies or yolk granules which are membrane- bound, they have to be dismantled and digested to release their subunits in an utilizable form. In general, yolk granules contain proteases that degrade their constituent proteins. Several proteases such as cysteine, aspartyl, and serine proteases have been identified in vertebrate and insect eggs. Such proteolytic processing release distinct set of peptides, which are then accessed by the embryo during development. In the insect, *Blattella germanica*, a cysteine protease is stored along the yolk proteins in its proprotease precursor form. The initiation of yolk degradation however starts after the acidification of the yolk granules, which activates the proenzyme to an active form. In insects, yolk accumulates centrally in the egg and is invaded by vitellophages, cells that originate from the cleavage nuclei that fail to migrate to the egg's surface during blastoderm formation. Vitellophages engulf yolk granules. However, during differentiation, the yolk mass becomes enclosed by the gut for digestion and utilization.

In insects, yolk proteins also play an important role in regulating embryogenesis. In many insects, the molting hormone, ecdysteroids, are stored as conjugates linked to the lipovitellin through non-covalent interactions. Interestingly, the release of free ecdysteroids coincides with the time of proteolytic cleavage of the yolk proteins. Thus, the timed release of ecdysteroids during the break down of yolk molecules during embryogenesis is helpful in the activation of embryonic cuticle formation and their subsequent moulting.

The crustacean egg, at the commencement of embryonic development, possesses a host of hydrolytic enzymes, not only to dismantle the vitellin complex, but also to release the component substrates in an utilizable form. In the brine shrimp, *Artemia salina*, the lipovitellin undergo proteolytic cleavage to give rise to smaller peptides by an alkaline lipovitelline- specific protease. Another protease, located in the lysosome, degrades a smaller lipovitellin fraction. A vitellin-bound trypsin-like protease plays an important role in the regulation of vitelline utilization during embryogenesis of this brine shrimp. During the encysted period of *Artemia* eggs, these proteins are not activated. Thus, in *Artemia* yolk utilization, protease activity appears to be a programmed developmental event with possible control mechanisms conferred by its association with lipovitellin.

Yolk acidification is also a necessary prerequisite for its degradation and the concomitant embryonic cell differentiation in the frog *Xenopus*. In *Xenopus*, the stored yolk platelets inside the egg have mildly acidic pH (5.6). However, this pH is not sufficient for yolk degradation. During embryogenesis, the yolk platelets become progressively much more acidic (pH <5). This acidification also occurs in various embryonic tissues. Yolk utilization was inhibited when acidification was inhibited with bafilomycin, which is an inhibitor of vacuolar ATPase. Interestingly, bafilomycin also inhibited cell differentiation in embryos. Thus, yolk platelet acidification is developmentally regulated in *Xenopus*. The yolk degradation due to acidification is also associated with cell differentiation and with the formation of the endosomal/lysosomal compartment, typical of adult cells.

With respect to the mode of yolk utilization in vertebrates, eggs could be classified into two types. They are cledoic and non-cledoic eggs. The term cledoic means a 'closed box', which could be penetrated by matter only in a gaseous state. According to Needham, the following are the characteristic features of cledoic eggs. 1) Cledoic egg is not dependent on the environment for water or ash; 2) Oxidation of stored protein is suppressed to a considerable extent and the end products of protein metabolism are accumulated in the form of non-soluble uric acid to save water for the developing embryo and to reduce the exchange of matter through the eggshell; and 3) oxidation of lipid is enhanced to meet the energy requirements of the developing embryo. The avian eggs as well as eggs of the terrestrial reptiles are all cledoic. In general, the cledoic eggs develop several extraembryonic membranous folds, such as amnion, yolk sac and allantois. They serve the function of gaseous exchange, yolk storage and utilization, waste disposal etc. On the other hand, the non-cledoic eggs, including the aquatic eggs, absorb water and minerals from the environment, and utilize both protein and lipids as energy sources for development. Also, they do not develop any extraembryonic layers to store the yolk separately or to accumulate metabolic waste materials.

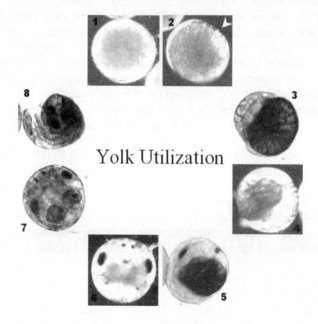

Fig. 17.1 (1)Freshly laid egg with dense yolk granules, (2) A yolk- free white streak (marked as arrow) makes its appearance at the animal pole, (3) One quarter of the yolk cleared, embryo in the late gastrulation stage, (4) One third of the yolk is cleared, beginning of organogenesis, (5) Two third of the yolk is cleared, yolk is found in the vegetal pole, eyes well developed, (6) Yolk is found as a single cluster in the centre, appendages in developing stage, (7) Yolk is found as two clusters in the centre, appendages well developed, (8) Hatched out larvae with little yolk. (For more detail see table 17.1)

In the marine crab, *Emerita asiatica*, lipovitellin I and II are progressively cleaved proteolytically by serine proteases into their constituent polypeptides during embryonic development until they are finally utilized as a source of amino acids. Intense esterase activity also occurs at the time of vitelline degradation (Tab.17.1). Glycosidases activity at this time of yolk degradation also helps in the release of bound glucose and galactose from the glycolipid and oligosaccharide components of the major yolk proteins, as well as to hydrolyze stored glycogen during embryogenesis. Enzyme activity, especially of esterases, during embryogenesis in this crab resulted in the release of vitellin-bound ecdysteroids into their free forms, such as 20-hydroxyecdysone and ecdysone (see Tab.17.1). The release of free active ecdysteroids from stored vitellin has important implications in their morphogenetic control over the synthesis of embryonic envelopes and larval cuticle. Gradual clearance of complex yolk along with appearance of embryonic structures is shown in Fig.17.1.

Table 17.1: Relationship between esterase and protease activity during yolk utilization and ecdysteroid fluctuation during embryonic development in the crab *Emerita asiatica* (From Subramoniam 2000)

Embryo Stage	Characteristics of embryonic development based on yolk clearance	Esterase activity (nmol naphthol/mg protein per min)	Protease activity in enzyme units (1μg leucine equivalent/30 min)	Total Ecd (ng/g egg wet weight)	Free Ecd (pg/mg lipid)	Conjugated Ecd (pg/mg lipid)	
						Polar	Apolar
I	Egg mass bright orange in colour; yellow yolk granules seen	ND	5.5	6.5	80.7	8.33	8.33
II	Cleavage has taken place and blastomeres are seen; egg mass bright orange in colour	ND	ND	ND	ND	ND	ND
III	A yolk free white streak makes its appearance at the animal pole	0.1198	8.69	15.2	146.7	613.8	25.0
IV	One quarter of the yolk cleared; red pigments are seen at the edge of yolk; Egg mass dull orange in colour	0.1983	ND	4.62	ND	ND	ND
V	One third of the yolk is cleared; two eye spots appear; red spot prominent and seen at the end of the animal pole	0.3086	12.6	6.92	83.33	20.33	25.0
VI	Eyes well developed; yolk is found in the vegetal pole; two third of the yolk is cleaved; red pigments seen all over the white space; egg mass brownish orange in colour	0.1585	ND	15.0	125.0	18.33	20.83
VII	Yolk is found as two clusters in the centre; appendages of the embryo are developed; heart beat seen; egg mass grayish orange in colour	0.115	10.1	6.15	50.0	12.5	16.67
VIII	Heart beat more prominent; embryo almost developed; egg mass pale grey in colour	0.0523	9.35	36.20	291.66	83.33	233.33

ND - Not determined

In a similar way, carotenoids conjugated to vitellin molecules of Emerita, as well as those existing in a free state undergo metabolic changes during embryonic development. Esterase activity in the embryo may also release carotenoids that are esterified to the long chain fatty acids of the lipovitellins. Several intermediate pigment compounds appear during the oxidation of beta-carotene into their final product, such as astaxanthin and isozeaxanthin. Most of the astaxanthin in the zoeal larvae is esterified, contributing to chromatophore formation and biosynthesis of visual pigments.

In the hematophagus arthropods such as the ticks, the yolk protein has developed into heme-binding protein to help these blood sucking arthropods for efficiently using the heme from their vertebrate hosts. In the cattle tick, *Boophilus*, vitellin degradation starts immediately after oviposition and by the time of hatching 40% of the yolk content is consumed. However, the total amount of heme present in the egg remained remarkably constant throughout embryogenesis and larval development. Most of the heme released during the yolk degradation remains bound to the undigested vitellin molecule. Free heme is a powerful generator of oxygen radical species and therefore the binding of heme by the tick vitellin might constitute an antioxidant mechanism. Because of the absence of de novo synthesis of the porphyrin ring in the ticks, yolk protein becomes the sole heme deposit for the free-living stages (eggs and larvae) of this ectoparasite. By virtue of the heme-binding ability, cattle tick vitellins are also capable of inhibiting heme-induced lipid peroxidation.

18

EXTRAEMBRYONIC MEMBRANES

When the land reptiles evolved from their aquatic ancestors, namely the amphibians, they had to adopt a reproductive strategy with complete emancipation from aquatic habitats for reproductive phenomena, like egg laying. The development of the eggs in the aquatic medium provides protection from extremes of environmental conditions such as heat and osmotic pressure. It helps in getting rid of toxic excretory metabolites, by way of simple diffusion into the water medium. On the contrary, laying of eggs on land requires protection from extreme heat and mechanical shocks from the terrestrial environment. When the eggs develop autonomously, provisions should also be made for gaseous exchange between the environment and the embryo as well as storage of excretory metabolic wastes.

In the embryos of reptiles, birds and mammals, the three primary germ layers participate in the formation of structures that are not part of the embryo itself, but nonetheless, are essential to the physiological functions of the embryo. These are termed extraembryonic membranes, which are discarded when embryonic development is completed. Although mammals have evolved intra-uterine development of embryos and their eggs totally lacking yolk, they still produce the extraembryonic membranes, just like the reptiles and birds, perhaps as a common feature of their phylogenetic relationship.

The extraembryonic membranes form four kinds of structures such as amnion, allantois, yolk sac and chorion. In the birds and reptiles, these membranous structures are an adaptation to development within the shelled eggs that are deposited on land. The amnion furnishes a fluid filled space in which the embryo can develop unhindered by physical restrictions and protected against the mechanical shocks. The yolk sac encloses the yolk and supplies the nutritional needs of the embryo. The allantois functions in gaseous exchange and storage of metabolic wastes. In mammals, the extraembryonic membranes have been adapted to intrauterine development. The amnion functions as a fluid filled cavity and supports the embryo without physical restrictions, but the yolk is vestigial in many mammals. The allantois, together with the part of the chorion of the egg, develops into placenta by means of which the embryo is connected to the maternal circulation and through which metabolic wastes are removed and nutrients and oxygen are provided.

All the four extraembryonic membranes develop from the mesoderm and one other germ layer (ectoderm or endoderm). The mesoderm is derived from the extraembryonic mesoderm (the lowest on the lateral side) which splits into an upper layer called as the somatic mesoderm and a lower layer called as the splanchnic mesoderm. The somatic mesoderm lies just ventral to the ectoderm and splanchnic mesoderm lies just dorsal to the endoderm. The ectoderm and the underlying somatic mesoderm are called as the somatopleure and the endoderm and the adjacent mesoderm is called as the

146   Molecular Developmental Biology

splanchnopleure. The amnion and the chorion develop from the somatopleure, whereas the allantois and the yolk sac develop from the splanchnopleure.

EXTRAEMBRYONIC MEMBRANES IN BIRDS

The formation of the four extraembryonic membranes in chick is depicted in Fig.18.1. They are yolk sac, amnion, chorion and allantois.

YOLK SAC

The yolk sac is the first extraembryonic membrane to be formed, as it mediates nutrition in developing birds and reptiles. It is derived from endodermal cells that grow over the yolk to enclose it. The yolk sac is connected to the midgut by an open tube called the yolk duct, so that the walls of the yolk sac and the walls of the gut are continuous. The blood vessels within the mesoderm of the splanchnopleure transport nutrients from the yolk into the body, for the yolk is not taken directly into the body through the yolk duct. The endodermal cells digest the protein into soluble amino acids that can be passed on to the blood vessels surrounding the yolk sac.

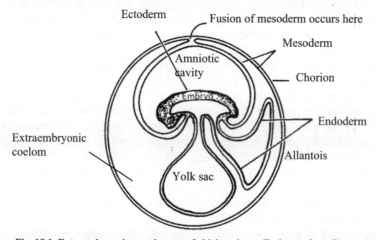

Fig. 18.1 Extraembryonic membranes of chick embryo (Redrwan from Davenport, 1979)

AMNION

The amnion and the chorion are developed together as upwardly projecting folds. The amniotic folds appear on the area pellucida just outside the body folds and eventually closing over the dorsal surface of the embryo. The rudiments of the amnion and chorion first appear anterior to the head as a transverse fold, which runs backwards over the anterior end of the head and covers it. The lateral ends of the folds are prolonged backwards along both sides of the embryo. The lateral folds meet together over the body of the embryo and fuse together enclosing the dorsal aspect of the embryo. Following this, a fold develops posterior to the embryo and ultimately, their free edge fuses together in such a way that the entire embryo is enclosed inside a cavity called amniotic cavity. Soon, a fluid is secreted into this cavity, enlarging it considerably. This enables the embryo to float freely in the fluid, with its umbilical cord connecting it to the extraembryonic parts. The amniotic folds are formed by the extraembryonic ectodermal layer. However the mesodermal cells penetrate into this layer subsequently.

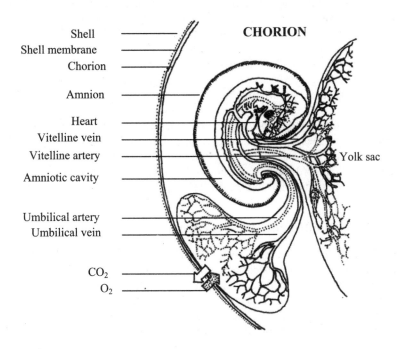

Fig. 18.2 Extraembryonic membranes of the chick embryo to indicate the circulation of the embryo and the exchange of gases through the chorion (Note : the umbilical artery takes waste products to the allantois and the umbilical vein brings oxugen to the embryo (Redrawn from Wolpert, 1998)

The amniotic fold has two surfaces. The inner surface facing the amniotic cavity is made up of extraembryonic ectoderm and the outer surface facing away from the embryo is made up of extra embryonic somatic mesoderm. The chorionic membrane is formed from the outer layer of the amnion by establishing a continuation with the outer ectoderm formed on the dorsal side of the embryo. The chorion, thus formed, grows downwards gradually and completely encloses the embryo along with the yolk sac. The cavity between the amnion and the chorion is called as exocoel or extra embryonic coelom. The chorion lies close to the shell. The chorion of the birds and reptiles are otherwise called as the serosa, so as to distinguish it from the chorion of the mammalian embryo with a well defined function.

ALLANTOIS

In birds, the allantois appears as ventral outgrowth of the endodermal hindgut. In this respect, it corresponds to the urinary bladder of amphibian. Allantois consists of endoderm with a layer of visceral mesoderm covering it from the outside. As it accumulates the metabolic wastes, allantois grows very fast and so penetrates into the extra embryonic coelom in the space between the yolk sac, amnion and chorion. The distal part of the allantois expands while remaining connected to the hindgut of the embryo by a narrow neck. By the middle of the incubation period the allantois spreads over the

148 Molecular Developmental Biology

complete surface of the egg underneath the chorion. This condition facilitates the allantois to subserve another function of supplying oxygen to the embryo. A network of blood vessels developed on the external surface of the allantois and converged towards the stalk of the allantois and through the umbilical cord established vascular connection with the embryo proper. This also accomplishes the oxygen transport to the embryo. The allantoic ciruclation continues until the chick breaks the egg shell and begins to breath the surrounding air. The allantois is essentially nonfunctional until it fuses with the outer chorion to form the chorioallantoic membrane. The membrane serves three major functions. First, it provides a large reservoir for the storage of nitrogenous wastes. Second, it transports calcium from the shell back to the embryo, where it is utilised in bone formation. Third, it functions in respiration by transporting CO_2 from the embryo through the pores in the shell to the outside environment and by transporting O_2 in the opposite direction (Fig.18.2). The chick embryo has turned on its side and has a beating heart. The yolk is surrounded by the yolk sac membrane. The vitelline vein takes nutrients from the yolk sac to the embryo and the blood is returned to the yolk sac via the vitelline artery. The umbilical cord takes waste products to the allantois and the umbilical vein brings oxygen to the embryo. The amnion and the fluid filled amniotic cavity provide a protective chamber for the embryo.

EXTRA EMBRYONIC MEMBRANES IN MAMMALS

We have seen that the extra embryonic membranes provide a microenvironment for the cledoic egg of reptiles and birds, affording physical protection and performing certain physiological processes such as respiration, excretion and nutrition. Although these functions are executed by the placental system in mammals, we come across the occurrence of these extra embryonic membranes during early embryonic development of mammals, thus reminiscing their phylogenetic relationship with reptiles and birds. Such common features among these higher vertebrate groups have also been seen in the pattern of cleavage and gastrulation in spite of the fact that the mammalian eggs are alecithal, unlike the meroblastic eggs of the reptiles and birds.

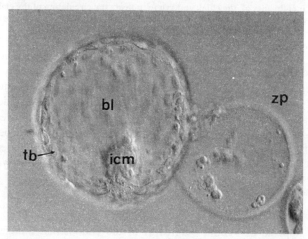

bl – blastocoel, icm – inner cell mass, tb – trophoblast, zp – empty zona

Fig. 18.3 Hatching of blastocyst in human embryo (Courtesy, Dr. Balaji Prasath, K.K. Womens and Childrens Hospital, Singapore)

The implantation of the blastocysts after hatching from the zona pellucida (Fig.18.3) in the uterine wall is followed by the progressive elaboration of the extra embryonic membranes. These membranes are homologous to those found in birds, but their functions have been modified to some extent. As these membranes expand, they fold in such a way that three cavities are formed; the amnion, the allantois and the yolk sac. Collectively, they form the extra embryonic coelom so designated because it is completely lined with mesoderm, known as the extra embryonic mesoderm. The origin and composition of the extra embryonic layers are strikingly similar to those of birds and reptiles. The amnion with its fluid filled amniotic cavity serves as a shock absorber to the developing embryo while preventing its desiccation. The yolk sac in mammals is vestigial and does not contain yolk. Similarly the allantois does not accumulate waste products as in birds but, together with chorion forms the placenta through which the nutrients, gases and metabolic wastes are exchanged (such a placenta is called chorioallantoic placenta to indicate its origin). In some of the more primitive mammals, the yolk sac and chorion may fuse to form the choriovitellin placenta.

19

PLACENTATION

PLACENTA OF MAMMALS

In the cledoic eggs of reptiles and birds, the extra embryonic membranes help the embryonic development in nutrition, gaseous exchange and excretion besides providing a safe aqueous medium for the embryo. In mammals, though the same extra embryonic membranes are present, their physiological functions are taken over by the placental system which provides an organic nexus between the foetus and its maternal tissues. In the absence of any yolk in mammalian eggs, the placenta transports nutrient substances from the tissues of the mother into the embryo. Nevertheless, the foetal and the maternal tissues in the placenta do not blend together, inasmuch as the blood of the mother and that of the embryo do not at any time mix with directly.

IMPLANTATION

In all the viviparous mammals, implantation of the embryo onto the uterus is a prerequisite to placentation. Implantation is a stepwise series of interactions between the hatched embryo and the uterus. These interactions occur in two phases, viz., adhesion and invasion. During adhesion, the embryo aligns with and attaches to the luminal surface of the uterine wall. This is achieved by the formation of junctional complexes between the endometrium of the uterus and the trophoblast of the embryo. The trophoblasts are the giant cells which originate from the mural trophectoderm and give rise to the placenta. Invasion consists of trophoblast- originated cytolytic activity that leads to mechanical disruption of uterine epithelial cells. This is followed by intrusion of trophoblast processes into the basal lamina and the underlying uterine stroma. Thus implantation is a complex and diverse process that accomplishes precise and specific attachment of the embryo to the uterine wall of the mammals. The formation of extra embryonic tissues by the foetal membranes to give rise to the placental system is unique to mammals only.

FORMATION OF PLACENTA IN HUMANS

After establishing contact with the uterine wall, the human blastocyst destroys the adjacent epithelial layer of the uterus and sinks into the underlying connective tissue. In humans, where there is an extremely intimate contact developed between the foetal and the maternal tissue, the trophoblast plays a vital role in the formation of the placenta and this type is termed as interstitial placentation. The trophoblast, which is in immediate contact with the maternal tissue starts proliferation and gives rise to a population of cells which are divisible into two categories. The first type is constituted by a layer of cytotrophoblast cells, whereas the other type is comprised of multinucleated syncytial type of cells called syncytiotrophoblast. The cytotrophoblast constitutes the deep lying portion of the trophoblast and are closer to the embryo, whereas the syncytiotrophoblast is more external and is closer to the uterine wall. The trophoblast now loses its compact cellular nature and becomes permeated by a system of cavities, called the trophoblastic lacunae. As a result of this change, the outer syncytioblast becomes converted into a meshwork of irregular strands with interstices in-between. The trophoblast now penetrates deeper into the uterine wall and reaches the maternal blood vessel. By their proteolytic activity the trophoblastic cells breakdown the endothelium of the maternal capillaries, thereby establishing a communication between the lacunae of the blood vessels and the trophoblastic lacunae. The blood flows into the lacunae, providing nutrient materials for the foetus. The maternal arterial and venous blood circulation is established with the lacunae, facilitating the pumping of the fresh blood and removal of the blood from the lacunae via. veins of the uterus. It must be pointed out that the strands of the outer trophoblastic syncytioblast are only bathed by the maternal blood, contained in the lacunae. As a next step, the underlying cytotrophoblast penetrates into the strands of the syncytioblast, forming a cellularised core within the syncytial mass. Following this, the underlying extraembryonic mesodermal connective tissue along with the ramified allantoic blood vessels penetrates into the cellular core of the cytotrophoblast. By these changes, the trophoblast layer is converted into a dendritic system, consisting of several finger-like ramifying branches called the chorionic villi. The chorionic villi make intimate connection with endometrium and the maternal circulation. However the fetal and maternal circulatory systems remains seprate. The chorionic villi appear uniformly around the chorionic sac, so that the decidual basalis and capsularis are penetrated by them (Fig.19.1). The human placenta thus consists of deciduas basalis (maternal portion) and the chorion frondosum (fetal portion).

Diffusion of soluble materials from the maternal blood collected in the lacunae can pass through the chorionic villi. In this manner, the mother provides the foetus with nutrients and oxygen, and the foetus sends its waste products (mainly CO_2 and urea) into the maternal circulation.

In addition to the above-mentioned functions, the placenta also becomes endocrine by secreting important hormones, which are helpful to the completion of embryonic development. The important among them is the human chorionic gonadotrophin produced by the syncytiotrophoblast portion of the chorion. This peptide hormone causes the other cells of the placenta as well as the ovary to produce a steroid, progesterone which is important for the maintenance and completion of embryonic implantation. The other placental hormones involved in the maintenance of and nutrition of the implanted embryo are listed in table 19.1.

Yet another function of the human placenta is the prevention of immunological rejection of the embryo by the mother. The fact that there is no graft rejection during normal pregnancy implies that the placenta enjoys privileged protection from the maternal immune system. The mother and fetus appear cooperatively to down-regulate the functions of immune cells that are important for graft rejection, while allowing inate host immune defenses to stay intact. The discovery of tumour necrosis factor-α (TNFα), and other members of the TNF-superfamily in trophoblast cells provide important insights into immunologic regulation at the maternal-fetal interface. TNFα is important for placental homeostasis, villous remodelling, cytotrophoblast cell differentiation, and defense from placental pathogens. The placenta can thus utilize these powerful cytokines to facilitate placental growth and development.

PHYSIOLOGY OF THE PLACENTA

Whereas in the ovioparous animals yolk forms the major nutritive source for embryonic development in the viviparous animals placenta helps to transfer the nutriments from the maternal source directly. In the first week of embryonic period, nutrition of the embryo depends upon the diffusion of materials from the endometrium. By the end of second week, the embryo is dependent upon transmission of metabolites to it through the syncytiotrophoblast from the maternal blood circulating through the intervillous spaces of the developing placenta. After 26 days of post-ovulation, movement of the blood through the embryonic heart becomes unidirectional. Thereafter, the nutrition of the embryo is dependent upon the placenta and the functional relationship between the fetal and maternal vascular systems is established.

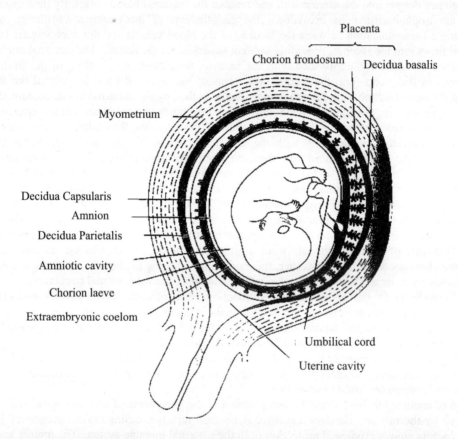

Fig. 19.1 Development of Human embryo and placenta at about 8 weeks. The deciduas capsularis is becoming smooth, and the placenta proper is beginning to take form. By trimester 3, the amniotic cavity has expanded to take up the uterine cavity completely. (Redrawn from Ramon Pinon 2002)

The placenta mediates the metabolic exchange between mother and fetus. It provides for the nutrition and respiration of the fetus. It also functions as an avenue for excretion of the fetal waste products, besides serving as a barrier between the fetal and maternal blood streams. At the end of the first trimester of pregnancy, the placenta occupies about one half the inner surface of the uterus.

Many substances traverse the placenta at different rates and in different ways. That includes the pinocytosis of whole droplets of maternal plasma into the fetal capillaries by the microvilli covering the syncytium. Water, gases, and simple solutes diffuse across the semipermeable placental barrier at varying rates. Substances other than water which transfer across placenta with great rapidity and by the mechanism of diffusion are urea, uric acid, simple amines, oxygen and carbon dioxide.

Fetal haemoglobin possesses the capacity to bind more oxygen than maternal haemoglobin at the same oxygen tension, and it combines with or releases oxygen at lower concentrations than does adult haemoglobin. The placental transfer of sugars, essential amino acids, and water soluble vitamins is accomplished by active or facilitated transfer by enzymatic carriers. Some enzymes produced by the placenta have a protective function preventing transplacental passage of materials potentially dangerous to the fetus. Similarly, cells and such large molecular materials as plasma proteins, antigens, antibodies, and various drugs are denied passage through the placental membrane. Thus placenta is one of the important organs of pregnancy. It not only maintains the fetus but is almost solely responsible for the intrauterine welfare of the fetus.

Table 19.1. Placental Hormones (from Ramon Pinon 2002)

Hormones	Function
Estrogens	Maintain estrogen dependant tissues
Progesterone	Maintains endometrial stability; inhibits maternal Hypothalomus- pituitary– ovarian axis; mammary gland development.
human chorioinic gonadotropin (hCG)	Maintains corpus luteum function during the first eight weeks of pregnancy; may stimulate embryonic and fetal pituitary.
human Placental lactogen (hPL) human Placental growth hormone (hPGH)	Regulate fetal growth, maternal and fetal energy utilization.
Corticotrophin - releasing hormone (CRH)	Stimulates hCG production; regulates placental estrogen production; initiates parturition

EVOLUTION AND CLASSIFICATION OF PLACENTA

Evolution of land vertebrates brought in changes in the embryonic development, which was hitherto occurring in aquatic medium. When the egg began to develop in the terrestrial habitat condition, a need to protect the developing embryos from the physical assaults as well as providing suitable physiological conditions became imminent. Thus in reptiles and birds, the evolution of extraembryonic membranes fulfilled the requirements of embryonic development in a shelled egg, laid on dry land. The class Mammalia represents a group of vertebrates in which the developmental adaptations of the embryo has reached the pinnacle, by way of internal incubation in the maternal body. This adaptation necessitated a contrivance with which the embryo could derive nutrients and their physiological protection from the mother. Thus evolved the placental system. Nevertheless, a trend of transition from the conventional foetal membranes giving rise to the typical placental system is evident among various mammalian groups.

Mammalian placentas are diverse in morphology, persistence, and tenacity of attachment between the foetal membranes and the maternal uterine tissues. The term placenta itself means the shape of a disc, referring to the human placenta (Latin placenta, flat cake). In spite of its morphological diversity, the mammalian placenta has a unifying function of physiological exchange between foetus and maternal tissue. A classification of placental types in Mammalia also reflects the evolutionary trend among them.

The origin of placenta may be traced to two primitive mammalian orders, where a placenta is either absent or is very rudimentary. Evidently, these two groups, viz., monotremes and marsupials exhibit a condition akin to that in reptiles and birds. The monotremes, represented by spiny ant eaters and the duck billed platypus, lay hard- shelled eggs, like the birds. The eggs develop a yolk sac, filled with yolk and a prominent allantois. However, a placenta is lacking as the embryonic development is independent of the mother. In marsupials, the young are born in comparatively underdeveloped condition and hence the placenta is found in a more primitive condition. In the opossum, there is a well- defined yolk sac, although no yolk is present. Hence, the nutrients should be derived from other sources. During implantation, the trophoblast becomes thrown into folds, which fit into the depressions on the uterine wall. However, the trophoblast has no inner mesodermal lining and hence no blood vessels. This is however lined inside by an ectoderm. The uterine wall secretes a viscous fluid, the uterine milk, which is absorbed through trophoblast and the endoderm, finally to reach the area vasculosa of the yolk sac. Such a kind of association between embryonic trophoblast and the uterine tissue is regarded as a precursor to the placenta proper. The allantois is insignificant and has no contact with the trophoblast. The mode of oxygenation in this placental type is unknown.

YOLK SAC PLACENTA

The first step in the development of a true placenta is the yolk sac placenta, found in the marsupial cats (Dasyuridae). Even here, the allantois is very small and the placenta-like structure is associated with the yolk sac. Furthermore, the trophoblast in contact with the non-vascular area of the yolk sac once more forms the connection with the uterine wall. However, in this instance, there appears for the first time uterine erosion, so noteworthy among the higher mammals. This erosion is caused by the trophoblast, which after becoming syncytial in certain regions eats into the uterine epithelium and engulfs some of the maternal blood vessels. The blood so obtained passes in between the trophoblast and yolk sac facilitating its digestion and absorption into the embryo. Respiratory exchange of gases is also possible by this arrangement. This type of placenta is called as yolk sac placenta. This type is also called non-vascular yolk sac placenta, as it is devoid of vascularisation in the yolk sac region. In most insectivores and metatherians, the blastocyst wall (also called omphalopleure) becomes vascularised and establishes close apposition to the uterine mucosa. This type is called choriovitelline placenta.

By far, the advanced placental types fall under two different categories, viz., choriovitelline placenta and chorioallantoic placenta. This division is based on the placental vascularisation, which primarily originates in the extra embryonic mesoderm, adhering to the yolk sac and the allantois respectively.

ALLANTOIC PLACENTA

It is present in yet another marsupial, the bandicoots (Perameles) in a primitive condition. Here, the yolk sac is still large and its area vasculosa could absorb some nutrients by way of the trophoblast. However, the allantois is also well developed and comes in contact with the mesoderm of the chorion. Implantation then occurs and the trophoblast in this area of contact becomes attached to the uterine wall, whose epithelium in this region is transformed into a vascular syncytium. The mesoderm of the allantois for the first time gives rise to foetal blood vessels, which comes into contact with the maternal blood vessels of the uterus because of the disappearance of the intervening trophoblast. This establishes the true allantoic placenta. In all the allantoic placental types, the foetal and the maternal blood actually do not mix. It is always separated by membranes through which the exchange of nutritive and waste materials as well as gases takes place.

Table 19.2 Classification of chorioallantoic placentas based on the intimacy between the foetal membrane (chorion) and uterine tissues (endometrium) (Based on Yolanda P. Creuz, In Embryology edited by Gilbert and Raunio, Saunders, 1997).

Placental type and corresponding morphological types	Characteristics	Placental shape	Found in
1. Epitheliochorial	The maternal tissues contribute endothelium connective tissue and epithelium, while foetus contributes, trophoblasts and endothelium	Diffuse	Horse, lemurs, whales and dolphins, ungulates such cattle, sheep, pigs and giraffe
2. Syndesmochorial	Like epitheliochorial excepting the absence of maternal epithelium	Cotyledonary	Ruminants
3. Endotheliochorial	Only maternal endothelium present. Foetal trophoblast, connective tissue and endothelium present	Zonary to discoid	Carnivores
4. Hemochorial	Consist mostly of all the three foetal tissue; No participation from the maternal tissue	Discoid	Insectivores, primates, bears and primitive rodents
5. Hemoendothelial	Only foetal endothelium participates	Discoid, cup-shaped, spheroidal	Modern rodents

Allantoic placenta assumes prominence only in the true placental mammals (Eutherians). Nevertheless, in different eutherian mammals, the placental structure shows variations with special reference to the arrangement of chorionic villi as well as by the degree of connection between the maternal and foetal tissues. In the most primitive eutherian mammals, the placentas are non-deciduous, which means that there is no erosion of the uterine epithelium. The chorionic folds simply fit in-between those of the endometrium from which they may be easily stripped away without causing bleeding. The best example is found in the pigs. In this case the chorionic epithelium is in contact with the uterine epithelium by fitting in the villi into the pocket like depressions of the uterine wall. This type is also known as epitheliochorial placenta. On the contrary, in the deciduate type, due to closer interaction with the uterine wall and their vascular system, there is tearing away of maternal tissue, causing bleeding, when the foetal part of the placenta separates from that of the mother. Humans are the best example for deciduate placenta. Since in this type of placenta, the chorionic epithelium comes into direct contact with the endothelial walls of the maternal capillaries, it is called as endotheliochorial placenta. In this type, since the chorionic villi are bathed in the maternal blood, it is also called hemochorial placenta (See Tab.19.2).

When the villi are scattered all over the surface of the chorion, as in the case of pigs, the placenta is known as diffuse placenta. However, in the cattle, the villi are formed in patches, while the rest of the chorionic surface is smooth. This type is known as cotyledon placenta. In the carnivores, the villi are developed in the form of a belt around the middle of the blastocyst and this type is called as zonary placenta. In humans and the apes, the chorion is initially covered with the villi, but later on the villi continues to develop only on one side of the embryo, giving the functional placenta the shape of a disc and hence this type is known as discoidal placenta. Similarly, when the placenta consists of two such discs as in the monkeys, it is called as the bidiscoidal placenta (Table 19.3).

Table 19.3. Classification of placental types based on their shape.

Placental type	Characteristics	Found in
1. Discoid placenta	Placenta, large flattened structure held in contact with uterine mucosa by numerous chorionic villi	Primates, humans, rodents, bats, rabbits and certain insectivores
2. Bidiscoidal	Two discs of placenta presents	Monkeys
3. Zonary placenta	Has a circumferential strip of chroionic villi called haemophagous organ contacting the surrounding mucosa	Carnivores like dogs and cats. Also elephants, dugong manatee with morphological variations
4. Cotyledonary placenta	Exhibits multiple coruncles or knob-like protrusions that establish contact with the uterine epithelia	Deer, goat, giraffe
5. Diffuse placenta	The villi are evenly distributed across the surface of the chorionic sac resembling a towel	Pig, horse, camels, lemurs, kangaroos, dolphins, opossums *etc.*

20

GROWTH

In all sexually reproducing forms, the developing embryos start life as zygote, the fusion product of male and female gametes. The single-celled zygote starts multiplying by a process of cleavage during which the enormous volume of the zygote cytoplasm is divided into numerous smaller cells without any intervening cell growth. Cleavage division is followed by gastrulation, which is characterized by extensive cell rearrangements to give rise to the establishment of three germ layers, viz., ectoderm, mesoderm and endoderm. Following the formation of the germ layers, the cells interact with one another and rearrange themselves by a process known as morphogenesis to produce tissues and organs. Growth begins in different animals at different stages during embryogenesis. For example, it starts after gastrulation in *Xenopus*, whereas in chick, growth occurs in the region anterior to the node, during primitive streak regression.

By the time the embryo is fully formed, most of the organ rudiments are completed. Different organs, their relative positions, and the structural features of each organ in the embryo are established according to the morphological plan of the animal. However, organ rudiments at this stage are not capable of performing their specific functions. Acquiring the functionality of the organs, therefore, requires further growth and differentiation.

Growth can be defined as a developmental increase in total mass that results in permanent enlargement of an organism. Growth can either be direct or indirect. In direct growth, the fully formed embryo is a miniature adult in the sense that they attain adulthood by growth without changing the morphology much. On the other hand, in indirect growth, the embryo hatches out as a larva, which is morphologically unlike the adult and attain adulthood by considerable body growth followed by a process called metamorphosis. The life cycle of a butterfly is an illustrative example to describe indirect growth.

TYPES OF GROWTH IN MULTICELLULAR ORGANISMS

In general, growth can result from cell proliferation, cell enlargement without division, or an increase in extra-cellular material, such as bone matrix or even water. By far, the most common mechanism of growth is by cell proliferation. This type of growth is referred to as multiplicative growth. The increase in number of all cells is brought about by mitotic division, while the average size of the cells remains the same. This type of growth is found mainly in embryos, and it is especially characteristic of the prenatal growth of the higher vertebrates. Under these conditions of growth, the growth of the animal's body is directly proportional to the number of its constituent cells. However, the actual size of the cellsdoes not remain constant. As the embryo develops and its tissues become differentiated, the cells increase in size. However, this increase in size of the individual cell is only limited and can account for only a very small proportion of the overall increase of the body. As the animal develops, the mechanism of growth also changes by introduction of other types of growth.

Growth by cell enlargement involves individual cells increasing their mass and getting bigger, without an increase in their number. For example, in the development of nematodes, cell divisions stop in the early stages of organogenesis. The number of cells in the fully grown nematode is thus the same as in a young one which has just emerged from the eggs. The number of cells in each rudiment may be definitely fixed in this type of development. Thus the entire excretory system of nematodes consists of only three cells. Example for this type of growth is also found during vertebrate development. Differentiated skeletal and heart muscle cells and neurons never divide again, but increase in size. Neurons grow by the extension and growth of axons and dendrites, whereas muscle growth involves an increase in mass, as well as fusion of satellite cells to pre-existing muscle fibers.

In other instances, growth occurs by a combination of cell proliferation and cell enlargement. Thus, lens cells are produced from a proliferative zone for an extended period, and their differentiation involves considerable cell enlargement. growth hormone.

The third strategy of growth, also known as accretionary growth, involves an increase in the volume of the extra-cellular space, which is achieved by cell secretion of large quantities of extra-cellular matrix. As an example, in bones and cartilage, most of the tissue mass is extra-cellular. The growth of the bones is dependent on the activity of the cells of the periosteum, which are capable of proliferation and can become osteocytes, while they secrete additional quantities of intercellular bone matrix. In other instances, such as the hematopoitic tissues and epithelia, there is continual cell renewal through out an animal's lifetime by cell division and differentiation from a stem cell population. The end product of this proliferative system, such as mature red blood cells and keratinocytes are themselves incapable of division and die.

GROWTH RATE

The growth rate of different organs and body parts of the animal or embryo is not same at a given time. In general, some organs grow at a faster rate and others at a lower rate. The result is that the proportions of the animal change with growth. For example, in the early embryo of the chick, the head is as large as the rest of the body, and a large portion of the head is made up by the relatively enormous eyes. However, at the time of hatching the head is much smaller than the body, and in the adult fowl the head is relatively small. Similarly, in human beings, growth of different parts of the body is not uniform, because different organs grow at different rates. At nine weeks of development the head of human embryo is more than a third of the length of the whole embryo, whereas at birth it is only about a quarter. After birth, the rest of the body grows much more than the head which is only about an eighth of the body length in the adult.

MEASUREMENT OF GROWTH RATE

A method for analyzing the unequal growth of organs has been proposed by Huxley. He found that if two parts of an animal grow at different rates, their sizes at any given moment are in a simple relationship to each other. This relationship of disproportionate growth of two different organs is known as allometric growth. In otherwords, allometry occurs when different parts of the organisms grow at different rates. The allometric growth can be expressed by the following formula; $y = bxk$ where y is the size of one of the organs, x is the size of the other organ, b is a constant, and k is known as the growth ratio. Growth ratio (k) shows the relation of the rates of growth of the two parts which are being compared. If the growth ratio equals 1 the two organs grow proportionally or isometrically. If the growth ratio does not equal 1 the growth is disproportional or allometric, the relative size of the two organs changing as growth proceeds. If the growth ratio is grater than 1, organ y grows at a quicker pace than organ x. In this case, organ y is called positively allometric to organ x. If the growth ratio is less than 1, organ y grows at a slower pace than organ x; it is negatively allometric to x. Examples for allometric growth are found in the growth of the appendages of crab. In the fiddler crab Uca pugnax, positive allometric growth occurs in the left chela. In adult male, the left chela grows disproportionality larger than that of the young male crab in which both chelae are of equal size. The left chela of the males may attain an extraordinary large size, upto 38 per cent of the weight of the body as a whole. Thus, the large chela is positively allometric with respect to the rest of the body, with a growth ratio of 1.62.

Another example for positive allometry is found in the development of vertebrate limb. During limb development, local differences in chondrocytes cause the central toe of the horse to grow at a rate 1.4 times that of the lateral toes. This means that, as the horse grew larger during evolution, this regional difference caused the five-toed horse to become a one-toed horse.

GROWTH CURVE

There is significant difference in the rate of growth among different organisms. But the generalized pattern of growth of an organism with respect to time follows an S-shaped curve. This is a universal expression of growth; when the abscissa is time and the ordinate represents a progressive scale of a unit of measurement of an organism or a population of organisms, an S-shaped growth curve results (Fig.20.1).

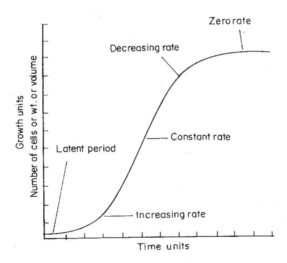

Fig. 20.1 A graph of growth patterned after the growth of microorganism in test tube cultures. When few cells are added to a fresh culture medium they show a latent period – that is, a period during little growth takes places. Then the number of cells increases rapidly with time (increasing rate). Then the growth may go along at a constant rate for a while but eventually it begins to slow down (decreasing rate) and final no growth occurs. Growth curves of higher organisms may deviate widely from this hypothetical curve (redrawn from Browder 1980)

A curve of this shape is known as the sigmoid curve. The organism starts with a definite size, which is maintained briefly without any apparent growth, then it increases very slowly for some time, and finally it begins to grow very rapidly. For a while, the increase may be at a rather constant rate and the slope of the curve does not change. But soon, certain factors cause a retardation of growth, a decreased increase in size with time. As a result, no further increase in the size of the organism occurs.

When this stage is reached, equilibrium is established between loss of material, which would tend to reduce the weight of the organism, and new material resulting from synthesis, which would tend to increase the weight. Thus in an adult animal with a stable weight, cellular growth continues in many organs of the body, but because of loss or death of cells, no over-all growth occurs. The time lapsed since the loss of growth is hence a measure of the age of the organism.

MECHANISM OF CELL PROLIFERATION

The eukaryotic cells go through a fixed sequence of events called cell cycle during cell duplication. The cell grows in size, the DNA is replicated, and these replicated chromosomes then undergo mitosis, and becomes segregated into two daughter nuclei. After this, the cell divides to form two daughter cells, which can go through the whole sequence again. However, during cleavage of the zygote, there is no cell growth, and hence the cell size decreases with each division. Only in other proliferating cells, the cytoplasmic mass must double in preparation for next cell division.

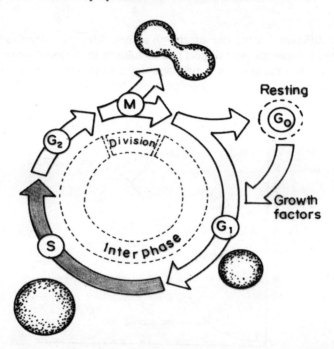

Fig. 20.2 The eukaryotic cell cycle. After mitosis (M), the daughter cells can either enter a resting phase (G0), in which they effectively withdraw from the cell cycle, or proceed through G1 to the phase of DNA synthesis (S). This is followed by G2, and then by mitosis. Cell growth occurs throughout G1, S, and G2. The decision to enter G0 or to proceed through G1 may be controlled by both intracellular status and extracellular signals, such as growth factors. Cells such as neurons and skeletal muscle cells, which do not divide after differentiation, are permanently in G0. (Redrawn from Wolpert, 1998)

The eukaryotic mitotic cell cycle is divided into distinctive phases (Fig.20.2). At the M-phase, the mitosis gives rise to two new cells. During the interphase, replication of DNA occurs during a defined period, called the S-phase. As shown in the figure, the S-phase is preceded by a period known as G1. The S-phase is followed by another interval, called G2 phase. The cell enters mitosis after G2.

During the interphase, the cell growth occurs through synthesis of protein, and the replication of their DNA. During embryonic cleavage, both G1 and G2 phases are lacking and also, during meiosis, there is no DNA replication at the second division. The timing of cell cycle events is controlled by a set of proteins called cyclins. Cyclin concentration oscillate from one phase of the cycle to the next. Cyclin acts by forming complexes with cyclin-dependent kinases, which in turn phosphorylate proteins that trigger the events of each phase, such as DNA replication in S-phase or mitosis in M-phase.

CONTROL OF CELL GROWTH

Cell growth through proliferation is under the control of both an intrinsic programme and external signals. Growth factors and other signalling proteins play significant role in controlling cell growth and proliferation. In cell cultures, growth factors are essential for cells to multiply, although different cell types require different growth factors. When somatic cells are not proliferating they are usually in a state known as G0, into which they withdraw after mitosis. Growth factors enable the cells to proceed out of G0 and progress through the cell cycle. Interestingly, growth factors are not only necessary for the cells to divide, but also for their survival. In the absence of all growth factors, the cells undergo programmed cell death, a process known as apoptosis (see chapter on apoptosis). There is a significant amount of cell death in all growing tissues, so that growth rate is dependent on the rates of both cell death and cell proliferation.

EFFECT OF GROWTH HORMONE AND MAMMALIAN GROWTH

Embryonic growth is dependent on growth factors. FGFs control cell proliferation in the progress zone of the developing chick limb, and are required for proliferation and outgrowth of the limb bud. Similarly, insulin-like growth factors 1 and 2 have a key role not only in post natal mammalian growth, but also in growth during embryonic development. Newborn mice lacking a functional Igf-2 gene develop normally, but weigh only 60% of the normal newborn body weight. Mice in which the Igf-1 gene is inactivated are also growth retarded, showing that Igf-1 has an important role in embryonic growth. Both Igf-1 and Igf-2 proteins act via Igf-1 receptor; the Igf-2 receptor by contrast reduces the concentration of Igf-2, by enabling it to be degraded.

Growth hormone synthesized by, pituitary gland is essential for the post-embryonic growth of humans and other mammals and is secreted throughout fetal life. A child with insufficient growth hormone grows less than normal, but if the hormone is administered normal growth is restored. Production of growth hormone is under the control of two hormones produced in the hypothalamus. One is the growth hormone-releasing hormone that promotes growth hormone synthesis and secretion. The other is somatostatin which inhibits its production and its release. Growth hormone produces many of its effect by inducing the synthesis of IGF-1 and IGF-2. In humans, there is a sudden spurt in growth after puberty which is caused by increased secretion of gonadotropins. This results in the increased production of sex steroids which, in turn, cause increased production of growth hormone.

21

METAMORPHOSIS

In most animal species the morphogenetic processes leading to organ formation are completed before the embryonic development is over. The young ones then grow to adult without much transformation in development except perhaps the gonadal development and maturity. This may be termed as direct development. In many other animal species, there is an intervention of a larval stage in the post-embryonic developmental cycle. These larval forms are embryologically incomplete but ecologically independent stages in the life cycle, necessitating a dramatic change in form to reach the adulthood. This phenomenon is called metamorphosis. The life cycle of the terrestrial frog with an aquatic larval stage, a benthic sea urchin with a planktonic pluteus larva, and a flying butterfly with a creeping caterpillar larva are some of the examples to illustrate metamorphosis. In all these forms, the morphogenetic or development potentials are kept in reserve even after the embryonic development is completed.

Cellular differentiation during embryogenesis occurs chiefly by inductive interaction of neighbouring cell types. However during the post embryonic morphogenesis, as it occurs during metamorphosis, cell differentiation is under the influence of the hormones. Under hormonal induction metamorphosis is accomplished in two major ways. One is by remodelling of the existing larval tissues into adult tissues as it happens in the amphibians. The other is by massive cellular destruction of larval tissues and the reformation of the adult tissues, as it happens in insects.

AMPHIBIAN METAMORPHOSIS

In amphibians, metamorphosis is generally associated with those changes that prepare an aquatic organism for a terrestrial existence. This is achieved by regressive changes including the loss of the tadpole's horny teeth and internal gills as well as the destruction of the tadpole's tail. Limb development and dermoid gland construction are concurrent events. In addition to these morphological changes, biochemical events are also associated with amphibian metamorphosis. For example, the liver starts producing the enzymes necessary for urotelic execretion in adults, in contrast to the ammonotelic tadpoles.

Amphibian metamorphosis is the developmental process, which transforms a tadpole into a frog. This transformation requires extensive remodelling of almost every tissue in the animal. For example, the small intestine of the tadpole requires a shortening in length as well as internal anatomical restructuring to function in the adult frog. The tadpole intestinal epithelial cells undergo a programmed cell death (apoptosis) and are replaced by a layer of newly formed adult epithelium. It has long been known that addition of iodine to the water or feeding with thyroid gland tissues causes metamorphosis to occur precociously in the tadpole; while the elimination of iodine from the diet postpones it. It is now known that the amphibian metamorphosis is brought about by the secretion of the hormone thyroxine (T4) and triiodothyroxine (T3) from the thyroid. T3 is more active than T4 as T3 causes metamorphosis in thyroidectomised tadpoles in much lower concentration than T4.

The release of T3 from thyroid is stimulated by the thyroid stimulating hormone (TSH) of the pituitary which in turn is controlled by the thyrotropin-releasing hormone (TRH) from the hypothalamus. Hypothalamus receives environmental cues, such as, temperature and light for their neurosecretory activity to release TRH. Interestingly, pituitary also secretes another hormone, namely the prolactin which is actually antagonistic to T3 in prohibiting metamorphosis (Fig.21.1). The prolactin secretion is inhibited by dopamine produced by the hypothalamus. Hypothalamus seems to finally control such a delicate balance between T3 and prolactin. The thyroid hormones also act on the hypothalamus and pituitary to maintain synthesis of TRH and TSH. Thus, metamorphic changes are controlled by the thyroid hormones on a threshold concept, which is regulated by prolactin.

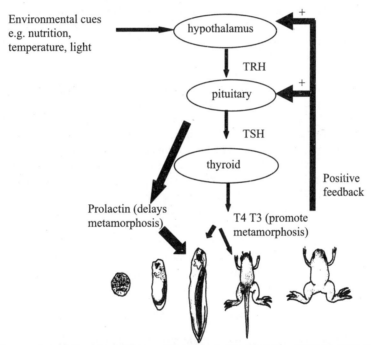

Fig. 21.1. Hormonal regulation of amphibian metamorphosis (Modified from Wolpert, 1998) TRH – Thyrotropin releasing hormone, TSH – Thyroid stimulating hormone, T3 – Triiodo thyroxin, T4 - Thyroxine

Prolactin upregulates the expression of type III iodothyronin 5-deiodinase, an enzyme that inactivates TH, in a highly tissue- specific manner in the tail but not in liver. By analogy, prolactin is the insectan juvenile hormone, as both hormones possess metamorphosis restraining properties.

Although thyroid hormone is the only obligatory signal for the initiation and completion of amphibian metamorphosis, other hormones and factors can modulate the onset and progression of metamorphosis. These include glucocorticoid hormones well as corticotrophin releasing factor (CRF) and adrenocorticotrophin(ACTH) which can accelerate TH- induced metamorphosis, both in intact tadpoles and isolated tissues. However, their control action depends on the environmental conditions, such as nutrition and developmental stage of the tadpole.

In Amphibia, the initiation and maintenance of an uninterrupted larval-adult transition requires the appearance and maintenance of high circulating level of thyroid hormone. Addition of T3 to organ cultures of early pre-metamorphic *Xenopus* tadpole tails caused them to undergo regression with the same morphological and biochemical criteria seen during natural metamorphosis. Evidently, the hormone merely serves to initiate a dormant developmental program laid down at an early stage in postembryonic development. In other words, the hormone does not determine the developmental program but serves to initiate it.

THYROID HORMONE RECEPTOR

Thyroid hormones work largely at the level of transcription of some genes and repressing the transcription of others. However, the earlier responses to thyroid hormone (T3) are the transcriptional activation of the thyroid hormone receptor (TR) genes. Thyroid hormone receptors are members of the steroid hormone receptor superfamily of transcription factors. There are two types of thyroid hormone receptor, namely, TRα and TRβ in amphibians as in all vertebrates. Expression of thyroid receptors in *Xenopus* tadpole is under developmental control. Although very small amounts of both TR transcripts can be detected in early embryos, substantial increase, particularly of TRα mRNA, occurs when the *Xenopus* tadpole first exhibits competence to respond to exogenous thyroid hormone. At this stage of development, several tissues which are programmed to undergo major changes later during metamorphosis show high concentrations of TR mRNA. These tissues include brain, liver, limb buds, small intestine and tail. Again there is good correlation between the accumulation of TR transcripts and the circulating level of thryroid hormone in *Xenopus* tadpoles.

The mechanism of TR expression in amphibians involves a direct interaction between TR proteins and the promoters of the genes encoding them. The promotor region of TR contains one or more thyroid hormone responsive element (TRE), to which the TR proteins bind. In *Xenopus*, the expression of TRβ- gene is modulated by T3 to a greater extent than that of TRα gene. Such direct interaction between the thyroid hormone receptor and the promotor of its own gene in *Xenopus* is termed as TR-autoinduction. TR has the unique property that it is a strong repressor of transcription which can only be relieved when it has T3 bound to it. So it is the liganded TR that enhances transcription, and which is responsible for the induction of the receptor. The extent of autoinduction of TR is dependent on the developmental stage of the tadpole. T3 is ineffective at early developmental stage, the autoinducibility increasing with development, the highest sensitivity being reached as metamorphosis progresses towards its climax. Furthermore, the magnitude and kinetics of TR mRNA upregulation varies from tissue to tissue, cell type to cell type and in different regions of the tadpole.

GENE REGULATION DURING AMPHIBIAN METAMORPHOSIS

The amphibian metamorphosis is initiated by a single molecular agent, the thyroid hormone. Many of the genes regulated by thyroid hormone during metamorphosis have now been identified and their influence on the biological process connected with metamorphosis has now been identified. The thyroid hormone regulated genes participating in the intestinal remodelling alone amounts to 20 genes. Among them, *Xenopus hedgehog* and *stromelysin-3* genes have been shown to participate in the apoptosis of the larval epithelium and development of the adult epithelium. The gene expression during the larval tail regression has also been identified to be influenced by the thyroid hormone.

NEOTENY

In general, metamorphosis results in the disappearance of larval characters and the acquisition of the adult characters such as maturity of gonads. However, in salamanders, juvenile characters are retained by retarded body development while the gonads achieve maturity at the normal time. This phenomenon is called neoteny. Since metamorphosis is normally induced by thyroid hormones, there are several different levels at which mutations could result in neoteny: the hypothalamus, the adenohypophysis, the thyroid gland, and the target tissues of the larva. Some neotenic amphibians can be experimentally stimulated to undergo metamorphosis. In *Ambystoma mexicanum*, the adenohypophysis apparently does not release thyroid stimulating hormone that promotes the synthesis of thyroid hormones by the thyroid gland. This salamander can be induced to undergo metamorphosis by transplanting the adenohypophysis of another species into it or by treatment with thyroid hormones.

A contrasting situation is obtained in certain arboreal salamanders such as *Bolitoglossa* of thyroid hormones by the thyroid gland. This salamander can be induced to undergo metamorphosis by transplanting the adenohypophysis of another species into it or by treatment with thyroid hormones. *accidentalis*, where gonadal maturation occurs precociously while the rest of the body develops normally. This condition is referred to as progenesis. In both cases, larval characters are retained when the gonads start functioning.

MOULTING AND METAMORPHOSIS IN INSECTS

Insects provide an interesting example for the occurrence of morphogenetic processes in the post embryonic developmental stages. These changes are caused by the massive destruction of larval tissues and the differentiation of adult tissues from the embryonic reserve cells called imaginal tissues. Like amphibian metamorphosis, insect metamorphosis involves the balance between a hormone necessary for continued larval growth and another hormone, capable of stimulating adult development.

Post embryonic growth in insects is discontinuous in view of the occurrence of a rigid exoskeleton called, cuticle. The chemical composition of the cuticle varies with stages in the life cycle. In order to facilitate body growth, the insects have to shed the cuticle periodically and then reform a new cuticle, which will be soft and flexible. This process of shedding the old cuticle and replacing it with a new cuticle by the underlying epidermis is called moulting or ecdysis. Moulting is initiated by the separation of the underlying epidermis from the cuticle by a process known as apolysis. The gap between the cuticle and the epidermis is filled with a fluid called moulting fluid, which contains several hydrolytic enzymes that will start digesting the inner part of the overlying cuticle. Simultaneously, the epidermis secretes a new cuticle, which is soft and extensible. Due to the resorption of the old cuticle by the moulting fluid, it becomes brittle and friable and finally shed away from the insects. Moulting facilitates not only body growth, but also brings about metamorphosis in insects.

In general, the metamorphic insects can be divided into two types, viz., hemimetabolous and holometabolous. In the hemimetabolous insects (e.g. cockroach), the metamorphosis is incomplete and the larval forms (nymphs) resemble the adults except for the absence of wings. The adult organs are formed from the already existing rudiments by way of repeated moulting. On the other hand, in the holometabolous insects (e.g. butterflies), the metamorphosis is complete and there is a dramatic and sudden transformation between the larval and adult stages. The juvenile larva after undergoing several moults becomes a pupa. Within the quiescent pupa, the larval tissues undergo histolysis. Concurrently new adult organs develop from undifferentiated nests of cells called, imaginal discs or histoblasts. In *Drosophila*, there are 10 major pairs of imaginal discs, which reconstruct the entire adult and a genital disc, which forms the reproductive structures. The differentiation of imaginal disc cells into adult tissues is strikingly under the control of ecdysone, which is the primary moulting hormone of insects. The imaginal disc is made of embryonic cells tightly compressed along the proximal distal axis. During metamorphosis, on exposure to 20-hydroxyecdysone, these cells differentiate and elongate. Such an elongation is brought about by peripheral microtubules in the leg disc cells. This conversion of an epithelium of compressed cells into a longer epithelium of non-compressed cells represents a novel mechanism for the extension of an organ during development.

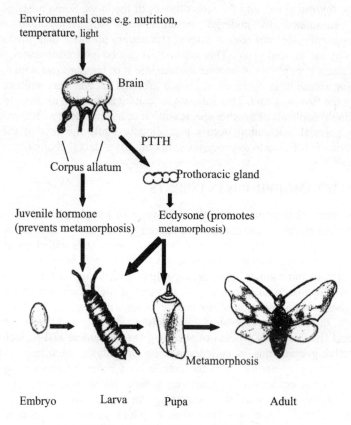

PTTH – Prothoracicotrophic Hormone

Fig. 21.2 Molting and metamorphosis in insects (Modified from Wolpert, 1998)

Hormonal induction of developmental processes during insect metamorphosis is better illustrated in the transformation of larval cuticle into the adult cuticle (Fig.21.2). Control of moulting and metamorphosis is under bihormonal regulation. Ecdysone, the so-called moulting hormone is secreted by the prothoracic gland. Ecdysone is however a prohormone that should be converted to an active form, 20-hydroxyecdysone, in the peripheral tissues like the fat body. The secretary activity of the prothoracic gland is controlled by a brain neurosecretory hormone called prothoracicotrophic hormone (PTTH). Corpora allata, another endocrine gland secretes a hormone called juvenile hormone (JH) which retains the larval characters. The secretary activity of corpora allata is again under the control of the restraining neuropeptide, allatostatin originating from the corpora cardiaca. During the larval stages, the titre of juvenile hormone in the blood far exceeds that of ecdysone. This results in the larva moulting into another larva. During the transformation of the larva into the pupa, the JH level drops below a critical threshold value, thus facilitating the ecdysone to commit the larval cells to pupal development. During pupation the corpora allata do not release JH and the 20-hydroxyecdysone-stimulated pupa will eventually metamorphose into the adult insect. Metamorphosis to the adult occurs when the insect is exposed to the 20-hydroxyecdysone in the absence of JH. As shown in the Figure 15.2, the main target tissue is the epidermis for the hormonal action to synthesize characteristic proteins, which will be responsible for the secretion of the larval, pupal and the adult cuticle respectively.

ECDYSTEROIDS ACTION DURING INSECT METAMORPHOSIS

An interesting aspect of insect post-embryonic development is that the steroid hormone, 20-hydroxyecdysone (20HE) triggers the larval to adult metamorphosis by inducing imaginal tissues to generate adult structures and at the same time larval tissues to degenerate. Such a dichotomous effect of ecdysone on larval tissues is due to differences in the hormone receptors. The target tissues of the metamorphic larval tissues for ecdysone action can be divided into three types: (1) the strictly larval tissues such as salivary glands, muscle, and the gut that undergo cell death in response to 20HE; (2) the imaginal tissues that divide and differentiate to produce adult structures on exposure to 20HE; (3) tissues that undergo extensive modification or remodelling, such as the fat body and the central nervous system. Members of the three types of tissues exhibit a wide variety of metamorphic responses to ecdysone. Different types of receptors for 20HE action are produced by different isoform combinations in a manner consistent with the metamorphic responses of the above three types of tissues.

In *Drosophila melanogaster*, three ecdysteroid receptor genes encoding three ecdysone receptor isoforms such as EcR-A, EcR-B1, and EcR-B2 have been discovered. These isoforms have common DNA- and hormone - binding domains with different N- terminal regions. The mode of action of 20 HE is like any other steroid hormone. The ecdysone enters the cell and binds with a cytosolic receptor. The hormone - receptor complex binds to the regulatory region of a number of different genes, inducing a new pattern of gene activity characteristic of metamorphosis.

ROLE OF JUVENILE HORMONE IN METAMORPHOSIS

As described above, insect moulting and metamorphosis are coordinated by the interplay of both ecdysone and juvenile hormone. Whereas ecdysteroids initiate and coordinate the moulting and metamorphosis, JH directs ecdysteroid action. Its presence during post embryonic development prevents the ecdysteroids from causing a switch in differentiative programme from larval to pupal or from pupal to adult. In other words, the switch to a new developmental program depends on the absence of juvenile hormone.

Recent studies have contributed to the receptor- DNA interactions during insect metamorphosis. For this interaction to occur, the intracellular ecdysteroid receptor requires to dimerize with another nuclear orphan receptor, called ultraspiracle (USP). USP is a structural homologue of the vertebrate retinoic acid X receptor, which also make dimerization partnership with vertebrate steroid hormones. The heterodimer of EcR and USP is the functional ecdysteroid receptor complex in all insects studied. However, the EcR and USP expression during post embryonic development is regulated differently by the molting hormone, ecdysone and the juvenile hormone. For example, in the insect, Manduca, EcR expression is stimulated by 20-OH-ecdysone, but this increase is inhibited by juvenile hormone. The expression of both USP mRNAs (USP exists as two isoforms, USP1 and USP2) varies during last two instars. Low levels of 20-OH-ecdysone increase USP concentration; whereas higher hormone levels inhibit USP1 expression. Juvenile hormone prevents the increase in USP1 specific mRNA, but USP2 is not affected. Evidently, tissue- and stage- specific in USP and EcR expression are responsible for ecdysteroid action in bringing about the metamorphic changes in insects. Recent studies have also indicated that JHs act as USP ligands and exhibit suppressive effects on ecdysone- dependent EcR transactivation.

The ecdysteroid receptor interacts with hormone response elements called ecdysteroid response elements (EcRE). The binding affinity between EcR/USP complex and various hormone response elements varies significantly. The wide range of EcREs recognised by the EcR/USP contributes to the complexity of the hormonal regulation of metamorphosis in insects Binding of EcR/USP to DNA is enhanced in the presence of 20-OH-ecdysone. Binding between EcR/USP and 20E directs the transcription of the early genes such as E74 and E75, most of which encode transcription factors. These factors in turn induce a secondary response by repressing some of the early genes and activating the expression of tissue-specific genes. These regulatory hierarchies consisting of a sequential series of transcription factors are induced and coordinated by ecdysteroids. Interestingly, JH modulates the ecdysone-regulated expression of the EcR and USP genes in an isoform –specific manner. Again, JH modifies the expression of the ecdysone –regulated transcription factors such as E75.

In the above described two examples of metamorphosis (amphibians and insects), there is a dramatic change in form that alters the physical appearance of the organism as it existed at the completion of the organogenesis. In a way, metamorphosis is a second period of organogenesis; the cell differentiation in this case is regulated by hormones produced by larval tissues.

22

REGENERATION

Regeneration can be defined as the capacity to replace parts of the body which are accidentally lost or which are willingly lost (autotomy). This is normally achieved by compensatory growth and differentiation of the tissues that compose the severed organ. In general, the ability to regenerate the missing parts is greater in soft-bodied invertebrates than in vertebrates. More specifically, those invertebrates which reproduce asexually by way of budding have a tendency to regenerate lost body parts spontaneously. For example, sponges, cnidarians, flatworms, nemertines, annelids and some echinoderms exhibit asexual reproduction by fission and regenerate lost body parts. In other words, regeneration and asexual reproduction are related events in these animals. In all these cases of restitution of body parts, original polarity of the animal is retained.

Not all animals are able to regenerate body parts, and not all tissues within a body can be equally repaired. For example, in the vertebrates, regenerative power declines, but surprisingly, the urodele amphibians possess tremendous ability to restore severed body parts like the limbs and eyes. In these cases, the damaged or amputated organs not only regenerate but are also able to build back an exact replica of the lost part.

TYPES OF REGENERATION

The simplest form of regeneration is the axonal outgrowth seen in severed nervous system. Regeneration can also occur as simple proliferation of cells that comprise the damaged organ. The example includes regeneration of cells in intestine, liver etc. Apart from this simple tissue regeneration, which is called 'passive regeneration', two major types of regenerative processes have been distinguished in animals. One is morphallaxis, which involves reorganisation of the remaining portion of the body to produce the missing structures. An example is the coelenterate, *Hydra*, which is able to reconstitute itself out of a bit cut from a whole animal. The other type is 'epimorphosis', which occurs by formation of an outgrowth of new tissue from the wound surface. Limb regeneration is the best example of epimorphosis.

REGENERATION IN *HYDRA*

Hydra is a freshwater coelenterate consisting of a hollow tubular body with a head region at the distal end and a foot, or basal disc with which it can stick to surfaces. The head consists of two parts, the conical hypostome, where the mouth opens, and below that is the tentacle zone, from which a ring of tentacles emerge. In between head and foot is the body column with tremendous capacity for regeneration. Throughout the animal, the tissue construction of the body wall is the same. It consists of two epithelial layers namely, the ectoderm and endoderm, separated by the basement membrane, called mesoglea (Fig.22.1). Each layer, which is a single cell deep, comprises a cell lineage, consisting of stem cells in the body column and differentiated cells in the extremities. All other cell types, namely,nematocytes (the cnidarian stinging cells), neurons, gametes, and secretary cells, lodged in the interstices among the epithelial cells of both layers, belong to the interstitial cell lineage.

CELL PROLIFERATION AND LOSS DURING *HYDRA* GROWTH

The *Hydra* grows continuously and the cells in the body wall proliferate steadily. The epithelial cells in the body column of both layers are continuously in mitotic cycle and, as the mass of the body column increases, the epithelial layers are continuously displaced along the body column. Tissues of the upper part of the column are displaced into the tentacle zone, and then eventually sloughed at the tentacle tips. Some of the endodermal tissue is displaced into the hypostome and sloughed off at the tip. However, the endodermal tissues in the hypostome are generated by dividing epithelial cells at the base of the hypostome, which are displaced apically and shed. Similarly, tissue of both layers displaced down the body column ends up on the foot and is sloughed. Loss of cells through these upper and lower extremities however accounts for only 15%, and the bulk of cell loss is through their displacement onto the developing buds, which is a form of asexual reproduction in *Hydra*. A bud is formed as an invagination in the lower middle part of body column, and then develops into a cylindrical protrusion. This is followed by the formation of head and foot at the ends of the protrusion.

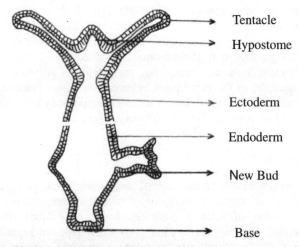

Fig. 22.1. Morphology of *Hydra*

The bud then detaches from the adult. The cells of the interstitial cell lineage are displaced along with their neighbouring epithelial layer cells into an extremity, or into a bud, and are also continuously lost. Thus, *Hydra* may be considered as an animal that is continually regenerating by replacing all its cells in three to five weeks. Hence, this organism is analogous in its growth to a vertebrate intestinal villus which is also getting continuously replaced by cellular proliferation.

By virtue of the tissue dynamics that take place in adult *Hydra*, the processes governing axial patterning are continuously active to maintain the form of the animal. Hence, regeneration in *Hydra* is closely related to these axial patterning processes. Regeneration in *Hydra* occurs by the process of morphallaxis, which does not require new growth by cell proliferation, but involves reconstitution of existing cells. As a consequence of this morphallactic regeneration, the *Hydra* is smaller in size but after feeding will grow eventually back to the normal size.

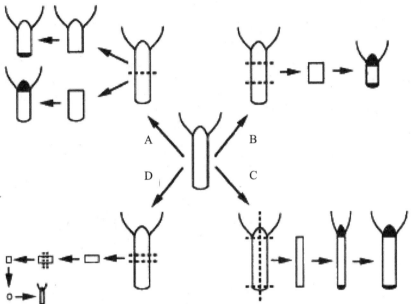

Fig. 22.2. Regeneration capacities of a *Hydra*. A: The regeneration of the missing extremity on each half after bisection. B: The regeneration of both head and foot after isolation of a piece of the body column. C: Regeneration of the extremities as well as the body column after isolation of a longitudinal half of the body column. D: Regeneration of complete animal from a small fraction of the body column. Dark head and foot regions arose by regeneration. (Redrawn from Bode, 2006)

The body column has an extraordinary capacity for the regeneration of a missing head or foot. When the body column of a *Hydra* is cut transversely, the lower piece will regenerate a head and the upper piece will regenerate a foot (Fig.22.2). Similar regeneration will occur when the animal is bisected anywhere along the upper 7/8ths of the body column. Similarly, when a piece of the body column is isolated, it will invariably regenerate a head at the apical end and a basal disc at the basal end, indicating that the tissue has a apical- basal or oral-aboral polarity which is maintained even after the excision. When an isolated body column is cut in half longitudinally, it rapidly heals into a narrow cylinder, and regenerates a head and foot, both with smaller diameters. The most dramatic form of *Hydra* regeneration is the ability to form complete animals from aggregates of cells. The cells of the aggregate sort out in such a way that in 24 hours, it will consist of a two-layered shell comprising ectoderm and endoderm of the adult animal. Thereafter, heads form, which subsequently organize surrounding tissue into the body column and foot of a *Hydra*, and detach from the remaining tissue. In these aggregates, the patterning processes for head formation, axial patterning, and foot formation start de novo, as there is no polarity in the tissue of the spherical shell.

However, the extensive regeneration capacity found in intact *Hydra* is only limited to the body column. Hence, an isolated head will not regenerate a foot, and an isolated foot will not regenerate a head. Similarly, an isolated tentacle disintegrates with time. Evidently, the regeneration capacity is limited to the body column, where the stem cells of all three lineages are undergoing division. This shows that there exist two organizing centres, one at each end. When regenerative growth occurs, the new cells must be derived either from a reserve population of previously undifferentiated, totipotent cells or by dedifferentiation from previously differentiated cells. In *Hydra*, there exists a pool of interstitial cells, from which the different classes of differentiated cells, such as the cnidoblasts are normally derived and constantly replaced. These interstitial cells migrate to the wound region and start differentiation into the lost part. These cells continue to proliferate and serve as a reserve for regenerative growth. Interestingly, the interstitial cells also congregate in the ectoderm prior to asexual reproduction by budding. Evidently, the mechanism underlying the repatterning of cells during regeneration and asexual reproduction is similar. The differentiation of the interstitial cells into the specialized cells is demonstrated during the mucous cell formation during the basal regeneration. When a *Hydra* is cut at the basal region, the interstitial cells from the gastric region migrate towards the basal region. As it moves along the body column to reach the base, they differentiate into the mucus cells characteristic of the foot. Interestingly, during normal life, the differentiation of the mucus cells of basal region also occurs through the interstitial cells, thus indicating that regeneration in *Hydra* follows the normal way of cell differentiation. In other words, regeneration is a way of life process in these lowly organized invertebrates.

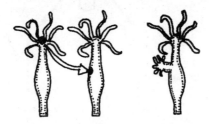

Fig. 22.3 Secondary axis formation by grafting an excised fragment of hypostome into the gastric region of an intact *Hydra* (Arrow indicates the grafting) (Redrawn from Wolpert, 1998)

ROLE OF EPITHELIAL CELLS IN *HYDRA* REGENERATION

In *Hydra*, the body column possesses formidable capacity for head and foot regeneration. This occurs in a morphallactic manner. Interestingly, only the two epithelial layers consisting of ectoderm and endoderm are required for regeneration in *Hydra*. Interstitial cell lineage is not needed for either head or foot regeneration.

AXIAL PATTERNING DURING REGENERATION

Adult *Hydra* is in a steady state of production and loss of tissue, coupled with continual displacement of tissue along the body axis. Hence, the processes governing the axial patterning need to be constantly active to maintain the morphology of the animal. Understandably, the same processes are also involved in head regeneration in *Hydra*. This axial patterning in *Hydra* is accomplished by the existence of a morphogenetic gradient. This gradient consists of two components namely, head activation gradient which is confined to the body column and a head organizer, located in the hypostome.

Transplantation experiments have revealed that the head organizer in the hypostome has the ability of typical organizer in that it can induce a second axis when transplanted to the body column of the host *Hydra* (Fig.22.3.). Similarly, when a piece of the upper part of the body column is transplanted into a lower axial location in another *Hydra*, it can also form a second axis consisting of head and body column. However, this axis is formed by self-organization of the transplanted tissue rather than by the induction of host tissue. On the contrary, a piece of column body tissue will not induce or form a second axis upon transplantation to a host body column. This indicated that the organizing capacity resides only in the hypostome.

The head organizer located in the hypostome transmits two signals to the body column. One is called the head activation gradient and the other is termed as head inhibition signal, the latter preventing the body column tissue from undergoing head formation.

MORPHOGENESIS DURING HEAD REGENERATION

By far, the morphogenesis as well as the molecular mechanisms underlying *Hydra* regeneration is well understood in head regeneration. After decapitation, the open wound is sealed by the stretching of the ectoderm and endoderm over the wound to form an apical cap. After about 30 to 36 hours, tentacle buds begin to emerge in the tentacle zone, which will subsequently elongate into tentacles as tissues from the body column is displaced apically through the tentacle zone onto the growing tentacles. Simultaneously, the flat to rounded apical cap changes its shape into a more dome-like or conical shape as it develops into the hypostome. The formation of the hypostomal dome will result in rings of cells with progressively smaller diameters and smaller numbers of cells within a ring as the dome or cone emerges from the cap. A similar phenomenon also occurs during bud formation. The epithelial cells of ectoderm and endoderm also change shape from a cylindrical shape to flattened shape during their transfer from tentacle zone onto the tentacle. During head regeneration, the head organizer expresses several molecules of *Wnt* signalling pathway. Just like the *HOX* genes of vertebrates, Wnt signalling may play vital role in the evolution of axial differentiation in early multicellular animals such as cnidarians (see chapter 28).

REGENERATION IN PLANARIANS

Freshwater planarians are another example of high regenerative power. They are triploblastic, acoelomate, unsegmented organisms that lack circulatory, respiratory and skeletal structures. Even a small fragment cut from the body can give rise to an intact animal. Transection of the planarian at various regions as well as removal of a fragment from the body would give rise to reconstitution of an entire planarian as in the case of *Hydra*. Furthermore, a head fragment of the planarian replaces the missing caudal region and the caudal fragment will replace the missing head (Fig.22.4).

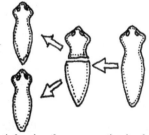

Fig. 22.4 showing the regeneration in planarians

The process of regeneration in planarians consists of two main events, namely, the onset of regeneration (blastema formation) and the pattern formation. When a planarian is cut, the epithelium around the wound rapidly closes up and a thin film of epidermal cells from the stretched old epidermis covers it. Below the wound epithelium, groups of undifferentiated cells aggregate to form a few layers of cells to give rise to an outgrowth which is termed as 'regeneration blastema'. The undifferentiated stem cells occupying the regeneration blastema are similar to the interstitial cells of *Hydra*, but are called as neoblast cells. In the adult planarians, the totipotent neoblast cells are the only cell types that can divide and proliferate. These cells in the intact planarian play a pivotal role as the stem cells to give rise to all specialized cells such as, muscle cells, nurse cells, gastrodermal cells and germ cells. The presence of neoblast cells in the body is inevitable for regeneration to occur as irradiation of these cells result in non-occurrence of regeneration in planarians. However, if a fragment from non-irradiated planarian is implanted into the mid region of the completely irradiated and beheaded planarian, a new head is formed with the neoblast cells derived form the grafted tissue. The neoblast cells, representing the multipotent stem cells are embedded in the parenchymatical tissues and constitute 20-30% of the total number of cells.

BLASTEMA FORMATION

In the normal planarian, the stem cells differentiate into all different kinds of cells. The neoblast cells after accumulating within the blastema finally differentiate into new structures following a distoproximal sequence, thereby restoring the lost structural pattern in proportion to their normal body size. The epithelium enclosing the blastemal mesodermal cells (neoblast cells) induces the expression of several homeobox-containing genes. These genes are important for maintaining the cell proliferation of the mesenchymal cells below the wound surface and also to certain early pattern formation. There are two theories put forward on the origin of blastema cells. They are the neoblast theory and dedifferention theory. However, the current view is that neoblast cells solely give rise to the specialized cells in planarian regeneration. There is some evidence from electron microscopic studies that during regeneration there appears to be dedifferentiation of intestinal and muscle cells. The resulting dedifferentiated cells look identical to neoblast cells. The gland cells lose their secretary granules and in the muscles breakdown of their fibre system occurs, thus transforming them into neoblast cells. Thus in planarians the regenerating cells include both the already existing neoblast cell and the transformed neoblast cells from other cell types. This condition is in contrast to the amphibian limb regeneration where only the dedifferentiated cells participate in the restitution of lost parts. Thus, regeneration in Planaria is considered to have both morphallactic and epimorphic properties.

URODELE LIMB REGENERATION

The formation of vertebrate limb or wing mainly involves interactions between the epidermal and mesodermal tissues, leading to tissue differentiation and pattern formation. Incidentally, this is a one time occurrence restricted to the embryonic phase. However, the events comparable to the embryonic limb development is also seen in the restitution or regeneration of the limb or other parts of the body in the later stages of the life in vertebrates such as the amphibians. These animals possess the unique potentiality of regenerating an amputated limb during either the larval or adult phase. Regeneration of limbs also involves tissue differentiation and pattern formation, mainly by epithelial mesodermal interactions. However, the major difference is that in regeneration, the mesodermal cells constitute the already differentiated cells whereas in limb bud formation, it is the undifferentiated embryonic mesenchyme cells. This necessitates the conversion of the differentiated cells to undifferentiated condition in the regenerating limbs. Yet another difference is that the ectoderm does not give rise to any internal tissues including those normally derived from the ectoderm (nerve cells) during regeneration.

After amputation of the urodele limb, the wound is quickly covered by a specialised epithelium, called wound epidermis which proliferates and forms into apical ectodermal cap (AEC). AEC is homologous to the apical ectodermal ridge of the embryogenic limb bud. The cells at the wound surface beneath the developing AEC include bone cell, fibroblasts, myocytes and neural cells. They all undergo a process of dedifferentiation by which their differentiated characteristics are lost.

The specialised cells at the wound site get loosened from the extracellular matrix by the hydrolytic action of the enzyme, matrix-metalloprotease. As a result of dedifferentiation all the cells assume the status of mesenchymal cells which are identical to one another regardless of their tissue of origin. Mesenchymal cells thus released during dedifferentiation accumulate below the AEC. These cells proliferate rapidly and cause the epidermis to bulge further. This mass of mesenchyme cells is called the regeneration blastema. Thus the blastema formation in regenerating limb produces an epithelial-mesenchymal reaction, much like the differentiating embryonic limb bud.

The wound epidermis provides the signals for the underlying tissues to dedifferentiate proliferate and form the blastema. The signals that are received from the wound epidermis to the underlying dedifferentiated cell include fibroblast growth factors (FGF1 and FGF2) as well as their receptors (FGFRs). In *Xenopus* limbs, FGFR1 and FGFR2 are expressed in the wound epithelium of the regenerating stumps (premetamorphic), whereas, the postmetamorphic limbs, which are able to regenerate, do not express them.

The presence of pedal nerve is important for the regeneration to occur. Limbs will not regenerate in the axolotl if the limb blastema is denervated. The severed nerve in the vertebrates releases a fibroblast growth factor (FGF2) that has a role in organising the apical wound epidermis in the axolotl regenerating limb. Fibroblast growth factors are growth promoting peptides that function as mitogens, motogens, and differentiation factors. FGFs are also the predominant outgrowth signal for normally developing vertebrate limb buds. FGF 2 is able to rescue denervated blastema in newts by replacing the function of wound epidermis and restoring the expression of an important developmental gene (*Dxl3*) in the epidermis. The apical epidermis of vertebrate regenerating limb buds releases other fibroblast growth factors (FGFs 4, 8, and 10) that serve to precipitate a cascade of organisational and positional signals, necessary for successful regeneration to occur.

MORPHOGENESIS AND REDIFFERENTIATION

As the blastema grows, cells showing the characteristics of muscles and cartilage reappear in the correct spatial relationship to each other and to the corresponding tissues in the stump. The first tissue to differentiate in the blastema is the cartilage as in the limb development. It first appears at the ends of the persisting bones which is completed by progressive addition to its distal end. When the cartilagenous reconstruction is completed, it is converted into the bone. Certain cell adhesion molecules such as N-CAM are known to enhance mesenchymal condensation and chondrogenesis. Muscle is formed around the cartilage as well as the terminal addition to persisting muscle. Restoration of vascularisation follows in the original pattern. The nerve cells, that are cut during amputation grow axons into the wound and reconstruct the original nerve pattern. The elbow bend as well as the digit formation follows the exact pattern formation characteristic of the forelimb development. In short, limb regeneration can be considered to be a reawakening of embryogenesis in an adult organism.

SOURCES OF BLASTEMAL CELLS

An important question concerning the blastema formation is the origin of mesenchymal cells. There is evidence from the preceding account that the mesenchyma cells originate from the dedifferentiation of already existing tissue cells in the stump. Alternative source of blastemal cells could be the embryonic reserve that are mobilised from elsewhere to the regenerating tip. If a portion of the limb is irradiated and amputated, that portion is incapable of regeneration indicating that all cells necessary for regeneration must be supplied by the region of the amputation. But then, the next question is with reference to the redifferentiative potentials of the dedifferentiated cells. Can a dedifferentiated muscle cell participate in the formation of alternate tissue types? In other words, are they pluripotent to differentiate into several cell types or totipotent if they can differentiate into all the cell types present in the organism. If we take the example of adult muscle cells, it is syncytial in the sense that a single muscle fibre is formed by fusion of many individual cells or myoblasts. The result of fusion is a single very long cell with many nuclei. Muscle cell nuclei do not divide in the adult organism. When the limb is amputated, the muscle cells lose their characteristic architecture (such as the actin-myosin arrangement) and then become mononucleated. These fragmented muscle cells with single nucleus is capable of replenishing their DNAs and dividing. The cytoplasm of these dividing cells still contain contractile fibres, but are broken down soon, thus making it impossible to distinguish themselves from mesenchymal cells. Interestingly, these cells in addition to redifferentiating into muscle cells also produce cartilage, thus providing evidence for the transdifferentiation in limb regeneration. A similar phenomenon also occurs during urodele lens regeneration, where the pigmented epithelial cells of the dorsal iris dedifferentiate and then transdifferentiate into a lens vesicle.

FACTORS STIMULATING REGENERATION OF LIMBS

When an adult newt is able to regenerate its lost limb, their close cousins frogs are incapable of doing it. Then, what could be the stimulating factor for regeneration that resides in the newt and not in the frogs (Fig.22.5). The ingenious experiments of nerve transplantation by Marcus Singer seem to provide a plausible answer for this. Regeneration of the limb blastema of vertebrates is dependent upon a critical supply of nerves at a very early stage. The nerve supplies trophic (nutritive like) materials to the cells, which are necessary for the growth and differentiation after amputation. Singer demonstrated that a minimum number of nerve fibres must be present for regeneration to take place. He believed that the neurones release a mitosis stimulating factor that increases the proliferation of the blastema cells. This neurotrophic substance is thought to be glial growth factor (GGF). This peptide, known to be produced by newt neural cells, is present in the blastema and is lost on enervation. When GGF is added to a enervated blastema, the mitotically arrested cells are able to divide again. Evidently, in addition to its well known function of integration, the nervous system also has a developmental role.

PATTERN FORMATION IN THE BLASTEMA

Regeneration always proceeds in a direction distal to the cut surface allowing the replacement of the lost part of the limb. If the hand is amputated at the wrist, only carpels are regenerated, whereas, if the amputation is through the middle of the humerus, everything distal to the cut is regenerated. Such an orderly formation of the missing structures is due to the occurrence of a morphogen, distributed as a gradient, in the blastema cells. Recent studies have indicated that retinoic acid acts as a morphogen, playing an important role in the respecification processes as the blastemal cells redifferentiate. Retinoic acid, synthesized in the regenerating limb wound epidermis, forms a gradient along the proximal to distal axis of the blastema. This gradient of retinoic acid can activate genes differentially in different regions of the blastema. One of the retinoic acid responsive genes is msx1 gene that is associated with the mesenchyme proliferation. Another set of genes is the *HOX* genes which are important in pattern formation. It is probable that during normal regeneration, wound epidermis produces retinoic acid, which activates those genes needed for cell proliferation, down regulating the genes that are specific for differentiated cells and, also activate a set of *HOX* genes that specifies the pattern formation. In short, regeneration of vertebrate limb is a well orchestrated developmental event governed by genes that control normal growth and differentiation.

AUTOTOMY AND LIMB REGENERATION IN CRABS

In arthropods, regeneration is limited to the renewal of lost appendages. But in most crustaceans, the limbs may regenerate at any stage of development. As a rule, the growth in all arthropods is discontinuous and is facilitated by a periodical shedding of the old cuticle followed by the formation of a pliable cuticle beneath the old exoskeleton. As regeneration is also a special type of growth, there is interdependence of regeneration and molting in arthropods.

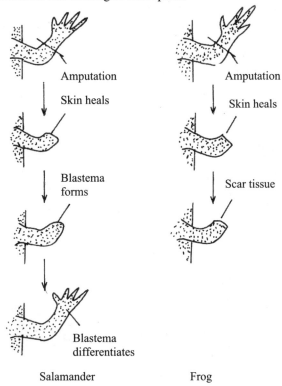

Fig. 22.5 Comparison of regenerative capacity between salamander and frog (Redrawn from Wolprt 1998).

Regeneration occurs in adult arthropods that continue to molt and grow as adults (primitive insects and crustaceans). In general, regeneration of limbs in crabs occurs after the process of autotomy. The legs of crabs are readily shed if seized by an enemy. The leg breaks off at a preformed breaking point, across the second leg joint. The leg is broken off by the violent contraction of the extensor muscle of the leg. This self-mutilation is known as autotomy. Regardless of where along the length of the limb an injury occurs, the autotomy response insures that the limb is cast off at the predetermined point between the coax and the basiischium. Different stages in the regeneration of crab limb after autotomy is illustrated in the fig.22.6. In the fiddler crab *Uca pugilator*, regeneration of the limb occurs in two phases following autotomy. The first phase follows soon after the loss of the limb and is called basal growth. Basal growth can take place at any time during the molt cycle. The second phase is called proecdysial growth which occurs only during the premolt stage.

In the normal limb, the preformed breaking point for autotomy is covered by a double membranous autotomy membrane, which leaves a central gap, through which traverses the large pedal nerve and a few blood sinuses. During autotomy, the pedal nerve is severed and the autotomy membrane seals off the opening left after the distal limb are cast off. However, the proximal portion of the autotomy membrane remains attached to the sheath of the severed nerve, thereby preventing the retraction of the nerve into the coxal stump and keeps it near the centre of the developing blastema. As in vertebrate limb regeneration, the presence of pedal nerve is essential for the regeneration of the crab leg. In the axolotl, the severed nerve releases a fibroblast growth factor (FGF2), which has a role in organizing the apical wound epidermis for its future inductive role. In the newt, FGF2 is also able to rescue denervated regenerates by restoring the expression of an important developmental gene (*Dxl3*) in the wound epidermis. Interestingly, two fibroblast growth factors namely FGF2 and FGF4 are expressed I the pedal nerve and in the epidermal cells in the crab *Uca limb* buds suggesting a regulatory role in the early blastema development in crustacean too.

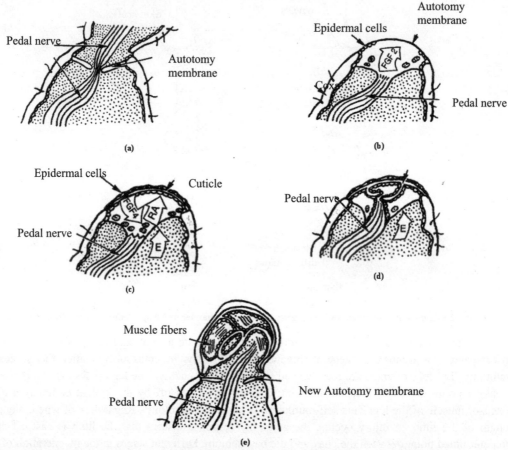

Fig 22.6 FGF 2 and FGF 4 are fibroblast growth factors; RA = Retinoic Acid; E = Ecdysteroid; (a) to (e) showing the sequential stages in the regeneration of the crab limb after autotomy. (a): is the longitudinal section of the limb before autotomy showing the preformed plane of autotomy (with double autotomy membrane). (b): two days after autotomy showing the extension of epidermal cells into the limb bud. (c): showing the completion of epidermal cell below the cuticle at the apex of the limb bud. (d): Eight days after autotomy infoldings from the cuticle are seen. This indicates the differentiation of leg into two segments. (e): After twelve days, developing muscle fibres are seen in the individual segments of the regenerating limb. Note the formation of new autotomy membrane. The regenerating pedal nerve is also seen. (Redrawn from Hopkins 2001)

During the basal growth, only differentiation occurs. Further growth takes place only during the proecdycial growth period. However, all of the leg segments are evident during basal growth.

Small segments of the limb bud are folded upon one another and encased in a flexible cuticular sac. Further proecdysial growth of these basal limb buds initiates hypertrophic growth of the myotubules formed during basal growth. Such growth is primarily due to muscle protein synthesis and water uptake. The cuticular sac surrounding the regenerating limb bud is flexible and can expand with rapidly growing bud during proecdysial growth.

HORMONAL CONTROL OF REGENERATION IN CRAB LEG

Since regeneration and growth of limb bud is intimately connected to molting process, the molting hormone, ecdysteroids also control the regenerative processes in the crab, Uca. Since the Y organ in crustaceans and the prothoracic gland of primitive apterygote insects are active even in adults, ecdysteroids are available to control regeneration of limbs after autotomy in these arthropods. Ecdysteroids are steroid hormones and hence affect their target tissues by interacting with nuclear receptors, namely, ecdysteroid receptors (EcR). These receptors have been found to be expressed in the regenerating limbs of the crab, Uca. The expression of EcR mRNA in the *Uca limb* bud increases slowly during the first 4 days of basal growth when the limb bud epidermis becomes organized and begins to secrete a thin cuticle. The basal growth occurs any time during the intermolt stage in crustaceans and the small cyclical ecdysteoid peaks that occur in the anecdysial intermolt period stimulate the basal growth of the regenerates. Further growth of the limb bud has to wait until the onset of premolt. Thus, proecdysial growth of the regenerate starts at the D zero stage of the premolt, but will be completed before D1 stage. The onset of premolt is signaled by two ecdysteroid peaks, one minor peak at the D zero stage and the other major peak appearing at D1 stage. Interestingly, the basal growth of the limb bud coincides with the minor peak at D zero stage and the apolysis as well as the new body cuticle synthesis corresponds to the major peak spanning D1 to D4 in the crabs.

In addition to ecdysteroids, retinoic acids are also involved in the control of limb regeneration in the crabs. In *Uca pugilator*, mRNA of the retinoic acid receptor is also expressed in the wound epidermis of the blastema of the regenerating limb at one and four days after autotomy. Endogenous retinoids have also been known to supply constitutive signals that play important roles in the normal development of the vertebrate limbs as well as the regenerating tissues from axolotls, frogs and chicks. Evidently, retinoic acid may have a synergistic effect on the basal growth of the crab limb along with ecdysteroids. It has also been suggested that the release of growth factors from the severed nerve and the production of FGF-like compounds by the wound epidermis are responsible for the basal growth in the crustacean limb regeneration. The effect of these growth factors may be modulated by the endogenous retinoids.

23

EXPERIMENTAL EMBRYOLOGY

It is well known that the growth of any scientific discipline is correlated with the development of techniques. In olden times, science was practised and nurtured by scientific philosophers, who propounded theories not just based on scientific discoveries, but by their conceptual ability. However, their understanding of the world phenomena was not always correct and hence, their theoretical advancements were challenged from time to time.

Any person cannot resist his curiosity to know how a chicken emerges from an egg, which when laid, had little resemblance to the young hatchling. The interpretations of the developmental process have varied throughout the history of embryology and the study of development has passed through different phases characterised by the methods of study. For centuries, observations and descriptions were the primary bases for the interpretation of the development. With the advent of microscope in the 17th century, there was a transformation in the descriptive embryology, to follow the embryological changes at the tissue and cellular levels. When the theory of evolution was proposed by Charles Darwin, comparative methods became the dominant approach to the study of development. Thus, similarities in the developmental patterns of the embryos of different organisms were used for establishing the phylogenetic relationships between them. These approaches in the study of organismal development, especially at the embryological level, did not focus on the mechanisms controlling developmental changes. Even later studies on the biochemistry and physiology of the embryo did not reveal the secret of embryonic regulation.

Experimental embryology thus became the necessary approach to analyse the factors controlling development by direct manipulation of the embryo. The operational skills of many embryologists in the early twentieth century, facilitated experimental studies involving nuclear transplantation, tissue interaction by way of induction, to mention a few. Two German embryologists who pioneered such embryological studies were Speeman and Mangold. Their studies explained the role of cellular interaction as the primary factor in influencing the cellular differentiation, which is so crucial in the developmental biology. More studies also emphasised the fact that the cell differentiation during embryogenesis should be accompanied by the integrated and co-ordinated development of the differentiated tissues leading to organ formation.

The chemical nature of influencing factors in cellular interactions remained elusive for a long time. The modern embryology however strives to find out answers in terms of molecular biology. It is well known presently that the cells undergoing differentiation produce proteins which characterise the type. Synthesis of distinctive protein molecules is based on selective expression of genes in certain cells at certain times. For example, 95% of the proteins that the red blood cells make are haemoglobin, which is not being synthesised by any other cell type during development.

Recent developments in biological techniques have also contributed much to our understanding of developmental processes. For example, the culture techniques applied to the organs, tissues and cells have greatly improved experimental embryology, by facilitating cell isolation and their subsequent culture under *in vitro* conditions so as to understand the factors regulating cell differentiation.

The understanding of developmental processes, especially during embryogenesis has led to the development of applied embryology by which we can control and correct developmental processes, by way of manipulative techniques. Our understanding of mammalian, including human, embryology has paved way for many technological developments, broadly categorised in the term, Artificial Reproductive Techniques (ART).

MAMMALIAN REPRODUCTIVE CYCLE

The reproductive cycle in mammals includes an ovarian cycle and a correlated uterine cycle. The ovarian cycle comprises of two important phases: growth of the ovarian follicle and following the ovulation, growth of the corpus luteum. These changes in the ovary are paralleled by the changes in the lining of the uterus. The uterus is composed of three layers: an outer peritoneal layer, subjacent to it a layer of smooth muscles, the myometrium and an inner layer the endometrium, which is composed of columnar epithelium composed of glands and an underlying stroma. It is the endometrium which undergoes changes correlated to the ovarian cycle.

Each maturing egg is contained in a special chamber called follicle. A ripening follicle grows in size and the follicle cells around the egg secretes a fluid. Soon there is so much fluid in the follicle that a mature follicle balloons outward from the surface of the ovary. At ovulation, the follicle ruptures and the egg escapes from the ovary. It is then swept by the ciliary action into the oviduct and reaches the uterus. Following the egg's expulsion, the ruptured follicle develops into an endocrine gland known as the corpus luteum. This gland becomes a major hormonal control agent during subsequent reproductive events.

The cyclical changes that occur in the uterus can be divided into two major phases. Preceding ovulation the endometrium thickens and some epithelial cells are shed from the lining of the vagina. This is the preovulatory or proliferative phase of the ovarian cycle. In lower mammals, the females become receptive to the males towards the end of this phase. This condition of heightened receptivity before ovulation is called estrus. Following ovulation is the second period, during which the endometrium becomes more thickened due to increased vascularization and secretions from the hypertropied glands. This is known as the post ovulatory, progestational phase of the uterine cycle. In lower mammals, the thickened endometrium regresses without breaking down, and another cycle begins if the ovum is not fertilized. In man, apes and old world monkeys, if the fertilization fails to occur, the endometrium breaks down, and is sloughed. The disintegration of the highly vascular progestational endometrium produces a bleeding phase, or menstruation. The above phases in the uterine changes are called as menstrual cycle. In other mammals, where the uterine changes are not accompanied by bleeding, it is called as estrus cycle.

HORMONAL REGULATION OF REPRODUCTIVE CYCLE

The changes occurring during ovarian and uterine cycles are controlled by subtle feedback relationships among an array of hormones (Fig.23.1). At the start of the ovarian cycle, hypothalamus stimulates the pituitary to secrete the gonadotropic hormone called the Follicle Stimulating Hormone (FSH), which in turn stimulates the growth of follicles. Now the cells of the follicles secrete estrogen, which triggers the release of the second hormone, Leutinising Hormone (LH) from the pituitary. It is the mid cycle peak of the FSH and LH levels that triggers ovulation. After ovulation, the high levels of

LH in blood stimulate cells in the ruptured follicles to differentiate into the corpus luteum and increase their rate of progesterone production. For the next 10 to 15 days, the corpus luteum secretes estrogen and progesterone.

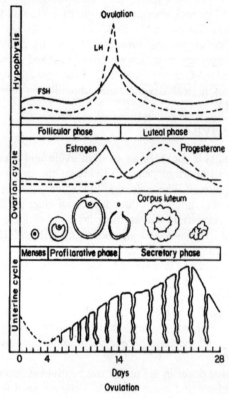

Fig. 23.1. Schematic representation of the estradiol, progesterone. LH and FSH profiles during human ovulatory cycle. (Redrawn from Ramon Pinon 2002)

The changing estrogen and progesterone levels throughout the cycle cause profound changes in the uterus. Estrogen causes the endometrium to thicken, whereas the progesterone stimulated development of glands that secrete various substances.

Progesterone accounts for the culminating events that prepare the estrogen influenced endometrium for pregnancy. In the event of the failure of fertilization, the progesterone level continues to rise and inhibits the pituitary's secretion of FSH and LH. This completes the feedback loop; without the stimulation of the FSH and LH, the corpus luteum stops secreting estrogen and progesterone. Without estrogen the blood vessels in the endometrium becomes constricted cutting off the oxygen and nutrition. Hence, the endometrium begins to disintegrate and is gradually cast off. Now the blood vessels open wide and cause the weakened endometrium in the uterus to haemorrhage. During the menstrual period the level of all sex hormones is low. Now the pituitary begins to secrete the FSH once more, because it is no longer inhibited by progesterone and the new follicle begins to develop and secrete estrogen. The estrogen shuts off the menstrual flow and stimulates the uterus to begin rebuilding for a new cycle.

ENDOCRINE CHANGES ASSOCIATED WITH NORMAL PREGNANCY

Pregnancy occurs by the implantation of the embryo at its blastocyst stage onto the endometrial wall of the uterus. After the formation of the placenta, the trophoblast cells (the outer syncytioblast) produce a hormone called chorionic gonadotropin (=human chorionic gonadotropin, hCG). Chorionic gonadotropin has no influence upon maturation of the ovarian follicles, but in the presence of the pituitary it causes ovulation and corpora luteum formation in immature mice. A high titre of the hormone in the maternal blood is hence important to prevent aborted pregnancy. It thus seems that in mammals and monkeys, the corpus luteum is only essential for implantation and early development of the embryo and not for the later stages of pregnancy.

In humans, the placenta becomes independent of the maternal hormones and forms a self differentiating endocrine unit. The human placenta synthesizes chorionic gonadotropin. Thus, as the pregnancy advances, the major production of these hormones is gradually transferred from the ovary to the placenta. The main function of the placental gonadotropin is the transformation of the corpus luteum of the menstrual cycle into the corpus luteum of the pregnancy, thereby maintaining gestation until the placental steroid production is sufficient to replace the ovarian secretions. In addition, the placenta is also known to produce another hormone, relaxin, which causes relaxation of the pubic symphysis and the pelvic ligaments, thus allowing the greater expansion of the birth canal at the time of the delivery of the foetus.

A basic understanding of the endocrine regulation concerning ovulation, implantation and embryo growth in mammals including humans has paved the way for correcting any infertility related problems with hormonal therapy. In addition, hormonal treatment can also be employed in increasing the reproductive potential of the commercially important species such as the cattle, horses etc. as well as controlling the fertility in the case of human population. The following is an account on the commonly employed methods in achieving the above goals.

INDUCED OVULATION IN HUMANS

The hormonal therapy in human induced ovulation consists of two steps, (i) induced maturation of oocytes and (ii) induced ovulation. In order to induce oocyte maturation, PMSG (Pregnant Mare Serum Gonadotropin) is given. This is followed by treatment with human chorionic gonadotropic hormone which acts in the place of LH in inducing ovulation. HCG is a placental hormone which can be extracted from placenta or from maternal blood in large quantities.

MULTIPLE OVULATION AND EMBRYO TRANSFER IN CATTLE

In the normal course of events, only one oocyte undergoes maturation. However, by increasing the hormonal dosage at appropriate time, a number of follicles can be induced to ripen and ovulate. In practice, well managed domestic cattle can yield an average of about 8 eggs when superovulated.

The donor females are frequently injected with prostaglandin F2α (PGF2α) to induce a synchronised estrus before treatment is started. Ten days after estrus the donor females are injected with FSH over four days followed by PGF2α to induce estrus. Estrus should occur within 2 days and the superovulated females are mated with quality breed males or artificially inseminated with superior quality semen within 24 hours. The fertilized embryos are recovered 6-8 days after insemination for cattle by inserting a Foley catheter into the oviduct. By microscopic examination the embryos are identified in saline solutions and recovered. They are immediately transferred into a recipient called "surrogate mother" for implantation in the uterus. This process is called embryo transfer. By these methods superior quality embryos can be developed in a cheaper and sturdy surrogate mother cow.

EMBRYO SPLITTING

Embryonic cells at a very early stage of cleavage are totipotent and hence each cell has the potentiality to give rise to the whole organism. Utilising these principles, embryo splitting has been routinely done to produce identical twins. The methodology consists of culturing the embryos recovered from inseminated cattle in a medium containing hypertonic sucrose and bovine serum albumin in a plastic Petri dish. The embryo sinks to the bottom of the culture dish which is placed on the stage of an inverted microscope. With a fine surgical blade controlled by a micromanipulator the embryo is oriented and bisected. The splitting of the embryo should be done only at the blastocyst stage. The blastocyst consists of inner cell mass and surrounding trophectoderm. The bisection should pass through the inner cell mass by splitting it into roughly equal halves. Then they are simply transferred into the oviduct of synchronised recipients as per the normal embryo transfer.

Using the principle of induced ovulation in mammals, aquaculturally important fish species can be induced to spawn under controlled conditions. In fish, the egg maturation and ovulation are controlled by gonadotropic hormone which is secreted from the pituitary. The secretory activity of the gonadotrophs, in turn is controlled by hypothalamic factors. In practice, an extract of pituitary as well induces ovulation in fish. A commercially available fish gonadotropin analogue called ovaprim is widely used in induced ovulation of fishes.

IN VITRO FERTILIZATION

In artificial insemination, the chances of fertilization are still limited in view of the problems related to sperm concentration and motility as well as intrauterine conditions favouring internal fertilization. In embryo transfer procedures, although the number of superovulated eggs are many, only a few of them get fertilized by artificial insemination. As regards humans, there are couples with a long duration of unexplained infertility due to a variety of disorders such as the tubal disease endometriosis, immune and cervical infertility and male infertility. For all these patients, IVF is the only answer. IVF is quite successful, resulting in about 70 to 80 % of fertilized eggs.

IVF IN CATTLE

Superovulation is induced in the second follicular phase of the estrus cycle by injection of the gonadotropins. Preovulatory follicles lie against the surface of the ovary and are quite large. Laparoscopic surgery is used to recover oocytes from these follicles. The selected ripe eggs with intact polar body are cultured in a suitable medium, before it is challenged with the sperm for in situ fertilization. Once fertilization occurs, the zygote is placed in the culture medium and the embryo is allowed to grow upto the blastocyst stage. This blastocyst is then introduced into the uterus for implantation.

IVF IN HUMANS

Human IVF clinics have mushroomed in the recent past, thanks to the perfection of the technique as well as due to the increased number of infertility cases. Follicular growth stimulation is induced by 100 mg of clominphene citrate on days 2-6 and follicle stimulating hormone (FSH) is given at 150 IU daily from day 1 to 5. The egg retrieval is timed by an endogenous LH surge, detected in the urine using radio immuno assay (RIA). The oocyte collection is performed laparoscopically or via transvaginal ultrasound using a suction pump. Follicles are flushed repeatedly with warm culture medium and the insemination is performed 3 -12 hours after egg collection.

In another recently developed IVF method, instead of challenging the egg with a mass of sperm, a single sperm is injected into a ripe egg to effect *in vitro* fertilization. After evacuation, the eggs are placed in the culture medium and treated with an enzyme hyaluronidase to disperse and remove the outer cumulus cells. The denuded oocytes are then injected with a single immobilised sperm, using a micromanipulator injection microscope. The injected sperm transforms into the male pronucleus and fuses with the female pronucleus, formed after the extrusion of the second polar body. The experimental procedure is depicted in Figs.23.2 to 23.4.

Embryo culture technique also offers opportunities for the production of transgenic animals by injection of the foreign genes into the zygote.

CRYOPRESERVATION

Cryopreservation is a biotechnique that deals with the freezing of live biological materials at subzero temperatures using liquefied gases such as nitrogen (-196°C). In recent years the technique has found enormous use in biological research and more specifically in the long term storage of the live cells, organs and even the whole organisms. Cryopreservation of mammalian embryos has found wide applications in IVF techniques in humans, embryo transfer in cattle and the other commercially important mammals as well as in transgenic animal production.

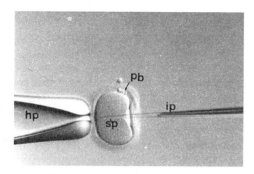

Fig. 23.2 Mature egg ready for sperm injection (op – ooplasm, pb – polar body, hp – holding pipette)

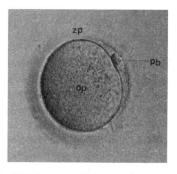

Fig. 23.3 Intracytoplasmic sperm injection (ICSI) (pb – polar body, sp – sperm, ip – injection pipette, zp – zona pellucida)

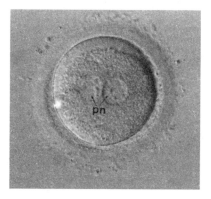

Fig. 23.4 Pronuclear fusion in the sperm injected egg (pn – pronuclei) (Courtesy, Dr. Balaji Prasath, K.K. Womens and Childrens Hospital, Singapore)

In living cells, water is the major constituent accounting for nearly 60-90% of the total volume. Liquid water within the cells is necessary for all the biological reactions to occur. When a cell or a tissue is cryopreserved at subzero temperatures (i.e. below its freezing point), water gets separated and begins to form ice crystals within the cells. If the cell is frozen in a condition which is not injurious to its life, then it can be kept under suspended animation. However, the intracellular ice crystal formation is injurious to the cell as the sharp ice crystals pierce through the internal membrane systems and kills the cells instantaneously. This mechanical damage is called cryoinjury. Another important consequence of subzero freezing is the increase in the solute concentration as pure water within the cytoplasm is separated in the form of ice crystals. This condition is called "solution effect". This condition brings about changes in the internal pH and causes destabilization of the cell membrane.

METHODS ADOPTED IN CRYOPRESERVATION

Cryopreservation is done in a liquid medium called extender medium. The cryoinjuries as well as the membrane damage due to increased solute or electrolyte concentration can be alleviated by using a cryoprotectant. A cryoprotectant is an organic or inorganic additive, which will protect the cell from freezing injuries during cryopreservation. A few examples are glycerol, dimethyl sulp*HOX*ide (DMSO) and ethylene glycol. A few characteristic features of the cryoprotectant are, i) it easily penetrates into the cells, ii) it should be a non-electrolyte, iii) freezing point depressant, and iv) it should be easily miscible with water.

The mechanism by which cryoprotectants afford protection to the freezing cells is not properly understood. However, a recent discovery of an 'antifreeze' protein in a Antartic fish by DeVries and associates in 1970 raised hope to understand the mechanism by which these proteins protect polar fishes inhabiting icy sea waters from freezing injury. In the polar regions, the seawater temperature remains at -1 to -1.9°C for all or most of the year. The freezing resistance found in the polar fishes is attributed to the antifreeze proteins. These proteins have the unusual property of lowering the freezing temperature of blood serum without affecting its melting temperature. The difference between the freezing and melting temperature is termed thermal hysteresis, which is a measure of macromolecular antifreeze activity. The antifreeze proteins lower the freezing temperature by binding to ice nuclei, thereby inhibiting further crystal growth. The binding occurs to the prism faces of ice crystals. The antifreeze protein of winter flounder is a small peptide consisting of 37 amino acids, with an amidated C-terminal arginine residue. This polypeptide is an alpha helix. It has amphipathic characteristics; most of the hydrophilic amino acids are aligned on one side of the helix, which presumably represents the ice-binding motifs. This hydrophilic side chains hydrogen-bond with the ice, then retarding its normal growth.

Cryopreservation can be done by two methods: i) slow cooling and (ii) fast cooling or vitrification. Slow cooling is done in a rate controlled programmable freezer (Kryo 10, Planer Biomed, UK). With this freezer a programmed slow cooling can be achieved, using vaporised liquid nitrogen. The principle involved in slow cooling is gradual removal of cellular water into the external medium, thereby precluding the intracellular ice crystal formation. The cryoprotectant added in the extender medium protects the embryo from other effects due to increased solute concentration etc. Invariably, in the slow cooling, the embryo is cooled to a temperature of -30 to -40°C, when all the cellular dehydration is supposed to be over and then the samples are plunged into liquid nitrogen for long term storage.

On the other hand, during vitrification, the sample is frozen in liquid nitrogen at ultrarapid rate. Vitrification occurs when a liquid solidifies without forming a crystalline phase, avoiding the damage associated with ice formation. In the above two methods, thawing of the frozen embryo is an important step to flush out all the cryoprotectants which are sometimes toxic to the embryos.

CRYOPRESERVATION OF HUMAN SPERMATOZOA

Among the earliest human cell types to have been successfully cryopreserved are the spermatozoa. Human semen is also today maintained in major commercial repositories around the world for various purposes. Most commonly, it is used for *in vitro* fertilization and also for intrauterine insemination. Human as well as many of the mammalian embryos are easier to cryopreserve compared to oocytes themselves. Most of the human embryo repositories from the donors are *in vitro* fertilized with cryopreserved semen from preferred male partners and stored.

A standard method of cryopreserving human semen outlined in many manuals consists of suspending the semen in 10-20% glycerol in egg yolk buffer, which normally contains egg yolk, glycin, citrate and antibiotics and freezing the semen in either straws or ampoules. Freeze rates of 1°C/min to 10°C/min have been found to be satisfactory. Prior to use, the semen has to be ridden of the cryoprotectant content as it may damage the embryonic metabolic status. The cryopreserved semen is centrifuged and pelletised. Then this pellet is overlayed with a culture medium and the spermatozoa are let to swim up and its concentration is adjusted to one million cells per ml.

The spermatozoan volume and water content are so less that they may not be often frozen at controlled rates and still permissible viability could be obtained. They may simply be frozen for 1 hour at 4°C and exposed to liquid nitrogen vapour phase for 5 min and plunged into liquid nitrogen. Cryopreservation of the sperm cells is also routinely done in animal husbandry and human IVF clinics. In animals, the cryopreservation of sperm is used for artificial insemination whereas human sperm banks are being established to ameliorate the infertility problems.

HUMAN EMBRYO CRYOPRESERVATION

Successful cryopreservation of the human embryos has been achieved for oocytes, zygote, 4- and 8-cell embryos as well as blastocyst stages. As per the procedure of the Australian scientist, Alan Trounson, 4 or 8 cell stage mammalian embryos were frozen in either 1.0 M glycerol or 1.5 M DMSO as cryoprotectants. The ampoules containing the embryos were cooled from room temperature to -6°C at 2°C/min and then seeded to initiate uniform ice nucleation in the medium. After holding for 30 min at -6°C, the cooling rate was changed to 0.3°C/min to reach -39°C. The embryos were then transferred to liquid nitrogen for storage. After storage the embryos were thawed rapidly by immediate transfer from liquid nitrogen to a water bath at 30-35°C. The thawed embryos were placed in culture for 1-12 hrs, prior to replacement in the uterus of the donor patient. An interesting aspect of embryo cryopreservation is that freeze/thawed embryos have the same chance of implantation as fresh embryos.

Cryopreservation of the zygote has also been found to be successful using 1,2 propane diol and sucrose as the cryoprotectants. These two cryoprotectants have not been shown to cause any chromosomal damage in mammalian studies. In other studies blastocystes were cryopreserved using the glycerol as the cryoprotectant. It has been shown that implantation rate was higher with blastocysts than the earlier stage embryos.

A few studies have also been made on the cryopreservation of oocytes of a variety of mammals such as mouse, rat, rabbit and the monkey and recently the humans. The oocytes were collected by laparascopy after ovarian stimulation with human chorionic gonodotrophin (hCG). The oocytes were normally frozen rapidly to -196°C and also thawed rapidly. After this, the oocytes were cultured in the medium and the viability tested. Then they are subjected to *in vitro* fertilization.

ETHICAL ISSUES IN CRYOPRESERVATION

Cryopreservation of human embryos raises ethical and legal problems in regard to preservation of human life. However, in some groups of society, freezing of unfertilized oocytes is ethically accepted, inasmuch as the oocyte does not represent a human life. However, freezing of embryos is equivalent to the manipulation of potentially new individuals. In this respect, it could be argued that the zygote is ethically different from the cleaving embryo, since the chromosome of the male and the female gametes are still grouped in separate nuclei in the zygote. A potential new individual has therefore not been genetically established. This consideration that the oocytes and zygotes are not life forms is like considering unfertilized hen's egg as a vegetarian item by some sections of the community.

24

STEM CELL BIOLOGY

Stem cells are defined functionally as cells that have the capacity to self-renew as well as the ability to generate more differentiated progeny. They can generate daughter cells identical to their mother (self-renewal) and produce progeny with more restricted potential (differentiated cells). The decisive, instructive, and permissive signals that decide the fate of self-renewal or differentiation of these cells are provided by growth factors in the microenvironment or "stem cell niche". The distinguishing feature of these pluripotent stem cells is that they exist in a transcriptionally permissive state. Potentially, this allows indefinite self-renewal until they are permitted to differentiate into diverse cell types.

MOLECULAR BASIS OF SELF-RENEWAL AND PLURIPOTENCY

Self- renewal is a cell division in which one or more of the resulting daughter cells remain undifferentiated, but retain the ability to give rise to another stem cell with the same capacity to proliferate as the parental cell. This kind of cell division, while maintaining pluripotency, is modulated by extrinsic factors such as a cytokine leukemia inhibitory factor (LIF) for mouse embryonic stem cells. Such external signals control stem cell gene expression by regulating transcription factors such as Oct-3/4, which acts as a pivotal player in determining self- renewal or differentiation. For example, differentiation of pluripotent stem cells is associated with down regulation of Oct4 levels and its upregulation or high level expression is related to their involvement in the self-maintenance process of stem cells. In the absence of Oct4, pluripotent cells in vivo (epiblast) and in vitro (ES cells) both revert to the trophoblast lineage. This implicates Oct4 once again as an important regulatory molecule in the initial cell fate decisions during mammalian development. In addition to its role in ES cells, Oct4 is also required for the maintenance of the germ cell lineage and thus appears to have an in vivo role in maintaining multipotency. The inner cell mass and then epiblast express Oct4 until gastrulation, when Oct4 remains expressed only in the posterior epiblast and the primitive streak, but not in cells that have undergone mesoderm induction. This expression pattern is consistent with Oct4 having a role in maintaining cells as multipotent until it is time for them to become committed to mesoderm, endoderm or ectoderm. Oct4 and the nuclear programmes it regulates might represent an older innovation designed to maintain a non-committed cell population in the early embryo until the appropriate time for induction of the germ layers. Maintenance of cells in a non-committed state prior to gastrulation is a fundamental aspect of trophoblastic organisms, and interestingly, the same genetic factors control this both in early embryo and in stem cells.

In addition, Nanog, a homeodomain containing protein, was identified as a factor that can sustain pluripotency in ES cells even in the absence of LIF. However, during development, Nanog function is required at a later point than the initial requirement for *Oct4*, but both are required for the maintenance of pluripotency. Yet another transcription factor that maintains pluripotency of the stem cells is SOX2.

The idea of stem cell existence in the adult organism came from the cell transplantation, which showed that mice could be protected against lethal irradiation by injection of spleen or bone marrow. A kind of seed cell or cells in the spleen or bone marrow preparations was thought to be responsible for regeneration of the repopulated marrow. These cells came to be called stem cells.

DISCOVERY OF STEM CELLS

The discovery of stem cells could be traced to 1953, with the discovery of teratoma in mice by Leroy Stevens. At that time, human teratoma has already been reported as a medical freak. Human teratoma contains a diversity of tissues and organs including synaptically connected neurons, muscles, beating heart tissue, bone, teeth and eye-like entities or whole limbs. The progenitor cell responsible for teratoma formation in mice has been traced to embryonic stage by Stevens in 1955. He noticed in the mice fetus that the genital ridge progenitor cells occasionally flowing off- course and develop like embryos, before forming into teratomas. Thus the concept of multipotent embryonic stem cells originated from the discovery that progenitor germ cells could give rise to tissues and organs of ectoderm, mesoderm and endoderm derivatives. Furthermore, ovarian cells of mice were shown to develop like embryos, but became disorganized into a tumour instead of a baby. This finally led to the extraction of embryonic stem cells from the mouse by Evans and Kaufmann in 1981. This technique was then perfected to extract the human embryonic stem cells in human beings. Human embryonic stem cells were first isolated and successfully propagated in 1998.

CULTURE OF STEM CELLS

Typically, pluripotent stem cell lines are isolated and maintained on mitotically inactive feeder layers of fibroblasts. The embryonal carcinoma cells, isolated from testicular tumours, maintained on feeder layers seem to retain developmental potency more readily than cells grown without feeders. Similar feeder- dependent culture conditions were used for the isolation of mouse and human ES and EG cells to maintain them in an undifferentiated state. The feeder cells provide a factor that suppresses the differentiation or promotes the self–renewal of pluripotent stem cells. Such property is termed as differentiation-inhibiting activity, which is found to be the same as leukaemia inhibitory factor (LIF), belonging to the family of cytokines. A molecular characteristic of stem cells is the capacity to sense a broad range of growth factors and signaling molecules and to express many of the downstream signaling components involved in the transduction of these signals. Thus, secreted growth factors play pivotal role in choosing differentiation pathways for stem cells.

STEM CELL NICHES

Stem cells are controlled by particular microenvironments known as 'niches'. A niche is defined as a subset of tissue cells and extracellular substrates that can indefinitely house one or more stem cells and control their self-renewal and progeny production *in vivo*. The stem cell niches are characterized by their expression patterns of signalling molecules and local environmental factors such as extracellular matrices. Heterologous cell types, matrix glycoproteins and the three dimensional spaces they form, provide ultrastructure for a stem- cell niche. The contact between these elements allows molecular interactions that are critical for regulating stem cell function. The concept of extracelular matrix regulating adult stem cells in mammalian stem cell systems is well known. For example, in the skin, β-1 integrins are known to be differentially expressed on primitive stem cells to interact with matrix glycoprotein ligands for establishing localization of a stem cell population. In the nervous system, absence of tenascin C alters neural stem cell number and function in the subventricular zone. Tenscin C modulates stem cell sensitivity to fibroblast growth factor 2 and bone morphogenetic protein 4, resulting in increased stem-cell propensity to generate glial cell offspring. In addition, tenascin C deletion in the hematopoitic system also affects the primitive cell populations. Another matrix protein belonging to sialoprotein family, termed as osteopontin (OPN), is known to contribute to the regulation of haematopoitic stem cells. OPN is chiefly produced by osteoblasts in the stem cell niche. It interacts with receptors on HS cells such as CD44, and α4 and α5β 1 integrins. OPN seems to serve as a constraint on HS-cell number, limiting the number of stem cells under homeostatic conditions or under stimulation. Taken together, matrix components contribute to the stimulatory or inhibitory influences on the stem cell pool within the niche.

Adult stem cells generally have limited function without the niche. For example, the haematopoitic stem cells, which regenerate the entire blood and immune system, have little function outside their niche. It is the specific cues from specific sites in the niche that allow stem cells to persist, and to change in number and fate. It is also the niche that provides the modulation in stem cell function needed under conditions of physiological challenge.

Various types of stem cells found in adult mammalian tissues interact with their respective microenvironments. They include stem cells found in skin, intestine, bone marrow and a few specialized regions of the brain where new neurons form. In each case, specific support cells seem to hold stem cells in place, and help to direct the production of daughter cells that differentiate to form specialized cells and tissues. For example, osteoblasts lining the inner bone surface are closely associated with haematopoitic stem cells and regulate their division and subsequent differentiation into the precursors of the blood cells. Similarly, in the brain region called hippocampus, neural stem cells take their cues both from astrocytes and from the endothelial cells that line the blood vessels. The endothelial cells help in the neural stem cells renewal, whereas astrocytes stimulate them to differentiate into neurons. The biochemical signalling between adult stem cells and their supporting cells involve the same molecular pathways that direct tissue generation during embryonic development. To name a few of the signal pathways are: Wnt, Notch and TGF- β / BMP.

TYPES OF STEM CELLS

Stem cells are of two different types, based on their origin. They are called adult stem cells or normal tissue stem cells and embryonic stem cells. Whereas embryonic stem (ES) cells can readily differentiate into essentially all cell phenotypes, adult stem cells have limited potential for differentiation. Again, most adult stem cells have a more limited life span in culture than ES cells have. There is also difference in their time of origin during development. The ES cells originate in the embryo before germ layer commitment. The adult stem cells are derived after somatic lineage specification, whereupon multipotent stem cells, progenitors, arise and colonize their respective cellular niches. The cellular niches provide a unique environment for the specialized functions of adult stem cells.

Figures (24.1) illustrate the putative developmental ontogeny of adult stem cells. The first view is that the development of stem cells occurs after the formation of germ layers. According to the second view, stem cells might develop similarly to primordial germ cells, in that they avoid the lineage commitments during gastrulation and subsequently migrate to specific tissue and organ nitches. In other words, the stem cells develop near the epiblast and subsequently migrate to positions inside the

192 Molecular Developmental Biology

embryo proper. Yet another alternative view is that adult stem cells are derived from primordial germ cells, although experimental evidence is lacking at present on this contention.

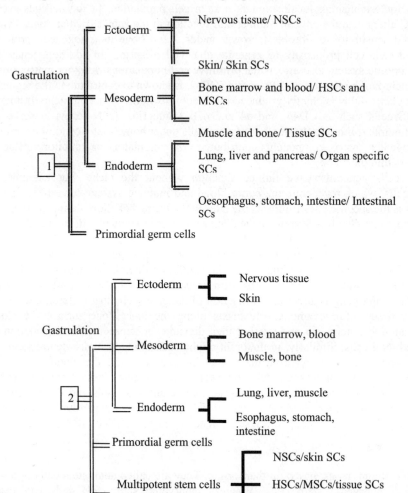

Fig 24.1. Putative developmental ontogeny of stem cells. In lineage tree 1, the development of stem cells occurs after the formation of germ layers. These stem cells are thus restricted by germ layer commitment to their respective lineage (e.g., mesoderm is formed giving rise to haematopoietic progenitors that become hematopoietic stem cells). Lineage tree 2 illustrates the idea that stem cells might develop similarly to PGCs, in that they avoid the lineage commitments during gastrulation and subsequently migrate to specific tissue and organ niches (Redrawn from Melton and Cowan, 2004)

ADULT STEM CELLS

Many tissues and organs in adult mammals contain reserves of stem cells to ensure their long-term anatomical and functional maintenance. Thus, accurate maintenance and homeostasis of mammalian tissues depend on the adult stem cells. Adult organs have, in general, closed cell communities, reflecting their distinct developmental origins. Nevertheless, they are composed of a cellular hierarchy in which, a small population of stem cells give rise to progenitor cells that regenerate mature tissue cells. Therefore, the stem cells are expected to produce mature cell types of respective tissues from which they originated. However, recent discoveries have challenged this dogma of 'tissue specificity', and have indicated the multipotency of the adult stem cells. Thus the current concept is that stem cells from adult tissues are not constrained by their ancestry, but have the 'plasticity' to alter their destiny (Table 24.1). Exposing stem cells to the extracellular developmental signals of other lineages could induce this plasticity. Alternatively, adult stem cells could also acquire lineage-specific determinants mediated through cell fusion.

Location of stem cells	Types of cells generated
Brain	Neurons, oligodendrites, skeletal muscle, blood cells
Bone marrow	Endothelial cells, blood cells, cartilage, bone, adipocytes, cardiac muscle, skeletal muscle, neuronal cells, skin, oval cells, gastrointestinal tract cells, thymus, pulmonary epithelial cells
Skeletal muscle	Skeletal muscle, bone, cartilage, fat, smooth muscle
Myocardium	Myocytes, endothelial cells
Skin	Keratinocytes
Liver	Liver cells
Testis and ovary	Gonads
Pancreatic ducts	Islet cells
Fatty tissues	Fat, muscle, cartilage, bone

Table 24.1 Potential plasticity of stem cells From Ruddon, 2007

Adult stem cells, otherwise called tissue stem cells, are defined by three common properties. They are: 1) the presence of an extensive capacity for self renewal that allows maintenance of the undifferentiated stem pool over the lifetime of the host; 2) strict regulation of stem cell number; 3) the ability to undergo a broad range of differentiation events to reconstitute all of the functional elements within the tissue, resulting in ordered tissue regeneration or repair. The maintenance of stem or progenitor cells in the adult tissues is regulated by *Hedgehog* (Hh) and *Wnt* signalling pathways. These pathways play central roles in directing embryonic pattern formation, and also function post-embryonically in stem cell renewal, tissue repair and regeneration.

In adult tissues, stem cells play an important role in tissue renewal and repair, especially in organs such as skin and blood where turnover is high. To effect repair of a tissue, the stem cell undergoes proliferation and differentiation, controlled and coordinated by stem cell specific growth factors acting on existing quiescent stem cells. Another mechanism by which the adult stem cells undergo transdifferentiation is by cell fusion (see below).

STEM CELL DIVISION

Defining characteristics of stem cells is their capacity to self-renew or make more cells, and generate differentiated progeny. Two types of cell divisions they obligatorily undertake accomplish these two properties of the stem cells (Fig.24.2). In the asymmetric cell division, the stem cells can accomplish the two tasks of self-renewal and differentiation. In each asymmetric cell division, the parent cell divides to generate one daughter with a stem- cell fate (self-renewal) and another daughter that differentiates. One difficulty of this type of division is that the stem cells cannot expand in number. Increase in stem cell number is important especially when the stem cell pools are first established during development and when they are needed in large numbers during regeneration after injury of an organ.

In the second type of cell division, namely, symmetric division, generation of more stem cells (self-renewal) or differentiated progenitor cells at each division is possible. Most stem cells can divide by either asymmetric or symmetric modes of division and the balance between these two modes is controlled by developmental and environmental signals to produce appropriate numbers of stem cells and differentiated daughters. Some mammalian stem cells seem to switch between symmetric and asymmetric cell divisions. For example, both neural and epidermal progenitors change from primarily symmetric divisions that expand stem-cell pools during embryonic development to primarily asymmetric divisions that expand differentiated cell numbers in mid to late gestation.

TYPES OF ADULT STEM CELLS

Most of the known adult stem cells originate from the bone marrow. In addition to the well-characterized hematopoietic stem cells, several of the mesenchymal stem cell types including the adherent mobile cell types such as reticular endothelial cells, macrophages, osteoblasts, adipocytes, fibroblasts and stromal cells also could be derived from bone marrow.

HAEMATOPOIETIC STEM CELLS (HSCs)

Haematopoietic stem cells (HSCs), the cells that give rise to the blood throughout our lifetime, are the best understood stem cells in our body. They have been successfully isolated and the stages through which they progress during development delineated. The path to the identification of hematopoietic stem cells began in response to the clinical need for cells capable of protecting humans exposed to minimum lethal doses of irradiation or chemotherapy. The first insight came with the observation that lead shielding of haematopoietic tissues prevented death from otherwise lethal doses of irradiation. Intravenous infusion of syngeneic marrow after irradiation also prevented death. Lymphohaematopoietic cells arising from such infused bone marrow also repopulated haematopoietic tissues. During fetal development, haematopoietic activity sequentially progresses from the yolk sac to the liver, to the spleen, and thymus. During adult life, haematopoiesis occurs in the bone marrow, spleen and thymus.

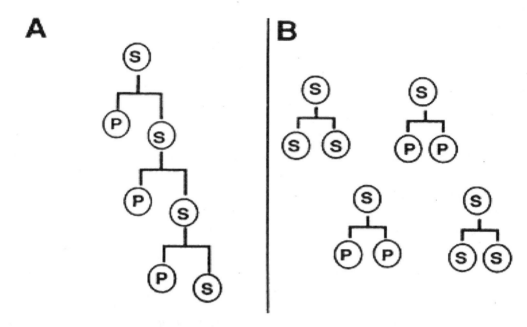

S= Stem cell; P= Progenitor cell (modified from Morrison et al, 1997).
Fig. 24.2 showing Asymmetric (A) and Symmetric (B) types of stem cell division (A) All divisions are obligatorily asymmetric and controlled by a cell-intrinsic mechanism. Note that no amplification of the size of the stem cell population is possible in this type of lineage. (B) A population of four stem cells is shown in which all divisions are symmetric, but half the time are self-renewing.

HSCs are non- adherent cells, which are derived from the bone marrow and are present in low numbers in circulating blood in adults and children. In humans, HSC are the only well-defined stem cells that are routinely in use therapeutically to regenerate the hematopoietic system following radiation therapy in childhood leukemia. A number of cell surface markers such as CD45 define HSC and help in their sorting out from other stem cell populations of bone marrow. Primarily, HSCs produce cells of blood and immune system (Table.24.2).

In addition, they also give rise to other stem cells of lymphoid and myeloid lineages from which are derived a number of lineage- restricted progenitors and precursor cells (See chapter on development of immune system). Donor HSCs are also known to give rise to non- hepatopoietic cell types such as hepatocytes by a process of transdifferentiation suggesting that tissue specific stem cells could generate the mature cell types of another tissue. Recent studies have claimed that bone marrow – derived stem cells could repair damaged liver in mice. The mechanism by which transplanted blood cells could transform into liver cells is cell fusion. Understandably, the blood cells change their identity through cell fusion with liver cells. In hybrid cells produced by such fusion, molecules from one fusion partner (in this case, liver cells), reprogramme gene expression in the genome of the other partner (blood cells).

Blood cells	Immune cells
Red blood cells	Natural killer cell
Neutrophil	T lymphocyte
Eosinophil	B lymphocyte
Basophil	
Monocyte/ Macrophage	
Platelets	

Table 24.2 Showing types of blood cells and cells of immune system derived from haematopoietic stem cells.

MESENCHYMAL STEM CELLS (MSC)

Mesenchymal stem cells are long established and follow lineage restriction rules laid down in the embryo. MSC is a pluripotent stem cell capable of generating a number of different tissue types of mesodermal origin. In the embryo, mesenchymal cells form blood and contribute to parenchymal organs such as the liver, lung and kidney. Hence, MSCs are thought to be the undifferentiated remnant of these fetal cells. However, they can have a separate adult origin. In effect, MSC can be derived not only from bone marrow, but also from a number of other mesodermal tissues including bone and skeletal muscles. In the adult, MSCs have been shown to generate mesenchymal cell types such as muscle, fat cells, bone, cartilage, tendon and other connective tissues. These cells secrete extracellular matrix and growth factors providing the bone marrow microenvironment. Hence, primary function of the MSC is to maintain the stem cell niche of the HSC and other precursor cells in the bone marrow.

MULTIPOTENT ADULT PROGENITOR CELLS (MAPC)

These are yet another pluripotent cells derived from the bone marrow. When injected into mouse blastocysts, they are shown to have the capacity to give rise to neural stem cells, and satellite cells, the latter having the capability to give rise to myogenic precursor cells.

SKELETAL MUSCLE STEM CELLS (SMSC)

Skeletal muscle contains a population of undifferentiated stem cells with the capacity to replace damaged muscle. The ability of a pool of undifferentiated SMSC to repopulate skeletal muscles following irradiation-induced damage or immune-compromised mice has been reported, although with limited success. The origin of SMSC is from the satellite cells occupying a region between the sarcolemma and basal lamina of the muscle fibres. Like other stem cell populations, the satellite cell population is heterogeneous and may contain a range of stem and precursor cell types. This is based on the fact that the satellite cells exhibit differential expression of a number of cell markers including CD34, MyoD, Myf5, pax7, m- cadherin, HGF and desmin.

The skeletal muscle stem cell populations have a well- defined function of muscle fibre regeneration. Upon physical stimuli such as injury and fibre degeneration, SMSCs undergo replication, migration and finally differentiation into new myofibrils. A number of growth factors serve to regulate skeletal muscle differentiation, proliferation and survival. Of these, the most important are the Insulin-like growth factor family (IGF), which promotes skeletal muscle stem cell differentiation (IGF-2), and survival and proliferation (IGF-1). Other growth factors that act on skeletal muscle stem cells include

TGF-β, PDGF, HGF and NGF. SMSCs have also been known to give rise to a variety of other cell types of mesodermal origin including osteogenic (bone) and adipogenic (fat) and myogenic lineages.

CARDIAC STEM CELLS

Till recently, heart was thought to lack the capacity to regenerate after injury that happens during heart attack. But the identification of cardiac stem cells has suggested that the heart has the inbuilt mechanism for repairing its own damaged tissues. Like many specialized cells, fully developed heart cells do not divide and so are unable to patch up damage. However, in patients who survive less severe heart attack, cells from the connective tissues called fibroblasts, which divide and migrate into the damaged area to form scar tissue, replace dead heart cells. In these areas, the ventricular valve becomes thin and no longer contracts properly. Recent discovery of both cardiac progenitor cells and cardiac stem cells in mouse adult heart has given hope for cardiac transplantation therapy in humans. Cardiac progenitor cells found in the left atrial wall of the adult mouse expresses islet-1 gene. Using this marker gene, these progenitor cells have also been identified in the developing mouse embryo. This may suggest that the islet-1 cells in the adult heart are remnants of a cardiac progenitor-cell population from the heart of the developing fetus. Postnatal cardiac progenitor cells in their niches have the potential to differentiate into endothelial cells for vessel formation, cardiac muscle cells for contractility, and cardiac conduction cells for coordinated electrical activity of the heart.

Additionally, cardiac stem cells have also been identified in the ventricular region. These primitive adult stem cells are capable of dividing and developing into mature heart and vascular cells. These cardiac stem cells are however distinct from cardiac progenitor cells as they express proteins such as c-kit and Sca-1, typical of stem cells from other organs. They do not express isl-1. Cardiac stem cells can also be grown from human biopsies as aggregates in suspension culture. These so-called cardiospheres contain a mixture of cardiac cell types. When transplanted into mice after an acute heart attack, they formed vascular cells and heart muscle cells. The problem faced in these transplantation studies is that whether these stem cells will do the contraction in resonance with the original cardiac cells in the heart. Irregular heart beats (arrhythmias) have occurred in patients who received their own skeletal muscle progenitor cells in an effort to repair heart muscle damage and this was attributed to poor communication between the transplanted cells and host heart muscles. Successful transplantation of cardiac tissues in humans depends upon further research in tackling this problem.

DIFFERENTIATION OF EMBRYONIC CELLS INTO THE CARDIAC LINEAGE

In the early embryonic cells, transcriptional factors such as NANOG, OCT4 and SOX2 maintain pluripotency. Decreased activity of these pluripotency factors is accompanied by increased activity of lineage-specific transcriptional activators such as *Brachyury* and MESP in the mesoderm lineage. Mesodermal cells that begin to differentiate into cardiac cells segregate into two distinct populations of cardiac progenitor cells (CPCs), marked by the transcription factor NKX2. These progenitors may be able to differentiate into several types of cardiac cell, and ISL1-expressing cells uniquely give rise to niches of postnatal CPCs.

NEURAL STEM CELLS (NSCS)

Adult mammalian brain has vastly reduced regenerative potential compared to developing brain. However, contrary to what was thought earlier, recent findings demonstrate that the mammalian brain contains a stem cell population that can give rise to differentiated cells, including mature neurons. Thus, cell proliferation does occur in the adult brain, which is capable of generating neurons and glia. The NSCs are defined by their ability to differentiate into cells of all neural lineages, to self-renew and to populate developing and/or degenerating CNS regions. In the normal brain, both neuronal stem cells as well as early progenitor cells are identified by their expression of CD133 marker; however, this cell surface marker is not expressed in late progenitor as well as the mature brain cells.

Neural stem cells and more restricted neuronal and glial progenitor cells are dispersed widely throughout the adult vertebrate brain. Long after fetal development, multipotent neural stem cells line the forebrain ventricles, whereas the committed neuronal progenitor cells remain within the ventricular wall, throughout its extensions to the olfactory bulb and the hippocampus. A large pool of glial progenitors also pervades both the ventricular zone and tissue parenchyma. The glial progenitors along with the neural precursor cells constitute the neural stem cell population of the adult human central nervous system.

A valuable feature of adult neural stem cells is that they are readily and extensively expandable when placed in culture and stimulated with the appropriate growth factors, such as epidermal growth factor (EGF) and fibroblast growth factor 2 (FGF2). This culture condition allows adult neural stem cells to be isolated, and their functional characteristics and developmental potentials to be investigated.

Neural stem cells are a focus of tremendous interest in recent years in view of their utility to replace neurons that have been damaged or lost as a result of injury or through neurodegenerative disorders such as Alzheimer's disease and Parkinson's disease. NSC has the characteristics of glial cells, and growth factors, neurotransmitters, and hormones control its stem cell activity in the adult brain. These stem cells can give rise to neurons as well as to two types of glial cells (astrocytes and oligodendrocytes). NSCs are derived from the embryonic, neonatal and adult rodent and human brains and can be maintained for long periods under the influence of EGF and FGF-2 growth factors in culture as suspension cell aggregates, known as neurospheres. Neurospheres contain a mixture of undifferentiated and differentiated cell types and can be induced to differentiate into functional neurons or astroglia by the addition of serum and plating onto a polylysine or laminin substrate. In mammals, there are two discrete regions in the brain that can generate new neurons. These are the ventricular subependyma (or subventricular zone) and the subgranular layer of the Dendate Gyrus. Precursors from these brain regions possess many characteristics of neural stem cells *in vitro* including self-renewal capacity.

NSCs are also shown to be able to differentiate into non- neuronal cell types. Recent studies have indicated that the mouse NSCs (committed to become neurons and glial cells), co-cultured with human endothelial cells, transformed up to 6% into endothelial cells by expressing their multiple endothelial markers, such as the endothelial cell adhesion protein, CD146. It is of interest to note here that NSCs are derived from ectoderm, whereas, endothelium is a derivative of mesoderm. In addition, NSCs co- cultured with pluripotent embryonic stem cells prior to injection into mouse blastocysts or chick embryos are reported to generate chimeric embryos with extensive NSC derived cell types. *in vitro*, these cells differentiated into myogenic and other cell lineages. Embryonic stem cells, being pluripotent, may produce instructive signals, which could induce NSC to differentiate pluripotently. However, the main difficulty in the replacement of injured neurons in the adult human brain with NSC is the remarkable resistance to accept such cells into a mature neuronal network.

DERMAL STEM CELLS

Skin is continually replacing its worn out cells. To allow replenishment of cells, a renewing tissue contains a stem cell population that provides a source of differentiating cells. The follicle (keratinocyte) stem cells occupy a position along the outer root sheath of the follicle. The multipotency of these stem cells is reflected in their ability to differentiate into skin epidermis and sebacious gland cells, in addition to making new hair follicles. The cell lineage choice in skin follicle stem cells is regulated by Tcf/Lef complex genes and mediated by *Wnt* and β - catenin signalling. The skin stem cells have been identified primarily in rodents. Recent experiments have also indicated their skin regenerative power in the mouse. Researchers have grown individual skin stem cells into hundreds of thousands of identical copies and then grafted them into a wound on the back of a hairless mouse. The cells grew to form patches of fur including skin, follicles, hair and oil- producing sebacious glands. If similar skin stem cells could be harvested from the dermis of human beings, they could also be grown up and re-implanted into a balding scalp or wound.

GUT STEM CELLS

Like other tissue- specific stem cells such as skeletal muscle satellite cells and follicular stem cells, gut stem cells are defined positionally. Stem cells in the gut reside near the base of the intestinal crypts. Gut stem cells give rise to a large number of 'clongenic' stem cells, which are involved in the regeneration of the intestinal crypts.

All the above tissue- specific adult stem cells are regulated by niche environment and by local growth factor action. Genes involved in the regulation of self- renewal in normal stem cells include Bmi-1, Notch, *Wnt* and *Sonic hedgehog* pathways. However, control mechanisms determining stem cell differentiation may require other genes expressing sets of transcriptional factors, which individually specify different lineages or combinations of lineages. For example, mutations in the ikaros gene, which encodes a zinc finger protein present in HSCs, prevent the development of multiple lymphoid sub-lineage derivatives. However, these adult tissue stem cells have more therapeutic value, as the extraction of these cells need not involve human embryos and hence will not raise any bio-ethical controversies.

STEM CELLS IN *HYDRA*

The adult stem cells of mammalian vertebrates are found in their niches in an inactive condition. Only on induction either by natural or artificial means, these stem cells will undergo division or differentiation into adult cell types. In *Hydra*, a member of the evolutionarily old metazoan phylum Cnidaria, the adult stem cells are active throughout its lifespan by maintaining the tissue homeostasis as well as taking part in budding, the chief form of asexual reproduction in *Hydra*. The tubular body of *Hydra* consists of two epithelial layers, the ectoderm and endoderm, separated by the basement membrane called mesoglea (Fig. 24.3).

Each layer is a single layer- deep and comprises a cell lineage. The third cell lineage is the interstitial cell lineage whose derivative cells are lodged in the interstices among the epithelial cells of both layers. The two layers of *Hydra* are dynamic and are in a steady state of production and loss of cells (see chapter on regeneration for details). Similarly, cell loss also occurs through the bud formation. Such constant replacement of differentiated cells is possible only through the continuous proliferation and differentiation of stem cells. The stem cells divide continuously as they are displaced apically or basally, and cease cell division and differentiate upon displacement into an extremity. Thereafter, they are displaced through the extremity to its tip and sloughed. Whereas the stem cells for ectoderm and endoderm are monopotent, giving rise to their respective epithelial cells, the interstitial stem cells are pluripotent in that it gives rise to four different cell types including the gametic cells. As shown in the Fig. 24.4, the differentiation pathway of the interstitial stem cells has both committed cells and differentiation intermediates. The derivation of gametes, especially, eggs, from the somatic interstitial stem cells is unique, because in all other higher metazoans, only primordial germ cells give rise to mature male and female gametes.

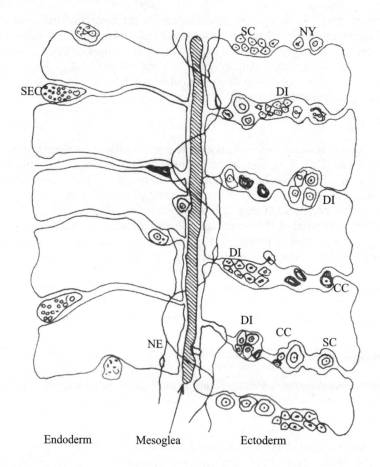

Fig. 24.3. Cross-section through the body column, indicating the location of cells of the interstitial cell lineage. SC – Stem cells; CC – Committed cells; DI – Differentiation intermediates; NE – Neurons; NY – Nematocytes; SEC – Secretory cells (Redrawn from H.R. Bode 1996).

In *Hydra*, pattern formation and morphogenetic processes are constantly active in the context of the tissue dynamics of the adult animal to maintain its morphology. Hence, the stem cells possess indefinite capacity for self-renewal. Unlike the stem cells of epithelia of more complex animals, whose sole functions are to proliferate and differentiate, the epithelial stem cells in the *Hydra* carry out several physiological functions, as they exist along the body column. However, the border between stem cells and differentiated cells is very sharp. For example, in the tentacle zone, the last ectodermal cell still undergoes division, while neighbouring cell across the border in the tentacle no longer divides, but undergoes differentiation. These differentiated cells in the tentacle also start expressing molecular markers such as TS-19 antigen on the cell surface and an annexin in the cytoplasm. Similarly, the change from dividing cells of endoderm to non-dividing cells also occurs at the same borders, and just as abruptly. However, the regulatory mechanism of the differentiation behaviour of the cells in the epithelial lineage is not properly understood.

As for the somatic differentiation products of interstitial cells, they also constantly move along with their epithelial neighbours and hence their population size gets diminished by sloughing at the extremities. Hence, there is a need for the interstitial stem cells to divide and generate different types of somatic cell lineages. Both intrinsic and extrinsic factors are supposed to control such differentiation pathways. The interstitial stem cells reside in the space between the ectodermal epithelial layer cells. A committed cell for neuron can differentiate in any region of the body column, but a particular type neuron will be formed only in response to the local environmental cues. Similarly, stem cells in the ectoderm committed to secretory cell differentiation, migrate across the mesoglea into the endoderm. Once in the endoderm, the migrated cell differentiates into the secretory cell type appropriate for the region: mucous cells in the head and gland cells in the body column.

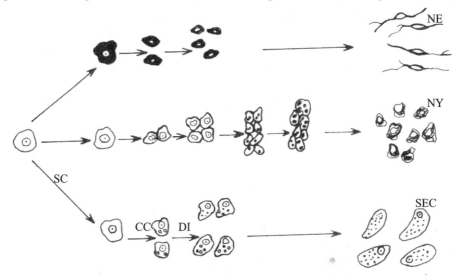

Fig. 24.4 Stem cell potentials of interstitial cells, indicating the differentiation pathways for the three classes of somatic differentiation products. Gamete differentiation pathway is not shown. Interstitial cells include stem cells and committed cells. Differentiation intermediates include dividing and differentiating intermediates. SC – Stem cells; CC – Committed cells; DI – Differentiation intermediates; NE – Neurons; NY – Nematocytes; SEC – Secretory cells (Redrawn from H.R. Bode 1996).

Molecular markers to distinguish stem cell commitment to different types of cells are not clearly understood. A small peptide of 11 amino acid lengths, the head activator, is known to stimulate mitosis in the epithelial and interstitial cells. The same peptide can also stimulate differentiation intermediates to traverse the pathway and thus affecting the commitment of stem cells to enter the neuron differentiation pathway. Vertebrate-type insulin growth factor, acting as signalling molecules may also enhance the rate of epithelial cell division in *Hydra*. Several marker genes, such as *Cnox-2*, a homologue of the *Drosophila HOM C/HOX* gene and *deformed* are also known to be expressed in a subset of interstitial cells that are most likely committed to nematocyte differentiation.

In *Hydra*, stem cells seem to have combined the functions of differentiated cells inasmuch as they do the physiological functions of the latter. Unlike vertebrates, the stem cells of *Hydra* are in constant state of proliferation and differentiation to meet the cell loss due to sloughing and bud formation.

CANCER STEM CELLS

And the last in the list is the cancer stem cells. Cancer is thought to arise either from normal tissue cells or committed progenitors. Just as stem cells are crucial for tissue development and regeneration, cancer stem cells play a pivotal role in tumour formation and maintenance. Normal stem cells are the only cells that self renew for the life time of the organism, and they produce the progenitor of mature cell populations required for tissue function and maintenance. On the other hand, cancer genes can subvert the stem cell population to turn it into cancer stem cells. These then produce cancer cells instead of normal cells. For example, some cancer genes involved in acute myeloid leukaemia turn progenitor cells into cancer stem cells by inducing expression of a specific set of stem-cell genes. In this way, the progenitor cells are not only subverted to self-renew, but also support tumour formation. In short, cancer stem cells are derived from normal stem cells that run amok, producing cancer cells instead of normal cells. Expectedly, many cancer-stem-cell populations express cell-surface proteins that are also found on normal stem cells. Cancer stem cells could also be derived from non-stemcell sources. For example, a leukaemia associated protein is found to transform non-stem cells to acquire stem cell behaviour, inducing them to form tumour. In the cancer stem cells some of the molecular signalling systems (such as *Wnt* and *Sonic hedgehog* which in their healthy manifestations play a role in embryonic development) may be aberrantly activated.

AGEING OF STEM CELLS

Adult stem cells ae otherwise known as tissue specific regenerative cells in the sense that they are required for tissue replacement throughout the human lifespan. Stem cells in several tissues are largely retained in a quiescent state but can be coaxed back into the cell cycle in response to extracellular cues, even after prolonged periods of dormancy. Once stimulated to divide, stem cells yield undifferentiated progeny, which in turn produce differentiated effector cells through subsequent rounds of proliferation. Under homeostatic conditions, there is, however, limited proliferative demand on the self-renewing stem cells themselves and so these cells divide infrequently. As stem cells appear to be less metabolically active in their quiescent state, they may be subjected to only lower levels of DNA-damage-inducing metabolic side products such as reactive oxygen species (ROS).

Nevertheless, recent studies have revealed the ageing of tissue specific adult stem cells along with mammalian ageing.

EMBRYONIC STEM CELLS

Embryonic stem cells (ESCs) are perpetually self-renewing, totipotent progenitors derived from the inner cell mass of blastocysts. Self renewal of embryonic stem cells is achieved by symmetrical cell division while maintaining pluripotency. Pluripotent ES cells have the ability to differentiate into derivatives of the three embryonic germ layers, even after being cultured continuously at the undifferentiated stage (Table 24.3). Depending on culture conditions, ES cells can differentiate into a variety of cell types. The ability to steer ES cell differentiation into specific cell types holds great promise for regenerative medicine.

The activity of ES cells can be modulated by extrinsic factors such as a cytokine leukemia inhibitory factor (LIF) in mouse. External signals control gene expression by regulating transcription factors like *Oct-3*/4, which acts as a pivotal player in self-renewal or differentiation. In the absence of Oct-4, the pluripotent cells *in vivo* (epiblast) and *in vitro* (ES cells), both revert to the trophoblast lineage. This implicates Oct-4 as an important regulatory molecule in the initial cell fate decisions during mammalian development. Furthermore, Nanog, a homeodomain-containing protein, was identified as a factor that can sustain pluripotency in ES cells even in the absence of leukaemia inhibiting factor (LIF). During development, Nanog function is required at a later point than the initial requirement for Oct-4, but both are required for the maintenance of pluripotency.

Self-renewal is defined as making a complete phenocopy of the stem cells through mitosis. In stem cell self-renewal, symmetric cell division results in one stem cell and another one, either a differentiated progeny or a stem cell with a restricted capacity for differentiation (see above). Self-renewal by symmetric cell division is often observed in transient stem cells appearing in early embryonic development. On the contrary, self-renewal by asymmetric cell division is found in permanent stem cells in embryos in later developmental stages and in adult to maintain the homeostasis of the established body plan. Another characteristic feature of ES cells is their clonogenicity. That means, each single ES cell has the ability to form a homogeneous line demonstrating all parental ES cell line features.

ES cell lines can contribute to the formation of all tissues, including germ and haematopoietic cells. ES cells have been described mainly from mammals including humans. Initially, ES cells were isolated from the inner cell mass (ICM) or germinal ridge of the mouse preimplantation embryo. The window of time for recovering ES cells from the embryo is narrow- about day 4 to 6 -for the inner cell mass and slightly later for the germinal ridge in the mouse embryo.

TRANSCRIPTIONAL CONTROL OF PLURIPOTENCY IN ES CELLS

Pluripotency is the hallmark of embryonic stem cells. The gene-expression programme of pluripotent ES cells is regulated by specific transcriptional factors, chromatin-modifying enzymes, regulatory RNA molecules, and signal-transduction pathways. Genetic studies have revealed that the homeodomain transcription factors Oct4 and Nanog are essential regulators of early development and ES cell identity.These transcription factors are expressed both in pluripotent ES cells and in the inner cell mass (ICM) of the blastocyst from which ES cells are derived. Disruption of Oct4 and Nanog causes loss of pluripotency and inappropriate differentiation of ICM and ES cells to trophectoderm and extraembryonic endoderm, respectively. However, recent evidence suggests that Nanog may function only to stabilze the pluripotent state rather than being essential for maintaining pluripotency of ES cells. Instead, Oct4 can heterodimerize with another transcription factor Sox2 in ES cells and Sox2 in turn contributes to pluripotency by regulating Oct4 levels. Oct4 is rapidly and completely silenced during early cellular differentiation. Oct4, Sox2 and Nanog collectively target two sets of genes, one that is actively expressed and another that is silent in ES cells but remains poised for subsequent expression of during cellular differentiation. Apparently, these regulators are central to the transcriptional regulatory hierarchy that specifies embryonic stem cell identity.

These ES cells are immortal, if cultured under appropriate conditions. However, in the embryo, after the window closes, these cells are no longer immortal and disappear as the embryo develops. An alternative explanation for the disappearance of these short- lived ES cells, in the *in vivo* condition, is that they probably become both the stem cells of the haematopoietic system and the most recently identified stem- like cells found in essentially every non-haematopoietic tissue of adult vertebrates (see above). Thus, ES cells appear to be the *in vitro* equivalent of the epiblast, as they have the capacity to contribute to all somatic lineages and in mice to produce germ line chimeras. However, ES cells cannot form 'extra embryonic' tissues, necessary for the development of placenta. Hence, a complete individual cannot be formed from ES cells.

ORIGIN OF EMBRYONIC STEM CELLS IN MOUSE

ES cells were isolated from the inner cell mass (ICM) of the mouse preimplantation embryo. These cell lines were generated by removing the ICM from the preimplantation blastocysts and serially passing these cells on irradiated mouse feeder layers consisting of mitotically inactivated mouse embryonic fibroblasts. The resulting cell lines showed unrestricted proliferative potential and were pluripotent. They could differentiate into derivatives of all three germ layers. Upon injection into blastocysts, they contributed to all tissue types in the resulting chimera.

Embryonic Germ Layer	Differentiated Tissue
Endoderm	Thymus
	Thyroid , parathyroid glands
	Larynx , trachea , lung
	Urinary bladder , vagina , urethra
	Gastrointestinal (Gi) organs (liver, pancreas)
	Lining of the Gi tract
	Lining of the respiratory tract
Mesoderm	Bone marrow (blood)
	Adrenal cortex
	Lymphatic tissue
	Skeletal , smooth and cardiac muscle
	Connective tissue (including bone, cartilage)
	Urogenital system
	Heart and blood vessels(vascular system)
Ectoderm	Skin
	Neural tissue(neuroectoderm)
	Adrenal medulla
	Pituitary gland
	Connective tissue of the head and face, Eyes, ears
Germ Cell	Sperm , Egg

Table 24.3. Development of different tissues from the embryonic stem cells

Culture methods for different types of ES cells have also been standardized in mouse embryos. Three types of permanent lines of pluripotent stem cells have been derived from embryos and fetal stages of mice: (i) embryonic stem (ES) cells isolated from blastomeres of the early mouse embryo from 8- cell upto blastocyst stage, (ii) embryonic germ (EG) cells isolated from primordial germ cells, the precursor cells of germ cells from 9.5-12.5-day fetal stages, (iii) embryonic carcinoma (EC) cells established from the stem cell population of teratocarcinoma (Fig. 24.5). In spite of differences in their origin, all three types share common characteristics, such as expression of alkaline phosphatase, the stage-specific embryonic antigen SSEA-1 and the germline – specific transcription factor Oct-4, a short G1 phase of the cell cycle and high telomerase activity. These cell lines have been largely used for the generation of genetically modified mice. They also provide a unique tool for studying the development of the embryo.

Proliferation of ES and EG cells is controlled by cytokines of the IL-6 family. Since EC cells are derived from malignant teratocarcinomas and proliferate independently of growth factors and cytokines, they have to be induced to differentiate by chemical agents. *In vivo*, ES and EG cells take part in embryonic development after injection into mouse blastocysts, resulting in chimaeric animals.

SURFACE ANTIGEN MARKERS AND IDENTIFICATION OF ES CELLS

Molecules on the surface of cells make them convenient markers for characterizing stem cell types. These surface antigens show highly restricted patterns of expression characteristic of particular cell types. Such antigens with restricted expression are called differentiation antigens. A good example of surface antigen to identify embryonic stem cell is stage- specific embryonic antigens (SSEA). SSEA1, a glycosphingolipid is found to be expressed by blastomeres of late cleavage-and morula- stage embryos by cells of the inner cell mass (ICM) and by trophectoderm. Mouse EC and embryonic stem cells characteristically express it. SSEA1 has therefore become one of the hallmarks for identifying these undifferentiated stem cells. However, as these ES cells start differentiation, there is disappearance of this antigen. On the other hand, SSEA3 and SSEA4 are not expressed in murine ES cells and embryonic cells of ICM, but they are expressed in human embryonic stem cells, making them surface antigens to identify human embryonic stem cells. Cell surface markers are also used to sort out distinctive adult stem cell populations. For example, CD45 marker molecule is used in identifying haematopoietic stem cells from the bone marrow. Other marker molecules used in the identification of both mouse and human embryonic stem cells are summarized in Table 24.4.

IN VITRO DIFFERENTIATION OF ES CELLS IN MOUSE

Figure(24.5) depicts the culture methods relating to the origin and differentiation potentials of mouse ES, EG and EC cells. In vitro, ES and EG cells differentiate spontaneously after cultivation as aggregates, called embryoid bodies (EBs; see figue). This cultivation system is, therefore, useful for investigating cellular differentiation and gene expression patterns during ES cell differentiation under in vitro conditions. Mouse ES cells were found to differentiate in vitro into many cell lineages including primitive ectoderm, visceral endoderm and early mesoderm. ES cells also differentiate into differentiated cell types such as cardiac, myogenic, chondriogenic, adipogenic, hematopoitic, epithelial, neuronal, endothelial and vascular smooth muscle cells. The differentiative capabilty of ES cells in vivo thus demonstrates the recapitulation of the early embryonic development. Similar to ES, EG cells also have the capacity to differentiate into many cell types when cultivated as EBs in vitro (See Fig.24.5).

GENE EXPRESSION DURING EB DIFFERENTIATION

Marker genes for early embryonic lineages, namely the neuroectoderm, endoderm and mesoderm, are expressed in regulated manner during EB differentiation (Table.24.4). During the first 2 days of EB development, genes such as Oct-3 and FGF-5 are transcribed. These genes are predominantly expressed during mouse pre-gastrulation stage before implantation. This is followed by expression of genes characteristic for early post-implantation stages, such as nodal and the early endodermal genes, *vHNF1, HNF3 β* and *HNF4* in 2-5 day-old EBs. In parallel, marker genes for early mesodermal differentiation such as *Brachyury, goosecoid* and *BMP-4*, which are expressed during gastrulation in mouse embryos, are expressed around day 4 in EBs. Genes involved in early neuroectoderm development *in vivo*, such as *Pax-6* and *Mash-1* are also expressed during early stages of ES cell differentiation at day 1 of EB development. Other genes coding for tissue- specific proteins, including ion channels and receptors, are expressed in the course of ES cell differentiation in a developmentally controlled time pattern closely resembling the pattern observed during mouse embryogenesis.

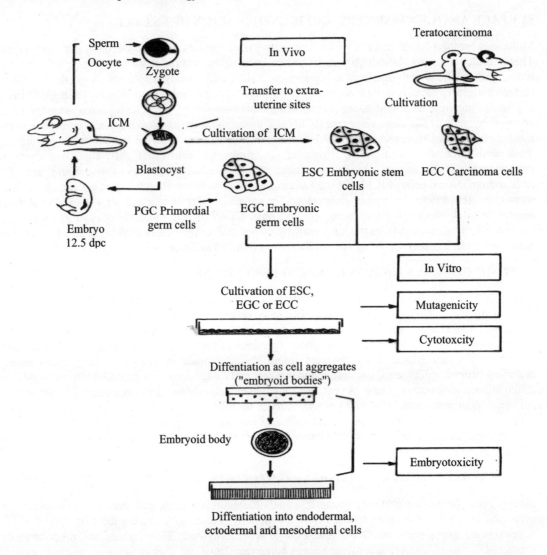

Fig. 24.5. Origin and differentiation potential of embryonic carcinoma (EC), embryonic stem (ES) and embryonic germ (EG) cells and the use of ES cell technology for mutagenicity, cytotoxicity and embryotoxicity analysis *in vitro*. This diagram also illustrates the protocol development for mouse embryonic stem cells.
(Redrawn from Rohwedel et al., 2001).

Morphogenetic development is, however, not possible within the EBs, because spatially controlled signals are lacking. Still, in EBs cellular differentiation is accomplished in a developmentally regulated and time-dependent manner, enabling analysis of developmental processes on a cellular level.

HUMAN EMBRYONIC STEM CELLS

Human embryonic stem (hES) cells are pluripotent cell line established from the explanted inner cell mass of human blastocysts. hES cells are maintained in media containing 20% fetal bovine serum or serum replacement media supplemented with basic fibroblast growth factor and require the presence of feeder cells or conditioned medium from feeders. hES cell lines have conventionally been characterized using surface markers and molecular markers used to characterize human embyonal carcinoma cells, mouse ES cells, and haematopoitic stem cells. As seen from the table 24.3, the overall expression of surface markers is similar in mouse and human embryonic stem cells. Like the mouse ES cell lines, they are able to self- renew as undifferentiated cells in culture indefinitely. Yet, when placed in the proper environment, hES cells retain the ability to form any cell type within the body. After undifferentiated proliferation *in vitro* for 4 to 5 months, these cells still maintained the developmental potential to form trophoblast and derivatives of all three embryonic germ layers, including gut epithelium (endoderm); cartilage, bone, smooth muscles and striated muscle (mesoderm); and neural epithelium, embryonic ganglia and stratified squamous epithelium (ectoderm). These dual capabilities make ES cells as an especially intriguing and informative source to understand early human embryology.

Marker genes	Mouse ES	Human ES
Alkaline Phosphatase	+	+
SSEA-1	+	-
SSEA-3	-	+
SSEA-4	-	+
Tra 1-60	-	+
Tra 1-81	-	+
Oct-3/4	+	+
Sox 2	+	+
Rex-1	+	+
Tert	+	+
Fgf4	+	+
Utf1	+	+
Foxd3	+	-
Cx45	+	+
Cx43	+	+
BCRP-1	+	+
LIFR	+	-
gp 130	+	+
Stat3	+	+
Nanog	+	+

Table 24.4 Comparison of mouse and human es cell expression of marker genes. For abbreviations, refer the text. Adapted from Carpenter, M.K. and Bhatia, M. (In Handbook of stem cells, Vol. 1, 2004, Academic Press).

To date, 71 independent hES cell lines have been developed worldwide. Of which, 11 are used currently for research purposes. Remarkably, all of the hES cell lines derived to date show similar expression of the surface and molecular markers.

TELOMERASE EXPRESSION

Telomerase is critical for the protection of germ line and stem cell chromosomes from fatal shortening during replication. The hES cell lines express high levels of telomerase activity. Telomerase is a ribonucleoprotein that adds telomere repeats to chromosome ends and is involved in maintaining telomere length, which plays an important role in replicative life span. Telomerase expression is highly correlated with immortality in human cell lines, and reintroduction of telomerase activity into some diploid human somatic cell lines extends replicative life span. Diploid human somatic cells do not express telomerase, have shortened telomeres with age, and enter replicative senescence after a finite proliferative life span in tissue culture. The high level of telomerase activity expressed by the human ES cell lines therefore suggests that their replicative life- span will exceed that of somatic cells.

Recently, telomerase activity has also been reported to be responsible for maintaining the pool of stem cells that self-renew throughout the 2-5 year colony life span of the ascidian, *Botryllus schlosseri*. In these colonial ascidians, these stem cells are involved in the asexual budding, which occurs continuously from parent body wall.

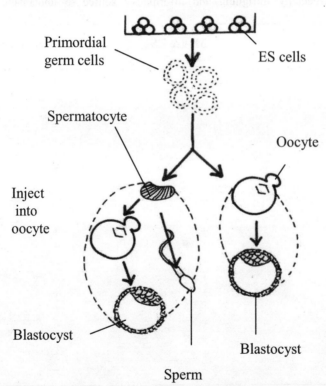

Fig. 24.6. Derivation of germ cells from stem cells. Embryonic stem cells (ES) cultured as aggregates, formed structures resembling early embryos (embryoid bodies). These aggregates underwent spontaneous differentiation to give rise to spermatocytes and oocytes. The sperm cells derived from these spermatocytes, when injected into the oocytes resulted in fertilization and further development upto blastocyst (modified from M.A. Surani, 2004).

GAMETE PRODUCTION FROM EMBRYONIC STEM CELLS

Zygote is a fusion product of egg and sperm, giving rise not only to a new individual, but also (theoretically at least) to an endless series of generation of life. The possibility to generate these gametes in a culture dish from mouse or human embryonic stem cells seems to be realistic from recent research. As mentioned above, embryonic stem cells, cultured as aggregates, will form cystic structures called embryoid bodies, resembling early embryos. Embryoid bodies, when cultured further in appropriate media, differentiated into primordial germ cells. Expression studies on genes implicated in ES cell pluripotency (*Oct4*) and germ cell development, including *stella* and *fragelis* (*Fgls*) have been employed in distinguishing the gametes from ES cells. These primordial germ cells further underwent maturation to give rise to functional haploid sperm and eggs. The differentiated gametes, particularly, the sperm expressed gamete-specific genes like, *Dazl, Piwil2, Rnf17, Rnh2, Tdrd1,* and *Tex14*. The oocytes when injected with the sperm developed into blastocysts (see fig.24.6). More importantly, these 'synthetic eggs' derived from ES cells could be used for nuclear transplantation experiments leading to enormous medical use. As shown in the Fig.24.6, the ES derived egg cells could be used as recipients for nuclei from adult somatic cells, after stripping off of their own genetic materials. The injected somatic nuclei could then be 'reprogrammed' by factors present in the oocytes, which are then allowed to develop to the blastocyst stage. Blatocysts contain epiblast cells, from which new ES cells can be derived. These ES cells, in turn, could be prompted to produce specific cell types for transplantation, to treat specific human conditions.

THERAPEUTIC USE OF STEM CELLS

Resident pools of somatic stem cells in many organs are responsible for tissue maintenance and repair. In regenerative medicine, the goal is to exploit these cells either by transplanting them from an exogenous source or by activating endogenous stem cells pharmacologically. For diseases caused by mutations in a single gene, the therapeutic goal is tissue replacement using stem cells engineered to correct the genetic defect.

Human embryonic stem cells have been regarded as an important source of cells and tissues for future transplantation and regenerative medicine therapies. Every year millions of people suffer and eventually die from serious and largely incurable degenerative diseases of the nervous system (e.g., Parkinson's disease, Alzheimer's disease, multiple sclerosis and stroke), heart (myocardial infarction), liver (hepatitis), pancreas (diabetes) and other organs. Stem- cell therapy offers hope for such diseases that result from a lack of a tissue reservoir of stem cells to replace defective and dying or ageing cells or tissues. The replacement of those cells could offer lifetime treatment.

PATIENT-SPECIFIC STEM CELL THERAPIES

Like organ transplants, stem cell transplants confront an immune barrier, which requires either that transplant be autologous (derived from 'self-tissues') or that patients take immunosuppressive drugs if the transplants are from unrelated donors (allogeneic). Hence, stem cell therapy also requires production of genetically equivalent (isogenic) cells. This can be achieved by producing pluripotent ESCs from adult somatic cells by a technique called somatic cell nuclear transfer (SCNT; in which an adult ssomatic cell nucleus is injected into an enucleated oocyte). Pluripotent ESCs created by SCNT (so-called ntESCs) can be used both to model diseases and as rejection-proof "autologous" tissue for cell replacement therapies. ntESCs sems to equivalent to ESCs derived from naturally fertilized embryos;however they have been produced so far from mouse and rhesus macaques. Due to ethical reasons, human ntESCs are not yet produced for therapeutic use.

INDUCED PLURIPOTENT STEM (iPS) CELLS

A somatic cell can be reprogrammed by transferring its nucleus into a eunucleated egg or oocyte or by fusion with an embryonic stem cell. This method is called somatic cell nuclear transfer (SCNT), demonstrating that pluripotency can be restored in a terminally differentiated cell and suggest that factors present in oocytes or ES cells can reprogram a somatic nucleus to a totipotent or pluripotent state. Following this, induced pluripotent stem cells (iPS) were first generated from mouse fibroblasts by retroviral-mediated introduction of four factors (Yamanaka factors) such as OCT4, SOX2, KLF4 and c-MYC or a different set of four factors (OCT4, SOX2, NANOG and LIN28). OCT3/4 and certain members of the SOX gene family have been identified as crucial transcriptional regulators involved in the induction process, whose absence makes pluripotency induction impossible. Additional genes including Klf family and Myc family, Nanog and LIN 28 have been identified to increase the induction efficiency. These four transcription factors, when ectopically expressed, enabled reprogramming of a somatic cell to a pluripotent state, circumventing SCNT. In fact, reprogramming of somatic cells can be achieved without the two oncogenic factors such as c-M and Klf4, but only with significantly lower efficiency. This is consistent with the notion that these oncognes are nonessential for the reprogramming process and may merely promote the epigenetic remodelling and activation of essential endogenous genes such as Oct4 and Sox2, which then establish the autoregulatory loop that maintains the pluripotent state. It is then possible that Oct4 is the only transcriptional factor that is indispensable for reprogramming and that other factors or signalling molecules can serve to facilitate activation of the "pluripotency circuitary".

In all cases, the retroviruses are silenced, and the pluripotent state of the reprogrammed cells ultimately hinges upon the activity of endogenous genes. Recently, researchers added fluorescent tags (Green fluorescent protein) to each of the reprogramming genes and made them fluoresce when expressed.As cells reprogram, they turn on their own pluripotency machinery and silence the introduced genes; the subsequent dimming of GFP is an important marker of reprogramming.

Mouse and human iPS cells closely resemble ES cells in morphology, gene expression, epigenetic status, and in vitro differentiation (pluripotency). Patient-specific iPS cells could be useful in drug discovery and regenerative medicine and creating in vitro disease model. Furthermore, mouse iPS cells give rise to adult chimeras and show competence for germline transmission. However, the chimeras and progeny mice derived from iPS cells frequently develop tumors, which may be due to reactivation of the c-Myc oncogene. Retroviral integration of the other three transcription factors may also activate or inactivate host genes, resulting in tumorigenicity. It has to be remembered that three of the reprogramming factors –Myc, Klf4, and Lin28-have been linked with oncogenesis. Although Myc is not essential for reprogramming, retroviral insertion alone can cause deleterious and cancer causing mutations. This suggest that reprogramming and tumorigenesis may entail a similar dedifferentiation process.

To reduce the potential oncogenic effect of genomic integration of viral vectors, different combinations or reduced numbers of factors have been introduced. Thus, it is now possible to reprogram both mouse and human somatic cells with the three- factor combination of OCT4, SOX2 and KLF4, without the potent oncogene c-MYC. Nevertheless, the three-factor reprogramming efficiency has been found to be very low. A major limitation of this iPS technology is the use of viruses that integrate into the genome and are associated with the risk of tumor formation due to the spontaneous reactivation of the viral transgenes. An alternative method is the use of adenoviruses, which do not invade the host genome, have been used to that allow for transient, high level expression of exogenous genes (Oct4, Sox2, Klf4, and c-Myc) without integrating into the host genome. These adenoviral iPS cells show DNA demethylation characteristics of reprogrammed cells, express endogenous pluripotency genes, form teratomas, and contribute to multiple tissues, including the germ line, in chimeric mice. Adenoviral reprogramming may be useful for generating and studying patient-specific stem cells and for comparing embryonic stem cells and iPS cells. Mouse iPS cells have also been derived from skin cells by multiple transfections with nonintegrating plasmids Repeated transfection of two expression plasmids, one containing the cDNAs of Oct3/4, Sox2, and Klif4 and the other containing the c-Myc cDNA, into mouse embryonic fibroblasts resulted in iPS cells without evidence of plasmid integration, which produced teratomas when transplanted into mice and contributed to adult chimeras.

In yet another attempt to produce the iPS cells, a new approach has been proposed to use adult stem cells instead of differentiated cells as the starting population. For example, mouse neural stem cells that express endogenous SOX2 could be reprogrammed to a pluripotent state with the addition of only two transcription factors. More recently, methods have been improvised to induce pluripotent stem cells from primary human fibroblasts with only OCT4 and SOX2. In this method, valproic acid, a histone deacetylase inhibitor, enabled reprogramming of primary human fibroblasts with only two factors, OCT4 and SOX2, without the need for the oncogenes c-Myc4 and Klf4. In the presence of valproic acid, OCT4 and SOX2 are sufficient for reprogramming human fibroblasts, providing further support for their central roles in the induction of pluripotency. The two-factor induced human iPS cells resemble human ES cells in pluripotency, global gene expression profiles and epigenetic states. These improvements in the methodologies with the use of small molecules not only enhance reprogramming efficiency in the differentiated human cells, but also make them safe and practical for therapeutic use.

In yet another method to boost the success rate of iPS formation, a recent approach used the silencing of the p53 pathway, which normally prevents mutations and preserves the sequence of the genome. After knocking out p53 pathway, iPS cells were produced with a success rate around a 100 fold, using 4-factor method or modified two or three factor method described above. Silencing of p53 has also been tried with siRNA which greatly improved the efficiency of human iPS generation. Though enhancing nuclear reprogramming, disabling the p53 invariably induced cancer in the iPS cells. With the p53 pathway intact, the cells that are reprogrammed are healthy and do not carry any DNA damage. Nevertheless, stifling of p53 to make DNA damaged iPS cells could provide useful cellular models to aid the understanding of many diseases.

FIRST PRODUCTION OF VIABLE MICE FROM iPS CELLS

The technique to produce iPS cells has become a popular means to reprogram somatic genomes into an embryonic-like pluripotent state, and a preferred alternative to somatic cell nuclear transfer and somatic-cell fusion with ES cells. In July, 2009, a team from China showed that the iPS cell lines produced from mouse fibroblast could go on to generate fertile live mouse pups. They generated 37 iPS lines that demonstrated Es-like characteristics and enhanced developmental potentials through retroviral infection with four Yamanaka factors. They injected these iPS cells into normal blastcysts to generate a chimera or inject them into a tetrapliod blastcyst. Tetrapliod complementation is a test for pluripotency and developmental potency, as any viable live-born animals resulting from the injection of diploid ES or iPS cells to create the tetrapliod (4N) embryos (blastocysts) will be predominantly from the diploid donor cells. The tetraploid host blastocyst primarily contributes to the extra embryonic lineages and not to the embryo proper. That the iPS cells can give rise to an entire viable adult through tetraploid chimera technology shows that in principle these cells can give rise to every cell type in the body.

REGENERATIVE THERAPY OF THE HEART

Recently, the possibility of repairing damaged heart tissue has been reported using stem cell therapy. The discovery of embryonic and postnatal cardiomyocytes, along with the finding that thymosin β 4 will influence the migration of embryonic endothelial and cardiac muscle cells, has paved the way for the regenerative therapy of heart muscles. Along with the self-renewing progenitor and stem cells from heart muscles, satellite cells from skeletal muscles, hematopoietic stem cells (HSCs), mesenchymal stem cells (MSCs), and side population (SP) cells from bone marrow, and endothelial progenitor cells (EPCs) from the blood stream have been used in such regeneration of cardiac muscles. During angiogenesis, the newly formed endothelial cells will arise from the proliferation of pre-existing endothelial cells. The recent finding of the transdifferentiation potential of NSCs into endothelial cells raises the hope that such stem cells could be used in the artificial vascularization of the surrounding tissues.

MUSCULAR DYSTROPHY

Stem cell therapy is valued for its potential to restore damaged or degenerating tissues. A new form of stem cell transplantation therapy to cure the muscular dystrophy has been reported recently in dog using canine mesoangioblasts. Duchenne muscular dystrophy is an untreatable crippling genetic disorder that severely limits motility and life expectancy in affected children. It primarily affects skeletal muscles, fibre degeneration,, progressive paralysis and death. It is caused by mutation in the gene that encodes the distrophin protein. Dystrophin is a protein that creates an elastic scaffold wich is able to absorb the stress during muscle contraction. Mesoangioblast is a stem cell which could be extracted from the aorta of the embryos as well as from small blood vessels of adult dogs. Unlike other stem cells, mesoangioblasts, when infused into the arteries, will cross the membrane of the blood vessels and distribute themselves evenly through the downstream muscles and rescue the damaged muscle cells with high efficiency. This property of stem cells passing through the blood vessel is not found in other cells such as satellite cells which would have to be injected into every two millimeter of muscle in the body, necessitating thousands of injections. The injected dystrophic dog showed dramatic improvement in structure and function of the transplanted muscles, increased dystrophin expression in muscle fibre and an unprecedented level of amelioration of spontaneous mobility of the dystrophic dog. In addition to the normal heterologous stem cells, autologous mesoangioblasts from the dystrophic dogs were genetically corrected by inserting a copy of the gene that encodes a micro-dystrophin has also been tried with limited success. Both autologous and normal mesoangioblasts have implications for their application in human patients. Particularly, the genetically corrected autologous mesoangioblasts would prove more useful for human treatment, because of the fact that the micro-dystrophin used to correct the mesoangioblasts was a human protein.

STEM CELL THERAPY OF THE HUMAN CENTRAL NERVOUS SYSTEM

Recent research has focused on the potential utility of neural stem and progenitor cells in the repair of human brain and spinal cord. In this regard, the pluripotent ES cells, neural-tissue derived stem cells, and phenotype-specified progenitor cells have all been investigated for their ability to generate neuron and glia. The therapeutic use of these stem cells has been investigated in the brain/ spinal cord disorders.

MULTIPLE SCLEROSIS

Multiple sclerosis (MS) is caused by the inflammation-induced destruction of myelin sheath that surrounds axons, leading to conduction deficits and a variety of neurological symptoms. Axonal loss as a consequence of chronic demyelination is an important cause of functional deterioration. Oligodendrocytes are the sole source of myelin in the adult CNS and their loss or dysfunction is the primary cause of diseases like MS in children and adults. The glial progenitor cells isolated from the adult and mid-gestation fetal brain have been used in generating myelin-producing oligodendrocytes phenotype to cure demyelinating and dysmyelinating diseases, such as multiple sclerosis.

PARKINSON DISEASE

Parkinson disease (PD) is characterized by progressive deterioration of dopaminergic neurons in the substantia nigra of the midbrain. The cell-based strategies aiming to replenish dopaminergic neurons involve implantation of fetal midbrain cells, which include those destined to become dopaminergic neurons, into the neostriatum. However, success is often limited because of the heterogeneity of cell types in the donor tissue. Hence, human ES cells serve as a potential source of dopaminergic neurons, as alternative to adult stem cells. In the use of human ES cells, fibroblast growth factor 8 (*FGF8*) and *Sonic hedgehog* were implicated as tandem initiators of dopaminergic neurogenesis *in vivo*. Even here, the ES derived dopaminergic preparations are not pure and may be contaminated with seratonergic and γ- aminobutyric acid (GABA) ergic neurons, as well as glia.

SPINAL CORD INJURY AND DISEASE

Axonal regeneration in the injured spinal cord using stem cell replacement therapy has gained immense importance in recent years, because of the dearth of available treatment options for spinal cord injury and trauma. Spinal cord injury is associated with the loss of both neurons and glia. Hence, the cell- based strategies for reconstituting the injured spinal cord must accommodate the need to replace multiple cell types. Neural stem cell implants into the injured spinal cord have been used under the premise that regionally appropriate phenotypes may be generated from undifferentiated cells in response to local signals competent to induce cell-type specification. Besides the loss of neurons and astrocytes in spinal cord injuries and infarcts, demyelination is also caused in the cord injuries. Hence, transplantation of myelinogenic glial progenitors has also been used in treating spinal cord injuries.

CELL BASED THERAPY OF MOTOR NEURON DISEASES

The major degenerative diseases of the spinal cord, including amyotrophic lateral sclerosis (ALS) and the spinal muscular atrophies of children, as well as the major viral infections of the spinal cord such as poliomyelitis are characterized by a selective loss of spinal motor neurons. Motor neurons have proved amenable to generation from human ES cells. Hence ES cell implants to treat experimental motor neuron disease have yielded some functional benefits. Alternatively, telomerase-immortalized cells of human fetal spinal neuro-epithelium with retroviruses overexpressing human telomerase reverse transcriptase have also been used in this transplantation therapy.

STEM CELLS AS CHAPERONS

Although cell replacement is an exciting future prospect, more work is needed before stem cells can be used reliably to replace damaged cells and tissues. Apart from replacing lost cells, stem cells have more subtle roles that could be exploited therapeutically. For example, in the nervous system, stem cells can act as 'chaperons' that nurse sick and injured neurons back to health. Neural stem cells secrete biochemicals that make the neurons function better, promote survival, decrease inflammation and encourage the growth of blood vessels. One of these factors is glial cell line- derived neurotrophic factor (GDNF). This growth factor protects both the cells that secrete the neurotransmitter dopamine, which is lost in Parkinson's disease, and the motor neurons that are destroyed in the paralyzing illness, amyotrophic lateral sclerosis (ALS). Thus, stem cells can be used as nursemaids to protect sick and dying tissues. Neural stem cells isolated from mouse fetuses secrete GDNF and promote recovery in mouse models of Parkinson's disease. Interestingly, engineered fetal neural stem cells, that pump out greater quantities of GDNF, when injected into the spinal cords of rats suffering from an ALS-like disease, survived well and continued to secrete GDNF and protected the motor neurons. Thus, genetically engineered stem cells have better prospects in the cure of neurodegenerative diseases like Parkinson's and ALS. Understandably, stem cell injections of the above therapeutic transplants are not intended to replace damaged nerve tissues. Instead, they are used to promote healing through the production of proteins called growth factors that stimulate cells to grow and divide, thus repairing the damaged nerves.

In yet another startling discovery, human neural stem cells derived from the brains of aborted fetuses, can migrate from one side of a mouse's brain to the other in response to distress signals issued by injured tissue. In these two instances, there is no replacement of the lost cells, but the stem cells protect the cells in the injured or diseased region.

In another instance, geneticists have successfully isolated human chromosome 21, and introduced it into the mouse embryonic stem cell. Any stem cells that absorbed human chromosome 21 were injected into three-day-old mouse embryos, which were then reimplanted into their mothers. The newly born mice carried copies of the chromosome and were able to pass it on to their own young. These mice showed nearly all of the characteristics of Down's syndrome in humans. Using knock out techniques for different genes on the transplanted chromosome, it will be possible to identify which gene or genes cause each of the symptoms common to people with Down's syndrome.

CONSTRAINTS IN STEM CELL THERAPY

One possible danger in using stem cells is that if the stem cells are not kept under proper control, they could give rise to cancer cells. That the stem cells are involved in the cancer development is evidenced by the fact that they are the only long-lived cells in a tissue that have the ability to replicate. Evidence indicates that many signal pathways that are classically associated with cancer may also regulate stem cell development. For example, several signalling pathways associated with oncogenesis such as the Notch, *Sonic hedgehog* (*Shh*) and *Wnt* also regulate stem cell self-renewal. Of these signal molecules, *Wnt* signalling molecules has a particular role in regulation of self-renewal. *Wnt* proteins are intercellular signalling molecules that regulate development in several organisms and contribute to cancer when dysregulated. The expression of *Wnt* proteins in the bone marrow suggests that they may influence HSCs as well. Over-expression of activated β-catenin (a downstream activator of the *Wnt* signalling pathway) in long-term cultures of HSCs expands the pool of transplantable HSCs. Conversely, ectopic expression of Axin, an inhibitor of *Wnt* signalling leads to inhibition of HSC proliferation and increased death of HSCs *in vitro*.

It is also possible that cancer-inducing mutations could accumulate in the long-lived normal stem cells. Recent research has also provided evidence that human embryonic stem cells grown in culture accumulate changes in their genetic material over time that might limit the usefulness of ageing stem cell lines. Especially, chromosomal rearrangements are identified in the long-stored stem cells.

Additionally, recent studies have revealed that the stem cells could be used in novel therapy. American cancer biologists have isolated the bone marrow mesenchymal stem cells and inserted a gene for interferon- β, a protein that can kill tumour cells. When these cells are injected into the carotid arteries of mice suffering from brain cancer, the cells migrated to the tumour and killed them. These animals lived significantly longer than those who received injections of normal cells. Nevertheless, the problem with embryonic stem cells is that they can form tumours called teratomas that contain all sorts of tissue types.

Another problem associated with stem cells in the treatment of human diseases is the possible histo-incompatibility with the existing tissues. Because human populations are genetically diverse, most types of transplantation have to overcome the problem of tissue rejection. Immune suppression and tolerance induction are two possible solutions. Since stem cells are amenable to genetic manipulation, the above problems of tissue rejections could be solved. The pluripotent embryo-derived stem cells can be made compatible with any individual by the so-called somatic cell nuclear transfer technique. This technique entails the isolation of the somatic cell nucleus from the patient and reprograms it in an oocyte cytoplasm (therapeutic cloning). The embryo is then allowed to develop and stem cells are derived from it. These stem cells are genetically identical to the patient. Alternatively, the existing stem cell lines are genetically modified by homologous recombination to create a stem cell that is compatible with the patient.

To overcome the histo-incompatibility of stem cell during tissue transplantation, recently stem cells have been harvested from the umbilical cord blood of the mother and then used in bone-marrow transplants for the same person. More recently, Canadian researchers have discovered that the connective tissue surrounding the blood vessels in the umbilical cord is rich in progenitor cells that generate bone, cartilage and other tissues and could be used in bone-marrow transplantations. The frequency of these progenitor cells in the human umbilical cord perivascular cells is one in 300, as compared to cord blood in which the frequency is only one in 200 million.

ETHICAL CONSIDERATIONS OF STEM CELL RESEARCH

Although cloning of human embryonic stem cells offer great hope in the alleviation of a wide variety of illnesses, the use of human embryos as a source of stem cells has proved morally and ethically contentious, and led to the bans on human embryo research in several countries, including United States, where it is even forbidden to use Government funds to make human embryonic stem cells. The reason given for this is that in order to isolate the stem cells from early human embryo, the embryo has to be broken up and effectively destroyed. The notion is that the fertilized egg is already a human being, and that the destruction of early embryo amounts to killing a person. Religious faiths differ in their contention on when life begins in the embryo. It could be immediately after the egg is fertilized, when the fetus gets a soul, when the fetus can live independently outside the mother, or when the mother delivers the baby. However, in the assisted reproduction, involving *in vitro* fertilization, more than one egg is produced by way of super ovulation. In such cases, there will be wastage of many unutilized embryos, which will be naturally destroyed. And hence there is no justification in banning the use of such embryos in generating human embryonic stem cells.

Furthermore, embryos derived from the transfer of a somatic cell nucleus (SCNT) into a donated oocyte from which the nuclear material has been removed should not raise ethical objections. The potential benefit of SCNT stem cells is that they could be made from the patient's own somatic cell nuclei, custom-built for immunological compatibility, and so transplant rejection should not occur. As this research also requires a supply of unfertilized human eggs, ethical issues could crop up. On the contrary, stem cells derived from adults (living or dead) should not raise any new ethical issues. In fact, bone-marrow transfusion and organ transplantation are widely practiced and regulated by guidelines. Yet again, the stem cells in the cord blood of babies are also stored for future use.

Recent discovery that mammalian eggs, including those of humans, can activate on their own without sperm (parthenogenesis) has thrown open certain possibility of obtaining human embryos for stem cell derivation without moral and ethical controversy. These mammalian parthenotes, like the fertilized eggs, undergo first few days of development, including the formation of pluripotent stem cells. But these parthenotes are not viable and do not have the potentiality to develop into offspring. Thus, there is a theoretical possibility for setting up parthenote stem-cell banks, in the place of the ethically controversial embryonic stem–cell banks.

25

THERAPEUTIC USE OF MESENCHYMAL STEM CELLS

INTRODUCTION

Discoveries in the field of stem cells have led to an avalanche of new knowledge and research tools that may soon lead to novel therapies for cancer, heart disease, Parkinson's disease, diabetes, lung diseases, and a wide variety of other diseases that afflict humanity. Embryonic stem (ES) cells can readily be shown to differentiate into essentially all cell phenotypes, whereas adult stem cells such as hematopoietic stem cells and mesenchymal stem cells have a more limited potential for differentiation (see chapter on stem cells). We are at a remarkable stage in research with both ES and adult stem cells, considering the unique possibility that these cells could be used to treat various human diseases. Use of ES cells is hindered by the tumorigenicity of the cells and the danger of immune responses, if they are used heterogeneously. On the contrary, adult stem cells have been shown to be promising, and several kinds of adult stem cells can be obtained in adequate amounts for autologous therapy. However, it is unlikely that one kind of stem cell will be ideal for all practical applications envisioned. ES cells may therefore prove to be ideal for creating new organs through new protocols that will circumvent the current ethical and technical minefields. Adult stem cells may be more useful for repairing damage to tissues by trauma, disease, or perhaps uncomplicated ageing. However, recent observations suggest that the adult stem cells are part of natural system for tissue repair. The initial response to tissue injury appears to be mediated through proliferation and differentiation of stem-like cells, endogenous to the tissue. After the endogenous stem-like cells are exhausted, non-hematopoietic stem cells such as mesenchymal stem cells from the bone marrow are recruited to the site of injury. The adult stem cells that home to injured tissues repair the damage by two or three mechanisms: by differentiating into appropriate cell type, by providing cytokines and other factors to enhance recovery of endogenous cells, and perhaps by cell fusion, a process that may provide a rapid mechanism for differentiation of stem cells.

WHAT ARE MESENCHYMAL STEM CELLS?

Two stem cell populations with distinct progenies are housed within adult bone marrow (BM), hematopoietic stem cells and mesenchymal stem cells (MSCs). MSCs are pluripotent adult stem cells, found in the bone marrow, have the capability of differentiating into multiple cell types of mesenchymal origin such as osteocytes, chondriocytes, fibroblasts, myoblasts, adipocytes and dermal epithelial cells. MSCs have been referred to in the literature by other names such as colony-forming fibroblastic cells, BM stromal stem cells, mesenchymal progenitor cells and BM stromal cells. MSCs are considered to be a potential source for cell and gene therapy.

LOCATION OF MSCS

MSCs must be found anywhere in which mesenchymal tissues turn over occurs. To date, mesenchymal progenitors have been isolated from marrow, muscle, fat, skin, cartilage, and bone. Moreover, every blood vessel in the body has MSCs on the tissue side of the vessel, although names like pericytes have been given to these multipotent cells.

USE OF MSCS IN MEDICINE AND TISSUE ENGINEERING

MSCs have been used to repair bone, cartilage, regeneration of bone marrow, muscle regeneration, tendon repair, a vehicle for gene therapy, lung repair, and as vascular support. MSCs can be generated in culture, stored frozen, or used directly for regenerative tissue repair. Culture-expanded marrow MSCs in a porous, calcium phosphate, ceramide- delivery vehicle are capable of regenerating structurally sound bone, whereas whole marrow or the vehicle alone can not satisfactorily accomplish this repair. Injected MSCs make it to the bone marrow to refabricate injured marrow stroma. MSCs could be injected into a specific muscle of the muscular dystrophy (mdx) mouse to cure it by providing newly synthesized dystrophin to the affected myotubules. Moreover, labeled MSCs injected into injured (infarct model) rat or pig heart appear to differentiate into cardiac myocytes. The studies with the mdx mouse cited previously establish the capacity of allo-MSCs to cure genetic defects. MSCs could be transfected with a lentiviral or retroviral construct without affecting their differentiation capacity. Molecules coded for by these gene inserts could be shown to be present many months after introduction into the *in vivo* model. Thus, the MSCs hold the potential for a variety of gene therapy applications. Autologous MSCs have been used to repair tendons. Although there are reports that MSCs can differentiate into neural cell types, the use of MSCs in neural environments as growth factor pumps that facilitate axonal reconnection and neuronal stem cell growth, migration, and differentiation may be clinically relevant.

USE OF MSCS IN LUNG DISEASES

Many lung diseases including chronic obstructive pulmonary disease (COPD) involve apparent bronchiolar and alveolar cell defects. It is likely that the delicate balance of stem, progenitor, and differentiated cell functions in the lung is critically affected in patients with COPD. Pulmonary emphysema is one of the major pathological abnormalities associated with COPD, which afflicts more than 650 million people in the world, and is expected to become the third cause of death worldwide in 2020. Lung parenchymal inflammation, protease/antiprotease imbalance, and oxidative stress are thought to be important processes in the development of emphysema. The damage to the lung in pulmonary emphysema is not reversible. Available treatments for moderate disease, such as inhaled bronchodilators, provide only limited, symptomatic relief. Therapies for advanced disease, such as long-term oxygen treatment and lung volume reduction surgery, yield modest extensions in life expectancy for selected group of patients, but none of these approaches can slow the progressive deterioration of the lung function that results in over 120,000 deaths in U.S each year. Hence, there is a strong need for new approaches that are capable of halting disease progression, and interventions that repair/regenerate alveolar structures.

Mesenchymal stem cells are prime candidates for the repair and regeneration of damaged tissues and organs. Bone marrow MSCs transplantation has been reported to ameliorate bleomycin-induced lung fibrosis, endotoxin mediated acute lung injury and ceccal ligation and puncture-mediated septic shock. Recent studies have demonstrated that bone marrow- derived MSCs can engraft in the injured lung and even differentiate into type I and type II alveolar epithelial cells, and bronchial epithelial cells. These studies raise the possibility that bone marrow MSC transplantation may be developed as an effective intervention in pulmonary emphysema.

HOW ARE MSCS ISOLATED AND CHARACTERIZED?

The basic characteristic of adult stem cell is that a single cell (clonal) would self-renew and generate differentiated cells. The most rigorous assessment of these characteristics is to prospectively purify a population of cells (usually by characterizing cell surface markers with flow cytometer), transplant a single cell into an acceptable host without any intervening *in vitro* culture, and observe self-renewal and tissue, organ, or lineage reconstitution. MSCs have been isolated and characterized from several species and tissues, but the most well characterized and probably the purest preparation is from human bone marrow. A number of techniques have been developed to fulfill this purpose, such as exposure of cultures to cytotoxic materials, cell sorting, low- and high-density culture and negative and positive selection, which permits the isolation of mouse mesenchymal stem cells (mMSCs) from mouse BM. The protocol for the isolation and culture of MSCs from mouse bone marrow is described below.

PROTOCOL FOR ISOLATION AND CULTURE OF MESENCHYMAL STEM CELLS FROM MOUSE BONE MARROW

MATERIALS

REAGENTS

DMEM medium with 2 mM L-glutamine and without ribonucleosides and ribonucleotides; FBS; 0.25% trypsin/1 mM ethylenediaminetetraacetic acid; Hank's balanced salt solution; Streptomycin; L-glutamine; Penicillin; $NaHCO_3$

EQUIPMENT

Hood for cell culture with vertical laminar flow and equipped with UV light for; decontamination; Water bath with temperature control; Centrifuge (no temperature control is needed);Incubator with both temperature and gas composition controls; Optical microscope with phase-contrast equipment; A procedure for material sterilization (e.g., Poupinel oven); 10-ml syringes with 27-gauge needles; Surgical forceps (straight and curved); Surgical scissors; Pipette; Filter mesh; 6-well culture dishes; 25-cm^2 flask.

REAGENT SETUP

HARVEST BUFFER

Mix Hank's balanced salt solution and 100 mg ml^{-1} streptomycin. Harvest buffer can be stored for 2 weeks at 4 °C.

COMPLETE DMEM MEDIUM

Adjust the volume to 100 ml by adding DMEM 'stock' to a solution containing 2 mm of L-glutamine, 100 μg ml^{-1} penicillin and 100 μg ml^{-1} streptomycin, 3.7 g liter^{-1} NaHCO3 and 15% of FBS. 'Complete DMEM medium' can be stored for 1 week at 4 °C.

POCEDURE

1. To isolate marrow, kill mice (Balb/c, 6–8 weeks old) by cervical dislocation. Then, rinse the animal skeleton freely in 70% ethanol, make an incision around the perimeter of the hind limbs where they attach to the trunk and remove the skin by pulling toward the foot, which is cut at the anklebone. Then, dissect the hind limbs from the trunk of the body by cutting along the spinal cord with care not to damage the femur. Store limbs on ice in DMEM supplemented with 1 ×penicillin/streptomycin while awaiting further dissection.

2. Perform further dissection of the hind limbs under the hood. Bisect each hind limb by cutting through the knee joint. Remove the muscle and connective tissue from the tibia and the femur by scraping the diaphysis of the bone clean, then pulling the tissue toward the ends of the bone. After cleaning, store the bones in DMEM supplemented with 1 ×penicillin/streptomycin on ice.

3. Harvest the BM in a hood using proper sterile technique. Cut the ends of the tibia and femur just below the end of the marrow cavity using a pair of sharp rongeur. Insert a 27-gauge needle attached to a 10-ml syringe containing complete media into the spongy bone exposed by removal of the growth plate. Flush the marrow plug out of the cut end of the bone with 1 ml of complete media and collect in a 10-ml tube on ice.

4. Filter the cell suspension through a 70-mm filter mesh to remove any bone spicules or muscle and cell clumps. Determine the yield and viability of cells by Trypan blue exclusion and counting on a hemocytometer.

5. Culture BM cells in 95-mm culture dishes in 1 ml of complete medium at a density of 25 ×10^6 cells ml^{-1} (on the basis of the cell count obtained in Step 4). Incubate the plates at 37 °C with 5% CO$_2$ in a humidified chamber without disturbing them. After 3 h, remove the non-adherent cells that accumulate on the surface of the dish by changing the medium and replacing with fresh complete medium.

6. After an additional 8 h of culture, replace the medium with 1.5 ml of fresh complete medium. Thereafter, repeat this step every 8 h for up to 72 h of initial culture.

7. Wash the adherent cells (passage 0) with phosphate-buffered saline, and add fresh medium every 3–4 d. The initial adherent spindle-shaped cells appear as individual cells on the third day in phase-contrast microscopy. Within 4–8 d, the culture becomes more confluent and reaches 65–70% confluence within 2 weeks. At this stage, the cultures typically exhibit two characteristics: first, plates may contain distinct colonies of fibroblastic cells that vary in size;

and second may contain very small numbers of hematopoietic cells interspersed between or on the colonies.

8. After 2 weeks of initiating culture, wash the cells with phosphate-buffered saline and lift cells by incubation in 0.5 ml of 0.25% trypsin/1 mM ethylenediaminetetraacetic acid for 2 min at room temperature. Neutralize the trypsin by adding 1.5 ml of complete medium, and culture all lifted cells in a 25-cm^2 flask. Discard the nonlifted cells.

9. Change the culture medium every 3 d (replacing with 6 ml of medium each time). Typically, cell confluence is achieved in 7 d

ANTICIPATED RESULTS

In the third week of culture, essentially all cells have spindle-shape morphology (Fig 25.1). These cells can be readily differentiated into osteocyte, adipocyte and chondrocyte cells by culturing in appropriate induction media. These mMSCs were negative for the endothelial markers [FLK1 (VEGF-R2), CD31 (PECAM), CD90 (Thy1) and CD117-APC (c-kit)] and hematopoietic markers [CD11b, Ter-119, CD45, CD45R/B220, Ly6C/Ly6D and CD3e]. They expressed higher levels of CD34 and Sca-1 antigens and moderate level of CD106 (VCAM-1). mMSCs isolated using this protocol are not contaminated with hematopoietic cell lineages, readily differentiated into mesenchymal lineages (more than 95%) and more than 70% of cells have colony-formation capacity.

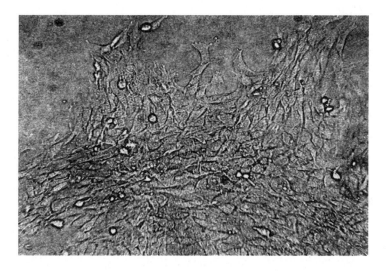

Fig 25.1: Culture of bone marrow derived mesenchymal stem cells in 10 CM culture dishes. Mouse MSC isolated from the bone marrows of C57 black 6 mice were cultured in DMEM complete medium at 5% CO2 for 14 days at 37° C using a cell culture incubator. The medium was replaced with the fresh medium for every 3 days. The photograph was taken using a Carl Zeiss light microscope (Germany) with 10 X lens. (Photograph courtesy: R.Thirumalai, University of Rochester, medical School, USA).

26

CANCER BIOLOGY

INTRODUCTION

Cancer is a complex family of diseases of higher multicellular organisms, particularly mammals. In humans, it has been treated as the most dreadful disease, and is a leading cause of death all over the world. Although cancer comprises at least 100 different diseases with multiple genotypic and phenotypic variations, all cancer cells have one common and very challenging characteristic: dysregulated cell cycle resulting in uncontrolled proliferation.

ORIGIN OF CANCER

It was Boveri in 1941 who proposed the somatic mutation hypothesis to explain the origin of cancer. He suggested that the origin of cancer cells was due to an inappropriate recombination of chromosome(s) in a somatic cell and that it caused abnormal cell proliferation. This chromosomal defect was transmitted to all cellular descendents of the original cancer cell. It is now well established that some human cancers have a familial distribution and that certain chromosomal rearrangements are associated with human malignant neoplasia.

On the other hand, the discovery that cancer cells arise due to alterations in the energy metabolism, especially the glycolysis was ground breaking. This discovery was proposed by the German scientist and Nobel laurate Otto Warburg who demonstrated in the 1920s that the most striking biochemical phenotype of cancers is their aberrant energy metabolism. This was the first biochemical evidence that showed "up regulated glycolysis or anaerobic respiration" as the cancer specific phenotype. He showed that cancer cells preferentially use glycolysis (anaerobic) as the principal pathway for the oxidation of glucose rather than oxidative phosphorylation (mitochondrial respiration). Warburg's seminal discovery was so overwhelming and compelling; his discovery was named as "Warburg's hypothesis". Later, it has been characterized that such a dramatic switch to glycolysis relies on the increased expression of the genes (corresponding to the enzymes) associated with glycolytic pathway. Thus, the pivotal role of genetic alterations (in terms of over expression) in cancer development is indispensable.

CHARACTERISTICS OF CANCER CELLS

Cancer results from a series of molecular events that fundamentally alter the properties of normal cells. In cancer cells the normal control systems that prevent cell overgrowth and the invasion of other tissues are disabled. These altered cells divide and grow in the presence of signals that normally inhibit cell growth; therefore, they no longer require specific signals to induce cell growth and division. Normal cells require mitogenic growth signals before they can move from a quiescent state into an active proliferative state. On the other hand, cancer cells show a greatly reduced dependence on exogenous growth stimulation by generating self-sufficiencey in growth signals. Molecular strategies for achieving this growth signal autonomy also involve alteration of extracellular growth signals, their transcellular transducers as well as the intracellular circuits that translate those signals into action. Other alterations in cell physiology that collectively dictate malignant growth include (i) insensitivity to growth-inhibitory (antigrowth) signals, (ii) evasion of programmed cell death (apoptosis), (iii) limitless replicative potential, (iv) sustained angiogenesis, and (v) tissue invasion and metastasis. Each of the above physiological changes acquired during tumor development represents the successful breaching of an anticancer defense mechanism hardwired into cells and tissues.

As these cells grow they develop new characteristics, including changes in cell structure, decreased cell adhesion, and production of new enzymes. These heritable changes allow the cell and its progeny to divide and grow, even in the presence of normal cells that typically inhibit the growth of nearby cells. Such changes allow the cancer cells to spread and invade other tissues. These changes are often the result of inherited mutations or are induced by environmental factors such as UV light, X-rays, chemical carcinogens in food products, tobacco, viruses etc. All evidence suggests that most cancers are not the result of one single event or factor. Rather, around four to seven events are usually required for a normal cell to evolve through a series of premalignant stages into an invasive cancer. Often many years elapse between the initial event and the development of cancer (latency period). The development of molecular biological techniques may help in the diagnosis of potential cancers in the early stages, long before tumors are visible.

The abnormalities in cancer cells usually result from mutations in protein-encoding genes that regulate cell division. Over time more genes become mutated. Often, the genes responsible for producing the proteins that normally repair DNA damage are themselves not functioning normally because they are also mutated. Consequently, mutations begin to increase in the cell, causing further abnormalities in that cell and the daughter cells. Some of these mutated cells die, but other alterations may give the abnormal cell a selective advantage that allows it to multiply much more rapidly than the normal cells. This enhanced growth describes most cancer cells, which have gained functions repressed in the normal, healthy cells. As long as these cells remain in their original location, they are considered benign; if they become invasive, they are considered malignant.

A malignant cancer, as different from a benign cancer, has the ability to invade locally, to spread to regional lymph nodes, and to metastasize to distant organs in the body. Malignant tumors invade and destroy adjacent normal tissues. On the other hand, benign tumors grow by expansion, are usually encapsulated, and do not invade surrounding tissues. Benign tumors may however secrete biologically active substances, such as hormones, that alter normal homeostatic mechanisms. Malignant tumors metastasize through lymphatic channels or blood vessels to lymph nodes and other parts of the body. Malignant tumors usually grow more rapidly with the involvement of surrounding tissues, whereas the benign tumors often grow slowly over several years.

GENOMIC ALTERATIONS AS THE BASIS OF TUMORIGENESIS

Cancer in general and carcinomas in particular, exhibit extensive modifications in genome composition. Alterations in gene expression are commonly witnessed in almost all types of cancers. Several lines of evidence substantiate the role of altered gene expression in uncontrolled proliferation and failure to respond to inhibitory signals. Hence, cancer is defined as "an abnormal growth of cells caused by multiple changes in gene expression leading to dysregulated balance of cell proliferation and cell death and ultimately evolving into a population of cells that can invade tissues and metastasize to distant sites, causing significant morbidity and, if untreated, death of the host" (Raymond Ruddin in "Cancer Biology"). In essence, cancer is a disease involving dynamic changes in the genome.

The genetic alterations range from subtle point mutations to dramatic gains and losses of genetic material. Though the role of mutations in tumorigenesis has been recognized ever since Hermann Muller proposed that multiple intragenic mutations can cause cancer, the contribution of aneuploidy in the pathogenesis of carcinoma has become more prominent in recent times. Gatekeeper genes that control cell growth and death and caretaker genes that maintain genome are the most common genes implicated in cancer. Members of either class can act as oncogenes when activated by gain of function mutations (e.g., ras, Flt-3, c-kit) or tumor suppressor genes when inactivated by loss of functions (e.g., p53, Rb, APC).

CANCER CELL TRANSFORMATION

Cancer cells behave as independent cells, growing without control to form tumors. Tumors grow in a series of steps. The first step is hyperplasia, meaning that there are too many cells resulting from uncontrolled cell division. These cells appear normal, but changes have occurred that result in some loss of control of growth. The second step is dysplasia, resulting from further growth, accompanied by abnormal changes to the cells. The third step requires additional changes, which result in cells that are even more abnormal and can now spread over a wider area of tissue. These cells begin to lose their original function; such cells are called anaplastic. At this stage, because the tumor is still contained within its original location (called *in situ*) and is not invasive, it is not considered malignant - it is potentially malignant. The last step occurs when the cells in the tumor metastasize, which means that they can invade surrounding tissue, including the bloodstream, and spread to other locations. This is the most serious type of tumor, but not all tumors progress to this point. Non-invasive tumors are said to be benign.

CHROMOSOMAL ABERRATION IN CANCER CELLS

Progression to increasingly malignant cellular phenotypes and to increasing drug resistance is the major characteristics of neoplastic cells. Underlying these characteristics is the property of genomic instability, a cellular hallmark of cancer. Genomic instability is manifested as a constellation of chromosomal aberrations including aneuploidy (gain and loss of genetic material), translocations, point mutations, deletions, duplications, and amplification of localized discrete chromosomal segments (see Fig. 26.1 for details).

These changes in the normal chromosome often act in combinatorial ways to bring about carcinogenesis. Certain rearrangements, deletions, and trisomies (3 copies of one chromosome) are characteristically associated with particular forms of cancer and are referred to as nonrandom chromosomal changes. In advanced neoplasias, both random and nonrandom rearrangements are found. In humans, such somatic mosaicism occurs in many cancer cells, including trisomy 12 in chronic lymphocytic leukemia and trisomy 8 in acute myeloid leukemia. Genetic syndromes in which an individual is predisposed to breakage of chromosomes (chromosome instability syndromes) are frequently associated with increased risk for various types of cancers, thus highlighting the role of somatic aneuploidy in carcinogenesis. Partial mono or trisomy describes an imbalance of genetic material caused by loss or gain of part of a chromosome.

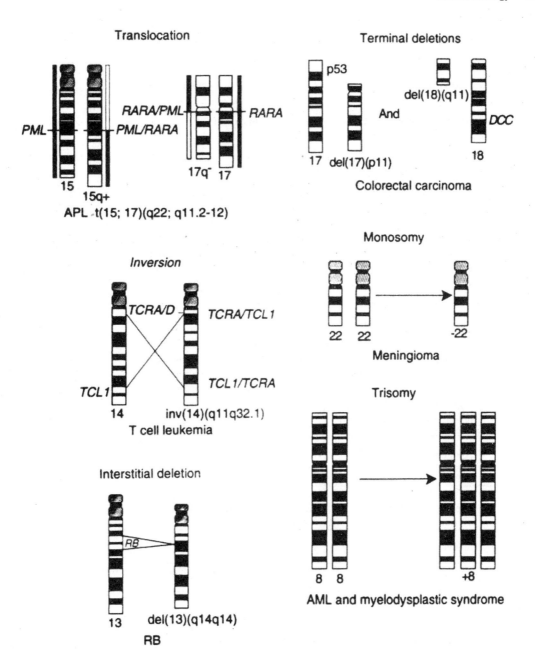

Fig. 26.1. Schematic representation of chromosomal aberrations observed in tumors (Redrawn from Ruddin, 2007)

NAME	FUNCTION	EXAMPLES of Cancer / Diseases	TYPE of Cancer Gene
APC	Regulation of transcription	Familial Adenomatous Polyposis	Tumor suppressor
BCL2	Apoptosis; pro-angiogenesis	Leukemia; lymphoma	Oncogene
BLM	DNA repair	Blood syndrome	DNA repair
BRCA1	May be involved in cell cycle control	Breast, Ovarian, Prostatic, & Colonic Neoplasms	Tumor suppressor
BRCA2	DNA repair	Breast & Pancreatic Neoplasms; Leukemia	Tumor suppressor
HER2	Tyrosine kinase; growth factor receptor-phosphorylation	Breast, Ovarian Neoplasms	Oncogene
MYC	Regulation of transcription	Burkitt's Lymphoma	Oncogene
P16	Cyclin-dependant kinase inhibitor – cell cycle regulator	Leukemia; Melanoma; Multiple Myeloma; Pancreatic Neoplasms	Tumor suppressor
P21	Cyclin-dependant kinase inhibitor– cell cycle regulator		Tumor suppressor
P53	Apoptosis; regulation of transcription	Colorectal Neoplasms; Li-Fraumeni Syndrome	Tumor suppressor
RAS	GTP-binding protein; important in signal transduction cascade	Pancreatic, Colorectal, Bladder Breast, Kidney, & Lung Neoplasms; Leukemia; Melonama	Oncogene
RB	regulation of cell cycle	Retinoblastoma	Tumor suppressor
SIS	Growth factor	Dermatofibrosarcome; Meningioma; Skin Neoplasms	Oncogene
XP	DNA repair	Xeroderma pigmentosum	DNA repair

Table 26.1. Important genes associated with cancer (Courtesy: http://www.learner.org)

As tumors develop and metastasize, continuing genomic changes provide the basis for tumor cell heterogeneity, drug resistance and the natural selection of even more malignant cell types. Thus, tumor progression is a highly accelerated evolutionary process, which occurs not over billions of years but within a single human lifetime. Localized amplification of chromosomal segments is a kind of genomic rearrangement seen in cells of advanced tumors undergoing malignant progression. The importance of amplification in tumor progression is underlined by the discovery that some DMs (double-minutes) consist of amplified copies of cellular oncogenes. DMs of unknown sequence and function have been identified in a large number of human and rodent tumors. Studies from murine carcinoma models suggest that expression of the cancer phenotype in carcinomas invariably involves gene dosage changes induced by chromosome instability described above as well as intragenic mutations. The genomic plasticity brought about by aneuploidy facilitates protumorigenic gene dosage changes and accelerate accumulation of oncogenes and loss of tumor suppressor genes. These conditions are furthermore favorable for loss of normal apoptotic mechanisms, and resistance to therapy.

TUMOR CATEGORIES

The type of tumor that forms depends on the type of cell that was initially altered. There are five types of tumors.
- Carcinomas result from altered epithelial cells, which cover the surface of our skin and internal organs. Most cancers are carcinomas.
- Sarcomas result from changes in muscle, bone, fat, or connective tissue.
- Leukemia results from malignant white blood cells.
- Lymphoma is a cancer of the lymphatic system cells that derive from bone marrow.
- Myelomas are cancers of specialized white blood cells that make antibodies.

TUMOR PROMOTION AND TUMOR INITIATION

Genotoxic carcinogens can induce damage in tumor suppressor genes or oncogenes, all of which contribute to transformation of normal cells to tumor cells. This is called tumor initiation stage in carcinogenesis. Some chemicals are also capable of promoting the outgrowth of those transformed cell clones into visible tumor cell masses. This stage of carcinogenesis is called tumor promotion stage.

TUMOR PROMOTION

Chemical compound like 2, 3, 7, 8-tetrachlorodibenzo-p-dioxin (TCDD) or benzo [a] pyrene (BP) result in tumor promotion through arylhydrocarbon receptor (AhR) - mediated signal transduction (fig. 26.2a). Binding of TCDD or BP to AhR leads to activation and translocation of the complex into the nucleus. This in turn heterodimerizes with AhR nuclear translocator (ARNT) and binds to xenobiotic response elements (XREs) and induces the expression of a variety of different genes involved in carcinogen metabolism, including CYP forms 1A1, 1B1, and 1A2 (see fig.26.2a). It also changes the expression pattern of several factors involved in cellular growth and differentiation, such as listed in the diagram. More importantly, pro-apoptosis factors such as tumor necrosis factor and heat shock protein 40 are down-regulated, and cell cycle genes can either be up-regulated (such as cyclin B2) or down-regulated (NEK2).

TUMOR INITIATION

Tumor initiation occurs through DNA adduct-derived mutations in cancer susceptibility genes. (DNA adduct is a piece of DNA covalently bonded to a cancer causing chemical (carcinogen) eg., acetaldehyde, a major component of cigarette smoke). DNA binding by genotoxic carcinogens such as activated BP leads to the induction of base pair or frameshift mutations in cancer susceptibility genes such as TP53 or RAS. The mutagenic activity of such DNA adducts involves DNA repair mechanisms. The mutagenic potency of such polycyclic aromatic hydrocarbon diol-epoxide-DNA adducts can be increased because of inhibition of nucleotide excision repair (NER) by metal ions (Me+; for example, Ni2+) or as a result of NER factor immobilization at repair-resistance DNA adduct sites, also known as decoy adducts (see fig.26.2b).

Noteworthy, only a small number of the approximately 35,000 genes in the human genome have been associated with cancer. Alterations in the same gene often are associated with different forms of cancer. These defective genes can be broadly divided into three categories. The first group, called proto-oncogenes, produces proteins that normally accelerate cell division or regulate normal cell death. The mutated forms of these genes are called oncogenes. The second group, called tumor suppressors that produce proteins that normally prevent cell division or promote cell death under undesirable situations. The third group contains DNA repair genes, which help prevent or correct gene mutations that lead to cancer. Some of the genes associated with cancer are listed in Tab. 26.1.

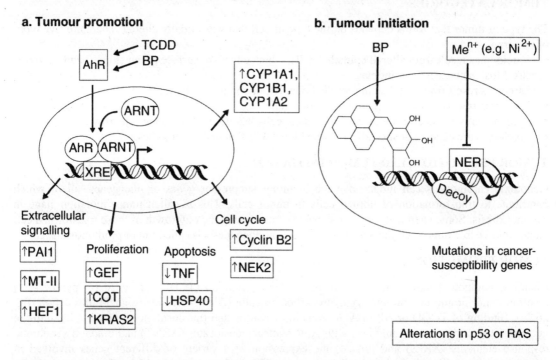

Fig. 26.2. Showing two stages in the transformation to cancer cells during carcinogenesis (Redrawn from Ruddin, 2007).

GENES INVOLVED IN CANCER DEVELOPMENT

Proto-oncogenes and tumor suppressor genes work much like the accelerator and brakes of a car, respectively. The normal speed of a car can be maintained by controlled use of both the accelerator and the brake. Similarly, controlled cell growth is maintained by regulation of proto-oncogenes, which accelerate growth, and tumor suppressor genes, which slow cell growth. Mutations that produce oncogenes accelerate growth while those that affect tumor suppressors prevent the normal inhibition of growth. In either case, uncontrolled cell growth occurs. Alterations in a proto-oncogene, resulting in an oncogene, can be inherited or caused by an environmental exposure to carcinogens.

SIGNAL TRANSDUCTION AND ONCOGENE ACTIVATION

Transcriptional factors are extremely powerful regulators of normal cells and cancer cells. In normal cells, proto-oncogenes code for the proteins that send a signal to the nucleus to stimulate cell division. These signaling proteins act in a series of steps called signal transduction cascade or pathway (Fig. 26.3).

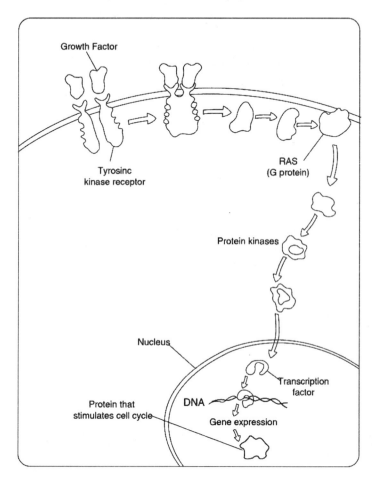

Fig.26.3. Signal transduction pathway in normal cell (Courtesy: http://www.learner.org)

A signal (growth factor) binds to a tyrosine kinase receptor on the cell membrane. This activates the membrane protein (through addition of phosphate groups), which in turn activates protein kinases in the cytoplasm. Several other proteins may be involved in the cascade, ultimately activating one or more transcriptional factors. The activated transcription factors enter the nucleus where they stimulate the expression of the genes. This is an example of the Ras pathway, which results in cell division.

This cascade includes a membrane receptor for the signal molecule, intermediary proteins that carry the signal through the cytoplasm, and nuclear proteins or transcription factors in the nucleus that activate the genes for cell division. In each step of the pathway, one factor or protein activates the next; although some factors can activate more than one protein in the cell. Oncogenes are altered versions of the proto-oncogenes that code for these signaling molecules. The oncogenes activate the signaling cascade continuously, resulting in an increased production of factors that stimulate growth. For example, *MYC* is a proto-oncogene that codes for a transcription factor. Mutations in *MYC* convert it into an oncogene associated with seventy percent of cancers. *RAS* is another oncogene that normally functions as an "on-off" switch in the signal cascade. Mutations in *RAS* cause the signaling pathway to remain "on," leading to uncontrolled cell growth. About thirty percent of tumors - including lung, colon, thyroid, and pancreatic carcinomas - have a mutation in *RAS*.

RAS is a family of genes encoding small GTPases that are involved in cellular signal transduction. Activation of RAS signaling causes cell growth, differentiation and survival. Since RAS communicates signal from outside the cell to the nucleus, mutation in *RAS* genes can permanently activate it and cause inappropriate transmission inside the cell even in the absence of extracellular signals. Because these signals results result in cell growth and division, dysregulated RAS signaling can ultimately lead to oncogenesis and cancer. RAS and ras-related proteins are often deregulated in cancers, leading to increased invasion and metastasis and decreased apoptosis. RAS activates pathways of mitogen-activated protein (MAP) kinase cascade. This cascade transmits signals downstream and results in the transcription of genes involved in growth and division. Mutations in the *RAS* family of proto-oncogenes are commonly found in 20% to 30% of all human tumors. Again, *RAS* point mutations are the single most common abnormality of human proto-oncogenes.

The conversion of a proto-oncogene to an oncogene may occur by mutation of the proto-oncogene, by rearrangement of genes in the chromosome that moves the proto-oncogene to a new location, or by an increase in the number of copies of the normal proto-oncogene. Sometimes a virus inserts its DNA in or near the proto-oncogene, causing it to become an oncogene. The result of any of these events is an altered form of the gene, which contributes to cancer.

Most oncogenes are dominant mutations; a single copy of this gene is sufficient for expression of the growth trait. This is also a "gain of function" mutation because the cells with the mutant form of the protein have gained a new function not present in cells with the normal gene. The presence of an oncogene in a germ line cell (egg or sperm) results in an inherited predisposition for tumors in the offspring. However, a single oncogene is not usually sufficient to cause cancer, so inheritance of an oncogene does not necessarily result in cancer.

TUMOR SUPPRESSOR GENES

These are protective genes. Normally, they suppress (limit) cell growth by monitoring the rate at which cells divide, repair mismatched DNA, and control cell death. When a tumor suppressor gene is mutated (due to heredity or environmental factors), cells continue to grow and can eventually form a tumor. Close to 30 tumors suppressor genes have already been identified, including *BRCA1, BRCA2,* and *p53*.

P53 TUMOR SUPPRESSOR PROTEIN

P53 (also known as protein 53 or tumor protein 53) is a transcription factor, which in humans is encoded by the *TP53* gene. It regulates cell cycle and thus functions as a tumor suppressor that is involved in preventing cancer. Originally p53 was thought to be an oncogenic protein. But it soon became clear that the transforming ability of p53 were due to a mutant protein and that the non-mutated, WT p53 (WT for Wilm's Tumor) negatively controlled cellular proliferation and suppressed cell transformation and tumorigenesis. The p53 is located on chromosome 17p13 in the human genome. It is now clear that WTp53 not only has anti-proliferative and anti-transforming activity but also possesses the ability to induce programmed cell death, after exposure of cells to DNA-damaging agents such as γ-radiation or anticancer drugs. Clearly, the biological effects of p53 include its ability to induce G1 arrest, to induce apoptosis following DNA damage, to inhibit tumor cell growth, and to preserve genetic stability. P53 becomes activated in response to several stress types, such as DNA damage (induced by UV, IR or chemical agents like hydrogen peroxide), oxidative stress and osmotic shock. In response to DNA damage, p53 elicits either cell cycle arrest to allow DNA repair to take place or apoptosis, if the damage is excessive. Aptly, p53 is termed as a protector or 'molecular policeman' monitoring the integrity of the genome. P53 acts as a transcription factor for several genes, including a number of genes involved in cell-cycle control, apoptosis, genetic stability, and angiogenesis, by stimulating the expression of genes that inhibit angiogenesis. The role of p53 in maintaining genetic stability appears to involve induction of genes that stimulate nucleotide excision repair, chromosomal recombination, chromosome segregation and induction of genes for ribonucleotide reductase. Thus mutations or inactivation of p53 function disrupt so many interconnecting pathways in such a way that, once that central control point is breached, numerous downstream regulators become dysfunctional, setting the stage for tumor progression. In fact, nearly 50% of all cancers involve a dysfunctional or impaired p53 at transcriptional or translational levels. Since the tumor suppressor gene *p53* is the most commonly mutated gene in human cancers, it is a good target for gene replacement therapy. Re-expression of p53 in human colon cancer cell lines bearing a mutated gene inhibits tumor cell proliferation. In a murine model of p53-mutated colon cancer, injection of an adenoviral vector encoding the *WT p53* gene into the tumors resulted in tumor regression and enhanced survival.

The proteins made by tumor suppressor genes normally inhibit cell growth, preventing tumor formation. Mutations in these genes result in cells that no longer show normal inhibition of cell growth and division. The products of tumor suppressor genes may act at the cell membrane, in the cytoplasm, or in the nucleus. Mutations in these genes result in a loss of function (that is, the ability to inhibit cell growth) so they are usually recessive. This means that the trait is not expressed unless both copies of the normal gene are mutated.

In some cases, the first mutation is already present in a germ line cell (egg or sperm); thus, all the cells in the individual inherit it. Because the mutation is recessive, the trait is not expressed. Later a mutation occurs in the second copy of the gene in a somatic cell. In that cell both copies of the gene are mutated and the cell attains uncontrolled growth. An example of this is hereditary retinoblastoma (RB), a serious cancer of the retina that occurs in early childhood. When one parent carries a mutation in one copy of the RB tumor suppressor gene, it is transmitted to offspring with a fifty percent probability. About ninety percent of the offspring who receive the one mutated RB gene from a parent also develop a mutation in the second copy of RB, usually very early in life. These individuals then develop retinoblastoma. Not all cases of retinoblastoma are hereditary: it can also occur by mutation of both copies of RB in the somatic cell of the individual. Because retinoblasts are rapidly dividing cells and there are thousands of them, there is a high incidence of a mutation in the second copy of RB in individuals who inherited one mutated copy. This disease afflicts only young children because only individuals younger than about eight years old have retinoblasts. In adults, however, mutations in RB may lead to a predisposition to several other forms of cancer.

Three other cancers associated with defects in tumor suppressor genes include familial adenomatous polyposis of the colon (APC), which results from mutations to both copies of the *APC* gene; hereditary breast cancer, resulting from mutations to both copies of *BRCA2*; and hereditary breast and ovarian cancer, resulting from mutations to both copies of *BRCA1*. While these examples suggest that heredity is an important factor in cancer, the majority of cancers are sporadic with no indication of a hereditary component. Cancers involving tumor suppressor genes are often hereditary because a parent may provide a germ line mutation in one copy of the gene. This may lead to a higher frequency of loss of both genes in the individual who inherits the mutated copy than in the general population. However, mutations in both copies of a tumor suppressor gene can occur in a somatic cell, so these cancers are not always hereditary. Somatic mutations that lead to loss of function of one or both copies of a tumor suppressor gene may be caused by environmental factors, so even these familial cancers may have an environmental component.

DNA REPAIR GENES

A third type of gene associated with cancer is the group involved in DNA repair and maintenance of chromosome structure. Environmental factors, such as ionizing radiation, UV light, and chemicals, can damage DNA. Errors in DNA replication can also lead to mutations. Certain gene products repair damage to chromosomes, thereby minimizing mutations in the cell. When a DNA repair gene is mutated its product is no longer made, preventing DNA repair and allowing further mutations to accumulate in the cell. These mutations can increase the frequency of cancerous changes in a cell. A defect in a DNA repair gene called XP (Xeroderma pigmentosum) results in individuals who are very sensitive to UV light and have a thousand-fold increase in the incidence of all types of skin cancer. There are seven XP genes, whose products remove DNA damage caused by UV light and other carcinogens. Another example of a disease that is associated with loss of DNA repair is Bloom syndrome, an inherited disorder that leads to increased risk of cancer, lung disease, and diabetes. The mutated gene in Bloom syndrome, BLM, is required for maintaining the stable structure of chromosomes. Individuals with Bloom syndrome have a high frequency of chromosome breaks and interchanges, which can result in the activation of oncogenes.

HUMAN GENETICS AND CANCER

If a person is born with a gene mutation that makes them more likely to develop cancer, he has inherited a cancer gene. This mutation can be passed on to his children. There are two types of mutations, germline mutation that is passed on from generation to generation, and somatic mutation, which happens during the life time of a person and are not passed on to the next generation.

Germline mutations are inherited at the moment of fertilization of mother's egg with father's sperm. All the cells of the body that develop from the zygote will contain the genetic change. You get half your genes from your father and half from your mother. If either of your parents has a genetic change that increases their risk of getting cancer (a cancer gene), you have a 50/50 (1 in 2) chance of inheriting it. This made the famous oncologist, Michael Bishop to state, "We carry the seeds of our cancer within us".

However, the cause of somatic mutation, which is acquired during the lifetime of a person, is not very well known. But factors such as radiation or the exposure to dangerous chemicals (for example, through smoking) can play a part. These somatic genetic changes are usually not passed on to the next generation.

The most common cancers through inherited mutation are breast, ovarian, bowel and womb (endometrial) cancer. Genetic tests can identify some of the genes responsible for these cancers. Other cancers such as prostate, pancreatic and testicular cancer may also be caused by an inherited gene mutation. Genetic tests for these cancers have not yet been developed.

Some very rare genetic disorders that can increase a person's risk of getting several types of cancer at young age are (1) Li-Fraumeni syndrome, (2) Multiple endocrine neoplasia type 1 (MEN1), (3) von Hippel-Lindau disease, (4) Neurofibromatosis, and (5) Retinoblastoma.

Germ line mutations are responsible for 5% to 10% of cancer cases. This is also called familial (occurring in families) cancer. However, most cancers are caused by a series of mutations that develop during a person's lifetime. These acquired mutations are caused by Tobacco, over-exposure to

UV radiation, and other toxins and chemicals cause acquired mutations. These mutations are not found in every cell of the body and are not passed from parent to child. Cancer caused by this type of mutation is called sporadic cancer.

It is believed that cancer happens when several genes of a particular group of cells become mutated. Some people may have more inherited mutations than others, and even with the same amount of environmental exposure, some people are simply more likely to develop cancer.

GENES KNOWN TO BE INVOLVED IN SOME CANCER TYPES

BREAST CANCER

Most breast cancers (about 90% to 95%) are considered sporadic, meaning that the damage to the genes occurs by chance after a person is born. Inherited breast cancers are less common (5% to 10%) and occur when gene mutations are passed within a family, from one generation to the next. Genetic mutations in certain types of genes are more likely to cause cancer. The most common gene mutations that can increase breast cancer risk occur in tumor suppressor genes. A tumor suppressor gene makes proteins that prevent tumor formation by limiting cell growth. Mutations in a tumor suppressor gene cause a loss of the ability to restrict tumor growth and, as a result, cancer can develop. A woman with an average risk of breast cancer has about a 12% chance of developing breast cancer. There are several genes linked to an increased risk of breast cancer. Some of the most common hereditary cancer syndromes associated with breast cancer risk is described below.

HEREDITARY BREAST AND OVARIAN CANCER (HBOC) SYNDROME

The two genes are associated with *HBOC, BRCA1* and *BRCA2* (BRCA stands for BReast CAncer), are tumor suppressor genes. Women who inherit a *BRCA* mutation have a 50% to 85% chance of developing breast cancer and a 15% to 40% chance of developing ovarian cancer. Men with *BRCA1* and *BRCA2* mutations may have an increased risk of breast and prostate cancers. Both women and men with *BRCA2* mutations may be at increased risk of breast cancer or other types of cancer. Approximately one in 40 women with Ashkenazi Jewish heritage carry a mutation in *BRCA1* or *BRCA2* genes that increases their risk of breast cancer to between 50% and 85%. Their risk of ovarian cancer is also increased to about 40%. About 80% of hereditary (inherited) breast cancer is caused by mutations in *BRCA1* or *BRCA2*.

ATAXIA TELANGIECTASIA (A-T)

Ataxia-telangiectasia (A-T) is a rare recessive disorder characterized by progressive neurological problems that lead to difficulty in walking. Signs of A-T develop in childhood. Children may begin staggering and appear unsteady shortly after learning to walk. Most people with A-T will eventually need to use a wheelchair. Slurred speech and difficulty with writing and other tasks develop over time. Red marks that are caused by dilated capillaries (tiny blood vessels), called telangiectases, often appear on the skin and eyes. People with A-T also have a weakened immune system and are prone to infections.

There is about a 40% risk of cancer for people with A-T. The most common cancers associated with A-T are leukemia and lymphoma. As the lifespan of individuals with A-T increases, other types of cancer, including melanoma, sarcoma, and breast, ovarian, and stomach cancers, are being reported in adults. However, it is too early to determine if there is truly an increased risk for these cancers that is caused by A-T, or whether these are sporadic cancers developing in adults who also happen to have A-T.

LI-FRAUMENI SYNDROME (LFS)

People with LFS have up to a 50% chance of developing cancer by age 40 and a 90% chance of developing cancer by age 60. Some of the cancers most commonly associated with LFS are osteosarcoma (a type of bone cancer), soft tissue sarcoma, leukemia, breast cancer, brain cancer, and adrenal cortical tumors. An adrenal cortical tumor begins in the adrenal cortex, which is the outer layer of the adrenal glands. The adrenal glands are located on top of each kidney and are a part of the body's endocrine system.

LFS is a rare condition. The gene associated with LFS is a tumor suppressor gene, *p53*. Testing for *p53* gene mutations is available for families who may have LFS. Another gene, *CHEK2*, may cause LFS for some families. Testing for mutations in the *CHEK2* gene is only available as part of a research study.

COWDEN SYNDROME (CS)

Women with CS have a 25% to 50% risk of developing breast cancer and a 65% risk of developing noncancerous breast changes. CS is a rare genetic condition caused by a mutation on the *PTEN* gene. People with CS also have a high risk of both noncancerous and cancerous tumors of the thyroid and endometrium (lining of the uterus). Also, people with CS often have small growths on the face or in the mouth and a larger than average head size. Genetic testing for the *PTEN* gene is available.

PEUTZ-JEGHERS SYNDROME (PJS)

Women with PJS have approximately a 50% risk of developing breast cancer. Women with PJS also have about a 20% risk of developing ovarian cancer. People with PJS often have multiple hamartomatous polyps, which are normal-appearing growths in the digestive tract that become a noncancerous tumor. These polyps cause an increased risk of colorectal cancer. People with PJS also have increased pigmentation (dark spots on the skin) on the face and hands. The increased pigmentation often appears in childhood and fades over time. Families with PJS also have an increased risk of uterine and lung cancers. The gene associated with PJS is called *STK11*. The *STK11* is a tumor suppressor gene, and genetic testing for the *STK11* gene is available.

Other genes may cause hereditary breast cancer. However, more research is needed to understand how gene mutations can increase breast cancer risk and to find other genes that may increase a person's risk of breast cancer.

PROSTATE CANCER

Most prostate cancer (about 75%) is considered sporadic, but the familial prostate cancer is less common (about 20%) and occurs because of a combination of shared genes and shared environmental or lifestyle factors. Hereditary prostate cancer is rare (about 5%) and occurs when gene mutations are passed on within a family, from one generation to the next. A man with an average risk for prostate cancer has about a 14% chance of developing prostate cancer by age 80. The risk of prostate cancer is slightly higher for black men than for white men. Researchers continue to find genes that may be associated with an increased risk of prostate cancer. However, more research is needed to better understand these genes before genetic testing can be used to determine a man's risk of developing prostate cancer.

One gene known to increase the risk of prostate cancer, by as much as three times the average risk, is located on chromosome 17. The normal function of this gene is not known, but men who inherit a mutated version of the gene have a 44% higher level of the prostate-specific antigen (PSA) protein, a protein in the blood used to help diagnose prostate cancer. However, not everyone with this gene will develop prostate cancer.

Other genes that may cause an increased risk of developing prostate cancer include *HPC1*, *HPC2*, *HPCX*, and *CAPB*. Research on these genes is still new and genetic tests are not yet available for routine screening because it is not clear that they definitely cause prostate cancer.

Men with *BRCA1* and *BRCA2* gene mutations have an increased risk of prostate cancer. *BRCA1* and *BRCA2* gene mutations are most commonly associated with hereditary breast and ovarian cancer (HBOC) syndrome. Men with *BRCA1* mutations have a slightly increased risk, while men with *BRCA2* mutations have about a 20% risk of developing prostate cancer during their life, usually before age 65. For this reason, men with *BRCA1* or *BRCA2* gene mutations are encouraged to begin annual prostate cancer screening at age 40. Men with *BRCA1* or *BRCA2* gene mutations also have an increased risk of breast cancer.

BRCA1 and *BRCA2* are tumor suppressor genes. Mutations in a tumor suppressor gene cause a loss of the ability to restrict tumor growth and, as a result, cancer can develop. Genetic testing for the *BRCA1* and *BRCA2* genes is available. However, mutations in *BRCA1* and *BRCA2* are thought to be responsible for only a small percentage of familial prostate cancer cases. Genetic testing may only be appropriate for families with prostate cancer that may also have HBOC.

COLORECTAL CANCER

A person with an average risk of colorectal cancer has about a 5% chance of developing colorectal cancer. Men have a slightly higher risk of developing colorectal cancer than women. Most colorectal cancers (about 95%) are considered sporadic and inherited colorectal cancers are less common (about 5%). There are several genes associated with an increased risk of colorectal cancer. Some of the most common hereditary colorectal cancer syndromes are described below.

HEREDITARY NON-POLYPOSIS COLORECTAL CANCER (HNPCC)

HNPCC, sometimes called Lynch syndrome, accounts for approximately 3 to 5% of all colorectal cancers. The risk of colorectal cancer in families with HNPCC is 80%, which is several times higher than the average risk. The average age for a person with HNPCC to be diagnosed with colorectal cancer is 45. For women with HNPCC, the risk of uterine cancer is about 20% to 60% and the risk of ovarian cancer is about 9 to 12%. In addition, people with HNPCC also have an increased risk of cancers of the stomach, small intestine, liver, bile duct, urinary tract, and the brain and central nervous system. There may also be some increased risk of breast cancer.

Mutations in the *MLH1*, *MSH2*, *MSH6*, and *PMS2* genes are the most frequent cause of HNPCC. The genes associated with HNPCC are mismatching repair genes. A mismatch repair gene makes proteins that repair DNA mistakes that occur as a cell divides. If one of these genes has a mutation, the mistakes cannot be repaired, leading to damaged DNA and an increased risk of cancer. Although multiple genes have been linked to HNPCC, most families with HNPCC have a mutation in only one of the genes. Genetic testing is available for the *MLH1*, *MSH2*, and *MSH6* genes. The *PMS2* gene is only tested as part of a clinical trial.

Because colorectal cancer is one of the most treatable forms of cancer, if found early, people diagnosed with HNPCC, or those who may have an increased risk based on their family history, often benefit from increased screening with annual colonoscopy examinations.

To prevent colorectal cancer, a prophylactic colectomy (the surgical removal of the entire colon) may decrease the risk of colorectal cancer for patients who have an increased risk. Drugs that may stop a tumor from forming are being tested to prevent HNPCC-related colon cancers.

FAMILIAL ADENOMATOUS POLYPOSIS (FAP)

FAP accounts for about 1% of colorectal cancers. People with FAP often develop hundreds to thousands of polyps (a growth in the colon or rectum) that are initially noncancerous, but there is almost a 100% chance that the polyps will develop into cancer, if untreated. Most people with FAP develop polyps by age 35, with the average age in the mid-teens. The risk of a person with FAP developing colorectal cancer by age 45 is 87%. People with FAP also have an increased risk of other types of cancers including stomach, small intestine, pancreas, thyroid, and hepatoblastoma (liver cancer that usually occurs in early childhood). Although FAP follows an autosomal dominant inheritance pattern, approximately 30% of people with FAP have no family history of the condition.

People with a variation of FAP, called attenuated familial adenomatous polyposis (AFAP), often develop fewer polyps, usually around 30 totals. For people with AFAP, colorectal cancer may not develop until around age 50.

FAP and AFAP are caused by a mutation on the APC gene, which is a tumor suppressor gene. Mutations in tumor suppressor genes result in a loss of the ability to restrict tumor growth. About 6% of people with Ashkenazi Jewish heritage have a specific mutation in the APC gene, called I1307K, which increases the risk of colorectal cancer. Genetic testing for the APC gene is available.

Since polyps can be found at an early age, people with FAP or who may have FAP should talk with their doctor about regular screening, such as a yearly sigmoidoscopy, beginning around ages 10 to 12. A colectomy may be recommended to decrease the risk of cancer. Continued screening for polyps in the rectal area and the upper digestive tract is recommended after the colon is removed.

MUIR-TORRE SYNDROME

Muir-Torre syndrome is a type of HNPCC. In addition to increasing the risk of colorectal cancer and the other types of cancer associated with HNPCC, people with Muir-Torre syndrome often develop skin changes in adulthood. These skin changes are usually not cancerous.

GARDNER SYNDROME

Gardner syndrome is a type of FAP. In addition to an increased risk of colon polyps and other FAP-related cancers, people with Gardner syndrome may also have osteomas (bony tumors) of the jaw, extra teeth, and soft tissue tumors including lipomas (fatty tissue) and fibromas (fibrous tissue).

TURCOT SYNDROME

Turcot syndrome is a type of both HNPCC and FAP. People with Turcot syndrome have an increased risk of colorectal cancer and brain tumors. Medulloblastoma, a type of brain tumor, is more common for people who have a genetic mutation associated with FAP. Glioblastoma, another type of brain tumor, is more common for people who have a genetic mutation associated with HNPCC.

MYH-ASSOCIATED POLYPOSIS (MAP)

People with MAP often develop multiple colon polyps that increase their risk of colorectal cancer. MAP is caused by a mutation on the MYH gene and follows an autosomal recessive inheritance pattern, in which a mutation needs to be present in both copies of the gene in order for a person to have an increased risk of getting that disease. Genetic testing for MYH gene mutations is available. Because many people with MYH gene mutations are thought to have FAP or AFAP, genetic testing may be recommended for people with multiple colon polyps who do not have mutations on the APC gene.

PEUTZ-JEGHERS SYNDROME (PJS)

People with PJS have a 40% risk of developing colorectal cancer. People with PJS often have multiple hamartomatous polyps, which are normal-appearing growths in the digestive tract that become noncancerous tumors. People with PJS also have increased pigmentation (dark spots on the skin) on the face and hands. The increased pigmentation often appears in childhood and fades over time. Families with PJS also have an increased risk of breast, uterine, ovarian, and lung cancers. The gene associated with PJS is called *STK11*. The *STK11* is a tumor suppressor gene, for which genetic testing is available.

JUVENILE POLYPOSIS SYNDROME (JPS)

People with JPS often have multiple juvenile polyps (a specific type of polyp) in the colon or other parts of the digestive system. Most people with JPS develop some polyps by age 20. Because of these polyps, people with PJS have an increased risk of colorectal, stomach, small intestine, and pancreatic cancers. Overall, people with PJS have a 9 to 50% risk of developing cancer. Two genes are associated with JPS; they are called *BMPR1A* and *SMAD4*, and genetic testing is available.

KIDNEY CANCER

Most kidney cancers (about 95%) are considered sporadic. Inherited kidney cancers are less common (about 5%). Kidney cancer may also be called renal cell carcinoma. A person at average risk for kidney cancer has less than a 1% chance of developing kidney cancer during his or her lifetime.

There are a growing number of genes thought to be associated with an increased risk of kidney cancer. Some of the most common genetic conditions that increase the risk of kidney cancer are described below. Most conditions are associated with a specific type of kidney cancer. Identifying a specific genetic syndrome in a family can help individuals and their doctors develop an appropriate cancer screening plan and, in some cases, help to determine the best treatment. Some genetic conditions also increase a person's risk for noncancer-related health problems for which screening and early detection may be beneficial.

VON HIPPEL-LINDAU SYNDROME (VHL)

People with VHL have an increased risk of developing several types of tumors. Most of these tumors are benign (noncancerous). However, people with VHL have about a 40% risk of developing kidney cancer. The specific type of kidney cancer associated with VHL is called clear cell kidney cancer. Other parts of the body where tumors can develop include the eye (retinal angioma), brain and spinal cord (hemangioblastoma), adrenal glands (pheochromocytoma), and ear (endolymphatic sac tumor). Tumors of the ear may lead to partial or complete hearing loss. People with VHL can also develop a cyst on their kidneys and pancreas, or on a man's testicles (epididymal cystadenomas). Symptoms of VHL usually develop in the age between 20 and 30, but can begin in childhood. Approximately 20% of people with VHL have no family history of the condition.

The gene associated with VHL is also called VHL. The VHL gene is a tumor suppressor gene. Mutations in a tumor suppressor gene results in a loss of the ability to restrict tumor growth and, as a result, cancer can develop. Genetic testing for VHL is available, and it is recommended that a person consults a doctor or a genetic counselor for more information based on his or her individual medical profile. Genetic testing is often recommended for anyone who may have VHL and for any family members, children or adults, at risk for VHL.

Screening for symptoms of VHL should begin early in families at risk and include:
1. Yearly eye examination to look for retinal tumors, beginning around age 2
2. Yearly physical examination
3. Yearly 24-hour urine test to screen for elevated catecholamines (a chemical produced by the body that may be at higher levels for people with VHL), beginning around age 2
4. Yearly abdominal ultrasound (the use of sound waves to create a picture of the internal organs) to look at the kidneys, pancreas, and adrenal glands beginning in the teenage years; change screening to abdominal computed tomography (CT or CAT) scan in adulthood
5. Magnetic resonance imaging (MRI) of the brain and spine every two years beginning in the teenage years

HEREDITARY NON-VHL CLEAR CELL RENAL CELL CARCINOMA

Hereditary non-VHL clear cell renal cell carcinoma is a genetic condition that increases a person's risk of developing clear cell renal cell carcinoma (CCRCC). A family may have hereditary non-VHL CCRCC, if multiple family members have been diagnosed with CCRCC.

A specific gene that causes non-VHL CCRCC has not been discovered. Some families who have hereditary non-VHL CCRCC may have a specific rearrangement of chromosome 3. In other families, the genetic change responsible for the increased risk of kidney cancer is unknown. Genetic testing is available to look for the specific chromosome 3 rearrangements. Yearly screening for kidney cancer with ultrasound, MRI, and CT scan beginning at age 20 has been suggested for families with this rare condition.

HEREDITARY PAPILLARY RENAL CELL CARCINOMA (HPRCC)

HPRCC is a genetic condition that increases the risk of type 1 papillary renal cell carcinoma. People who have HPRCC have an increased risk of multiple kidney tumors and an increased risk of developing tumors on both kidneys.

HPRCC is suspected when two or more close relatives have been diagnosed with type 1 papillary renal cell carcinoma. In these families, people are typically diagnosed with HPRCC while in their 40s (or sometimes at an older age). People with HPRCC may develop multiple kidney tumors in one or both kidneys. Some doctors suggest that individuals who have HPRCC, or a family history that suggests HPRCC, should have yearly screening beginning at age 30.

HPRCC is caused by a mutation on the *MET* gene. The *MET* gene is a proto-oncogene. Proto-oncogenes make proteins that promote normal cell growth. Mutations in proto-oncogenes result in too much cell growth and can lead to cancer. Genetic testing for mutations in the *MET* gene is available.

BIRT-HOGG-DUBÉ SYNDROME (BHD)

BHD is a rare genetic condition associated with multiple benign skin tumors, lung cysts, and an increased risk of both benign kidney tumors and kidney cancer. People with BHD have about 15% risk of developing kidney cancer. Most of the kidney cancers associated with BHD are classified as the chromophobe type (a rare type of kidney cancer) or oncocytoma (a slow growing type of kidney cancer that rarely spreads), but clear cell and papillary kidney cancers may also occur. Because of the increased risk of kidney cancer, people with BHD should consider yearly screening with ultrasound, MRI, or CT scan beginning at age 25.

BHD is caused by a mutation on the *FLCN* gene, and genetic testing is available for families who may have BHD.

HEREDITARY LEIOMYOMATOSIS AND RENAL CELL CARCINOMA (HLRCC)

People with HLRCC develop skin nodules called leiomyomata. The nodules are found mainly on the arms, legs, chest, and back. Women with HLRCC often develop uterine fibroids known as leiomyomas, or, less commonly, leiomyosarcoma. People with HLRCC also have about 20% risk of developing type 2 papillary renal cell carcinoma.

The most common screening options for people who may have HLRCC are regular skin examinations, abdominal/pelvic CT scans, or MRI every two years. Also, a doctor may recommend that women have regular gynecological examinations and ultrasounds to look for uterine fibroids. HLRCC is caused by the fumarate hydratase (FH) gene, and genetic testing for this gene is available.

Other genes may be associated with an increased risk of kidney cancer, and finding other genetic causes of kidney cancer is an active area of research. In addition, other genetic conditions may be associated with an increased risk of kidney cancer.

BECKWITH-WIEDEMANN SYNDROME (BWS)

Children with BWS have an increased risk of developing Wilms' tumor (a type of kidney cancer). BWS is a growth disorder associated with large body size, large tongue, abdominal wall defects, an increased risk of childhood tumors, kidney abnormalities, and low blood sugar when a baby is a newborn, and unusual ear creases or pits. Children with BWS may also have body parts that are larger on one side of the body than on the other.

In most people, the genetic changes that cause BWS occur by chance, but others may have inherited changes. Screening for people who may have BWS includes MRI or CT scans of the abdomen, abdominal ultrasound, blood tests, and regular physical examinations.

LI-FRAUMENI SYNDROME (LFS)

People with LFS have up to 50% chance of developing cancer by age 40 and a 90% chance of developing cancer by age 60. Some of the cancers most commonly associated with LFS are osteosarcoma (a type of bone cancer), soft tissue sarcoma, leukemia, breast cancer, brain cancer, and adrenal cortical tumors. An adrenal cortical tumor begins in the adrenal cortex, which is the outer layer of the adrenal glands. The adrenal glands are located on top of each kidney and are a part of the body's endocrine (hormonal) system. Wilms' tumor has occasionally been reported in such families, but the risk of developing Wilms' tumor is not known.

LFS is a rare condition. The gene associated with LFS is *p53*, which is a tumor suppressor gene. Testing for *p53* gene mutations is available for families meeting the diagnostic criteria for LFS. More research is needed to better understand LFS. Another gene, *CHEK2*, may be responsible for LFS in some families. Testing for mutations in the *CHEK2* gene is only available as part of a research study.

TUBEROUS SCLEROSIS COMPLEX (TSC)

TSC is a genetic condition associated with changes in the skin, brain, kidney, and heart. People with TSC have about a 4% risk of developing kidney cancer. Suggested screening to detect kidney cancer for people who have TSC includes an ultrasound of the kidneys every one to three years, more frequently if necessary, followed by MRI or CT scans of the kidneys, if more detail is needed. TSC is caused by a mutation on either the *TSC1* or the *TSC2* gene. Approximately 60% of people with TSC do not have a family history of the condition.

COWDEN SYNDROME (CS)

People with CS have developed kidney cancer. Although, it is not known whether having CS increases a person's risk of developing kidney cancer. Cowden syndrome is a rare genetic condition caused by a mutation on the *PTEN* gene. People with CS have a high risk of both benign and cancerous tumors of the breast, thyroid, and endometrium. Also, people with CS often have small growths on the face or in the mouth and a larger than average head size.

MELANOMAS

Most melanomas (about 90%) are considered sporadic. An increased risk of melanoma occurs when specific gene mutations are passed on within a family from generation to generation. Melanoma itself is not inherited; it is an increase in risk of developing melanoma that is inherited. Many people who have an increased risk of melanoma never develop the disease. Although 10% of melanoma is familial, known genetic variations only account for 1% of all melanoma diagnoses.

A person at average risk for melanoma has a 2% chance of developing melanoma during his or her lifetime. More specifically, the average risk for women is 1.6% and the average risk for men is 2.3%. There are a growing number of genes thought to be associated with an increased risk of melanoma. However, more research is needed to better understand how these genes affect the risk of melanoma.

HEREDITARY MELANOMA

At least three genes have been linked to hereditary melanoma. Families with mutations in these genes may have multiple dysplastic nevi. Although dysplastic nevi are likely to be related to altered genes, the specific genes involved have not been identified. The association of familial melanoma and multiple dysplastic nevi is also sometimes called familial atypical multiple mole melanoma (FAMMM) or atypical nevus syndrome.

Mutations in the *CDKN2A* gene (also called *p16* and *MST1*) are thought to account for approximately 25% of hereditary melanomas. People who have mutations in the *CDKN2A* gene have about a 70% risk of developing melanoma during their lifetime, but the risk varies by geographic location. Some families with these mutations also have an increased risk (up to 17%) of developing pancreatic cancer. Pancreatic cancer, however, is much less frequent than melanoma in these families. Pancreatic cancer also varies by geographic location, with much higher rates in Europe than in the United States, and extremely low rates in Australia.

The *CDKN2A* gene is a tumor suppressor gene. *CDKN2A* produces two different proteins, p16 and p14ARF, depending on how the gene is "read." Mutations affecting p16 are associated with an increased risk of melanoma and pancreatic cancer. The cancer risk associated with mutations affecting p14ARF is not as well understood, although there may be some increased risk for melanoma, and possibly other cancers, in a small number of families.

There is also growing evidence that variations in another gene, *MC1R*, alter the risk of melanoma for people with *CDKN2A* mutations and for people without *CDKN2A* mutations. MC1R is important in regulating pigment; variations have been associated with freckling and red hair.

Genetic testing for mutations in *CDKN2A* is available, but the American Society of Clinical Oncology's (ASCO's) Task Force on Genetic Testing recommends that such testing not be routinely performed outside of a research study, since researchers do not fully understand the implications of certain types of mutations. In addition, several types of melanoma that run in families do not yet have clearly defined genetic causes. ASCO bases its clinical recommendations for identifying melanoma risk on family history, skin type, and environmental factors, regardless of the presence or absence of known mutations.

Mutations in the *CDK4* gene are responsible for an increased risk of melanoma in a very small number of families. The cancer risks associated with mutations in *CDK4* are thought to be similar to those in families with mutations in *CDKN2A*. The *CDK4* gene is a proto-oncogene. Genetic testing for mutations in the CDK4 gene is not available because this genetic mutation is thought to be rare. Based on limited data, the characteristics of melanomas occurring in families with *CDK4* mutations are similar to those in families with *CDKN2A* mutations.

A third area on chromosome 1 (1p22) has been shown to possibly contain another melanoma susceptibility gene that has not yet been identified. There is also evidence that additional genes associated with hereditary melanoma exist, and this topic is an active area of research.

People at risk for hereditary melanoma should examine their skin carefully each month to look for changes in the appearance of moles. Screening should begin by age 10 in children at risk. Skin examinations by a trained health-care provider should be performed yearly or more frequently, if necessary.

MELANOMA-ASTROCYTOMA SYNDROME

People with this rare condition have an increased risk of melanoma and astrocytoma (a type of brain tumor). The specific gene for this condition is thought to be located on chromosome 9. Families with both melanoma and brain tumors have been shown to have alterations in *CDKN2A* that affect the p14ARF protein.

Other genetic conditions that are associated with an increased risk of melanoma include xeroderma pigmentosum, retinoblastoma, Li-Fraumeni syndrome, ataxia-telangiectasia, and Werner syndrome.

XERODERMA PIGMENTOSUM (XP)

XP is a group of rare conditions whereby a person's cells cannot repair DNA damage caused by ultraviolet (UV) exposure, such as sunshine. Signs of XP are caused by the increased sensitivity to UV light, and include dry skin, abnormal pigmentation, severe freckling, and blistering after minimal sun exposure. There is a high rate of skin cancer, including melanoma (over a thousand times more frequent than in the general population), and noncancerous skin abnormalities in people who have XP. People with XP also have an increased risk of other types of cancers including leukemia, brain tumors, and stomach, lung, breast, uterine, and testicular cancers.

There are several different forms of XP, but each is inherited as an autosomal recessive condition. In autosomal recessive inheritance, a person needs to have two copies of a gene mutation in order to have the disease. This means that each parent must have one mutated gene and one normal gene. If both parents carry a gene mutation for XP, each of their children will have a 25% chance of inheriting both gene mutations and having XP. Several genes have been linked to the different forms of XP, but genetic testing is not available.

RETINOBLASTOMA

Retinoblastoma is a childhood eye tumor. Tumors may develop in one or both eyes, but inherited retinoblastoma most often occurs in both eyes. Nearly all cases of retinoblastoma are found in children under age 5, and inherited retinoblastoma occurs most often in children younger than one. Approximately 30% to 40% of retinoblastoma cases are hereditary. Children with hereditary retinoblastoma have an increased risk of developing other cancers, as they grow older, including osteosarcoma (a type of bone cancer), other sarcomas, leukemia, lymphoma, melanoma (50 times more than the general population), lung cancer, and bladder cancer.

Retinoblastoma is inherited as an autosomal dominant condition. The gene associated with hereditary retinoblastoma is called *RB1*. However, many children with hereditary retinoblastoma have new mutations in the *RB1* gene, meaning that there is no other family history of the condition. People with new gene mutations can pass the mutation to their children. Genetic testing for the *RB1* gene is available.

LI-FRAUMENI SYNDROME

People with Li-Fraumeni syndrome (LFS) have up to a 50% chance of developing cancer by age 40 and a 90% chance of developing cancer by age 60. Some of the cancers most commonly associated with LFS are osteosarcoma (a type of bone cancer), soft tissue sarcoma, leukemia, breast cancer, brain cancer, and adrenal cortical tumors. Melanoma has occasionally been reported in such families, but the risk of developing melanoma is not known.

LFS is a rare condition. The gene associated with LFS is called p53. The p53 gene is a tumor suppressor gene. Testing for p53 gene mutations is available for families meeting the diagnostic criteria for LFS. More research is needed to better understand LFS. Another gene, *CHEK2*, may be responsible for LFS in some families. Testing for mutations in the *CHEK2* gene is only available as part of a research study.

WERNER SYNDROME

Werner syndrome is a rare condition characterized by premature aging and an increased risk of cancer. The first sign of Werner syndrome may be the lack of a growth spurt in puberty. People with Werner syndrome develop the signs of aging in early adulthood, including graying hair and hair loss, cataracts, and osteoporosis. The risk of developing cancer is approximately 10%. The most common types of cancer seen in people with Werner syndrome are osteosarcoma (a type of bone cancer), other sarcomas, melanoma, and thyroid cancer.

Werner syndrome is inherited as an autosomal recessive condition. The gene associated with Werner syndrome is called WRN. Genetic testing for the WRN gene is currently available only as part of a research study.

HEREDITARY BREAST AND OVARIAN CANCER

An increased risk of melanoma has also been identified in people with hereditary breast and ovarian cancer (HBOC). HBOC is a genetic condition that follows an autosomal dominant inheritance pattern. Two genes are associated with HBOC: *BRCA1* and *BRCA2*. A mutation in either of these genes gives a woman an increased lifetime risk of developing breast and ovarian cancers. Men with these gene mutations have an increased risk of breast cancer and prostate cancer. A genetic mutation on the *BRCA2* gene is more commonly associated with an increased risk of melanoma.

COWDEN SYNDROME

Melanoma has been reported in association with Cowden syndrome (CS). Cowden syndrome is a rare genetic condition that follows an autosomal dominant inheritance pattern. It is caused by a mutation on the *PTEN* gene. People with CS have a high risk of both noncancerous and cancerous tumors of the breast, thyroid, and endometrium. Also, people with CS often have small growths on the face or in the mouth and a larger than average head size. Although people with CS have developed melanoma, it is not known whether having CS increases a person's risk of developing melanoma.

PANCREATIC CANCER

Most pancreatic cancers (about 90%) are considered sporadic. However, inherited pancreatic cancers are less common (about 10%). A person with an average risk for pancreatic cancer has less than a 1% chance of developing pancreatic cancer. There are several genes linked to an increased risk of pancreatic cancer. The most common hereditary cancer syndromes associated with pancreatic cancer risk are described below.

HEREDITARY PANCREATITIS (HP)

People with HP have a 40% risk of developing pancreatic cancer. Pancreatitis is inflammation or swelling of the pancreas that causes severe abdominal pain. HP can begin in childhood and become worse over time. The gene most often associated with HP is called *PRSS1*, the role of which is not yet known. Mutations in two other genes, SPINK1 and CFTR, have also been linked to HP; however, it is not clear whether mutations in these genes can cause an increased risk of pancreatic cancer. Genetic testing for mutations in the *PRSS1, SPINK1,* and *CFTR* genes is available for people who may have HP.

PEUTZ-JEGHERS SYNDROME (PJS)

People with PJS have a 35% risk of developing pancreatic cancer. People with PJS often have multiple hamartomatous polyps, which are normal-appearing growths in the digestive tract that become noncancerous tumors. These polyps cause an increased risk of colorectal cancer. People with PJS also have increased pigmentation (dark spots on the skin) on the face and hands. The increased pigmentation often appears in childhood and fades over time. Women with PJS have about a 20% risk of developing ovarian cancer. Families with PJS also have an increased risk of breast, uterine, and lung cancers. The gene associated with PJS is called STK11. The STK11 is a tumor suppressor gene, and genetic testing for the STK11 gene is available.

FAMILIAL ATYPICAL MULTIPLE MOLE MELANOMA AND PANCREATIC CANCER (FAMMM-PC)

FAMMM-PC is also known as melanoma-pancreatic cancer syndrome. People with FAMMM-PC have up to a 17% risk of developing pancreatic cancer. FAMMM-PC is associated with mutations in the CDKN2A gene. People who have mutations in the CDKN2A gene also have about a 70% risk of developing melanoma during their lifetime. The CDKN2A (also called p16 and MST1) gene is a tumor suppressor gene. Genetic testing for mutations in CDKN2A is available, but testing is currently not recommended outside of clinical trials.

HEREDITARY BREAST AND OVARIAN CANCER (HBOC) SYNDROME

The two genes are associated with HBOC, BRCA1 and BRCA2 (BRCA stands for BReast CAncer), are most often associated with an increased risk of developing breast and ovarian cancer for women and breast and prostate cancer for men. However, mutations in both genes have been linked to an increased risk of other types of cancer, including pancreatic cancer. The risk of pancreatic cancer for families with BRCA2 mutations is estimated to be up to 10 times greater than the average risk. There may also be a small increased risk of pancreatic cancer for families with BRCA1 gene mutations.

HEREDITARY NON-POLYPOSIS COLORECTAL CANCER (HNPCC)

HNPCC (sometimes called Lynch syndrome) causes less than 5% of hereditary pancreatic cancer cases. HNPCC is most commonly associated with an increased risk of colorectal cancer. The risk of colorectal cancer for families with HNPCC is 80%, which is several times greater than the average risk. People with HNPCC also have an increased risk of cancers of the stomach and small intestine. Women with HNPCC have about a 9% to 12% risk of developing ovarian cancer and a 20% to 60% risk of uterine cancer. There may also be an increased breast cancer risk for families with HNPCC.

Mutations in the MLH1, MSH2, MSH6, and PMS2 genes are the most frequent cause of HNPCC. The genes associated with HNPCC are mismatch repair genes. Although multiple genes have been linked to HNPCC, most families with HNPCC have a mutation in only one of the genes. Genetic testing is available for the MLH1, MSH2, and MSH6 genes. The PMS2 gene is only tested as part of a clinical trial.

There may be additional hereditary cancer syndromes that increase the risk of pancreatic cancer, and anyone concerned about their family history should talk with their doctor or genetic counselor. As research continues, doctors may learn more about the causes of inherited pancreatic cancer.

LI-FRAUMENI SYNDROME (LFS)

People with LFS have up to a 50% chance of developing cancer by age 40 and a 90% chance of developing cancer by age 60. Some of the cancers most commonly associated with LFS are osteosarcoma (a type of bone cancer), soft tissue sarcoma, leukemia, breast cancer, brain cancer, and adrenal cortical tumors. An adrenal cortical tumor begins in the adrenal cortex, which is the outer layer of the adrenal glands. The adrenal glands are located on top of each kidney and are a part of the body's endocrine (hormonal) system. Pancreatic cancer has occasionally been reported in such families, but the risk of developing pancreatic cancer is not known.

LFS is a rare condition. The gene associated with LFS, called p53, is a tumor suppressor gene. Testing for p53 gene mutations is available for families who may have LFS. Another gene, CHEK2, may cause LFS for some families. Testing for mutations in the CHEK2 gene is only available as part of a research study.

Familial adenomatous polyposis (FAP) People with FAP often develop hundreds to thousands of polyps (a growth in the colon or rectum) that are initially noncancerous, but there is almost a 100% chance that the polyps will develop into cancer if not treated. Most people with FAP develop polyps in their 20s or 30s, although polyps can be found as early as the teenage years, and colorectal cancer often develops by age 40. People with FAP also have an increased risk of other types of cancer including stomach, small bowel, pancreas, thyroid, and hepatoblastoma (liver cancer that usually occurs in early childhood).

Although FAP follows an autosomal dominant inheritance pattern, approximately 30% of people with FAP have no family history of the condition. FAP is caused by a mutation on the APC gene, which is a tumor suppressor gene. Genetic testing for the APC gene is available.

THYROID CANCER

Most thyroid cancers are considered sporadic. Inherited thyroid cancers are less common (about 10%) and occur when gene mutations are passed within a family, from one generation to the next. A person with an average risk for thyroid cancer has less than a 1% chance of developing thyroid cancer during his or her life. Women develop thyroid cancer more often than men.

There are some genes associated with an increased risk of thyroid cancer. Some of the most common hereditary cancer syndromes associated with thyroid cancer risk is described below.

MEDULLARY THYROID CANCER

Medullary thyroid cancer (MTC) accounts for about 5% to 10% of all thyroid cancers. Approximately 25% of MTC are caused by an inherited cancer risk. Three conditions are responsible for hereditary MTC: multiple endocrine neoplasia (MEN) 2A, MEN 2B, and familial medullary thyroid carcinoma (FMTC). These conditions are caused by mutations in the RET gene, a type of gene called a proto-oncogene. A proto-oncogene makes proteins that promote normal cell growth. Mutations in proto-oncogenes result in too much cell growth and can lead to cancer. Genetic testing for mutations in the RET gene is available.

MEN 2A

MEN 2A causes 20% to 25% of all medullary thyroid cancers. People with MEN 2A have a 95% to 100% risk of developing MTC. Families with MEN 2A also have an increased risk of developing pheochromocytoma, a tumor in the adrenal glands. The adrenal glands are located on top of each kidney and are a part of the body's endocrine (hormonal) system. Families with MEN 2A also have an increased risk of noncancerous parathyroid tumors or an increase in the size of the parathyroid gland. The parathyroid glands are located in the neck near the thyroid gland and are part of the endocrine system. MEN 2A is diagnosed if a single person or close family members have at least two of the characteristic features. MEN 2A follows an autosomal dominant inheritance pattern, and less than 5% of people diagnosed with MEN 2A have no family history of the condition.

MEN 2B

MEN 2B causes less than 2% of all medullary thyroid cancers. People with MEN2B have an increased risk of developing MTC, pheochromocytoma, and physical abnormalities, such as a long face and long arms and legs. They also have thick, lumpy lips caused by noncancerous tumors called mucosal neuromas. Some people with MEN 2B have bowel problems. Signs of MEN 2B, including MTC, can be begin in childhood. MEN 2B is also an autosomal dominant condition, but approximately 50% of people with MEN 2B have no family history of the condition.

FAMILIAL MEDULLARY THYROID CANCER (FMTC)

FMTC causes less than 2% of all medullary thyroid cancers. FMTC is suspected if a person has four or more family members with MTC and no cases of pheochromocytoma or hyperparathyroidism (increase in hormone produced by the parathyroid). However, people with multiple family members with MTC should talk with their doctor about thyroid screening.

PAPILLARY AND FOLLICULAR THYROID CANCER

Papillary and follicular thyroid cancer make up 80% to 90% of all thyroid cancers. It is estimated that less than 5% of papillary thyroid cancer is inherited.

FAMILIAL PAPILLARY THYROID CANCER

Familial papillary thyroid cancer is thought to follow an autosomal dominant inheritance pattern. However, more research is needed to identify genes associated with papillary thyroid cancer risk. For this reason, genetic testing is not currently available.

FAMILIAL ADENOMATOUS POLYPOSIS (FAP)

People with FAP have a 2% risk of developing papillary thyroid cancer. The average age at thyroid cancer diagnosis is 28, and women with FAP appear to have a greater risk than men. People with FAP often develop hundreds to thousands of polyps (a growth in the colon or rectum) that are initially noncancerous, but there is almost a 100% chance that the polyps will develop into cancer if not treated. Most people with FAP develop polyps in their 20s or 30s, although polyps can be found as early as the teenage years, and colorectal cancer often develops by age 40. People with FAP also have an increased risk of other types of cancer including stomach, small bowel, pancreas, and hepatoblastoma (liver cancer that usually occurs in early childhood).

Although FAP follows an autosomal dominant inheritance pattern, approximately 30% of people with FAP have no family history of the condition. FAP is caused by a mutation on the APC gene, which is a tumor suppressor gene. A tumor suppressor gene makes proteins that prevent tumor formation by limiting cell growth. Mutations in a tumor suppressor gene cause a loss of the ability to restrict tumor growth and, as a result, cancer can develop. Genetic testing for the APC gene is available.

COWDEN SYNDROME (CS)

People with CS have up to a 10% lifetime risk of follicular or papillary thyroid cancer; follicular thyroid cancer is most common. Approximately 70% of people with CS have noncancerous thyroid changes, including multinodular goiter (enlarged thyroid gland with multiple growths or nodules), adenomatous nodules (growth on the thyroid gland), and follicular adenomas (another type of growth on the thyroid gland). CS is a rare genetic condition caused by a mutation on the *PTEN* gene. People with CS also have an increased risk of both benign and cancerous tumors of the breast and endometrium (lining of the uterus). Also, people with CS often have small growths on the face or in the mouth and a larger than average head size. Genetic testing for the PTEN gene is available.

CELL CYCLE

Normal cells grow and divide in an orderly fashion, in accordance with the cell cycle. (Mutations in proto-oncogenes or in tumor suppressor genes allow a cancerous cell to grow and divide without the normal controls imposed by the cell cycle.) The major events in the cell cycle are described in Fig. 26.4

Several proteins control the timing of the events in the cell cycle, which is tightly regulated to ensure that cells divide only when necessary. The loss of this regulation is the hallmark of cancer. Major control switches of the cell cycle are cyclin-dependent kinases. Each cyclin-dependent kinase forms a complex with a particular cyclin, a protein that binds and activates the cyclin-dependent kinase. The kinase part of the complex is an enzyme that adds a phosphate to various proteins required for progression of a cell through the cycle. These added phosphates alter the structure of the protein and can activate or inactivate the protein, depending on its function. There are specific cyclin-dependent kinase/cyclin complexes at the entry points into the G1, S, and M phases of the cell cycle, as well as additional factors that help prepare the cell to enter S phase and M phase.

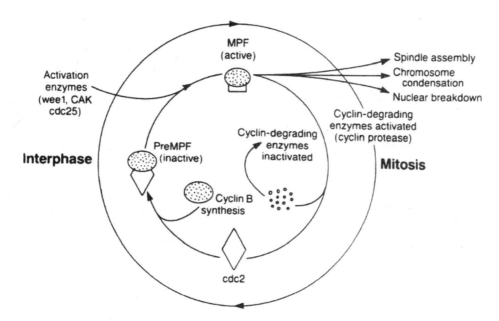

Fig 26.4 The cyclin-cdc2 cycle. During interphase, cyclin B accumulates and associates with cdc2 to form pre-MPF. This is then sequentially phosphorylated by wee1 (a cyclin-cdc2-specific protein tyrosine kinase) on Tyr15 of cdc2, and then by CAK (cdc2-activating kinase) on *Thr* 161:cdc25, a protein tyrosine phosphate then dephosphorylates *Tyr* 15, leaving active MPF. This triggers mitosis and activates cyclin protease. As cyclin is broken down, MPF disperses and the cyclin-degrading enzymes are inactivated. Thus, cyclin begins to accumulates once more. (Adapted from Ruddon, 2007).

One important protein in the cell cycle is p53, a transcription factor that binds to DNA, activating transcription of a protein called p21. P21 blocks the activity of a cyclin-dependent kinase required for progression through G1. This block allows time for the cell to repair the DNA before it is replicated. If the DNA damage is so extensive that it cannot be repaired, p53 triggers the cell to commit suicide. The most common mutation leading to cancer is in the gene that makes p53.

Li-Fraumeni syndrome, an inherited predisposition to multiple cancers, results from a germ line (egg or sperm) mutation in p53. Other proteins that stop the cell cycle by inhibiting cyclin dependent kinases are p16 and RB. All of these proteins, including p53, are tumor suppressors.

In respect of cell division, normal cells differ from cancer cells in the following four ways.

(1) Normal cells require external growth factors to divide. When synthesis of these growth factors is inhibited by normal cell regulation, the cells stop dividing. Cancer cells have lost the need for positive growth factors, so they divide whether or not these factors are present. Consequently, they become independent cells and do not behave like part of the tissue.

(2) Normal cells show contact inhibition; that is, they respond to contact with other cells by ceasing cell division. Therefore, cells can divide to fill in a gap, but they stop dividing as soon as there are enough cells to fill the gap. This characteristic is lost in cancer cells, which continue to grow after they touch other cells, causing a large mass of cells to form.

(3) Normal cells age and die, and are replaced in a controlled and orderly manner by new cells. Apoptosis is the normal, programmed death of cells. Normal cells can divide only about fifty times

before they die. This is related to their ability to replicate DNA only a limited number of times. Each time the chromosome replicates, the telomeres get shorten. In growing cells, the enzyme telomerase restores replaces these lost ends (see chapter on ageing). Adult cells lack telomerase, limiting the number of times the cell can divide. Conversely, telomerase is activated in cancer cells, allowing an unlimited number of cell divisions.

(4) Normal cells cease to divide and die when there is DNA damage or when cell division is abnormal. Cancer cells continue to divide, even when there is a large amount of damage to DNA or when the cells are abnormal. These progeny cancer cells contain the abnormal DNA; so, as the cancer cells continue to divide, they accumulate even more damaged DNA.

ANGIOGENESIS

Although tumor cells are no longer dependent on the control mechanisms that govern normal cells, they still require nutrients and oxygen in order to grow. All living tissues are amply supplied with capillary vessels, which bring nutrients and oxygen to every cell. As tumors enlarge, the cells in the center no longer receive nutrients from the normal blood vessels. To provide a blood supply for all the cells in the tumor, it must form new blood vessels. In a process called angiogenesis, tumor cells produce growth factors, which induce formation of new capillary blood vessels. The cells of the blood vessels that divide to make new capillary vessels are inactive in normal tissue; however, tumors make angiogenic factors, which activate these endothelial cells to divide. Without the additional blood supplied by angiogenesis, tumors can neither grow nor spread, or metastasize, to new tissues. Tumor cells can cross through the walls of the capillary blood vessel at a rate of about one million cells per day. However, not all cells in a tumor are angiogenic. Both angiogenic and non-angiogenic cells in a tumor cross into blood vessels and spread; however, non-angiogenic cells give rise to dormant tumors when they grow in other locations. In contrast, the angiogenic cells quickly establish themselves in new locations by growing and producing new blood vessels, resulting in rapid growth of the tumor.

How do tumors begin to produce angiogenic factors? An oncogene called BCL2 has been shown to greatly increase the production of a potent stimulator of angiogenesis. It appears, then, that oncogenes in tumor cells may cause an increased expression of genes that make angiogenic factors. There are at least fifteen angiogenic factors and production of many of these is increased by a variety of oncogenes. Therefore, oncogenes in some tumor cells allow those cells to produce angiogenic factors. The progeny of these tumor cells will also produce angiogenic factors, so the population of angiogenic cells will increase as the size of the tumor increases.

How important is angiogenesis in cancer? Dormant tumors are those that do not have blood vessels; they are generally less than half a millimeter in diameter. Several autopsy studies in which trauma victims were examined for such very small tumors revealed that thirty-nine percent of women aged forty to fifty have very small breast tumors, while forty-six percent of men aged sixty to seventy have very small prostate tumors. Amazingly, ninety-eight percent of people aged fifty to seventy have very small thyroid tumors. However, for those age groups in the general population, the incidence of these particular cancers is only one-tenth of a percent (thyroid) or one percent (breast or prostate cancer). The conclusion is that the incidence of dormant tumors is very high compared to the incidence of cancer. Therefore, angiogenesis is critical for the progression of dormant tumors into cancer.

VIRUSES AND CANCER

Many viruses infect humans but only a few viruses are known to promote human cancer. These include DNA viruses and retroviruses (RNA viruses). Viruses associated with cancer include human papillomavirus (genital carcinomas), hepatitis B (liver carcinoma), Epstein-Barr virus (Burkitt's lymphoma and nasopharyngeal carcinoma), human T-cell leukemia virus (T-cell lymphoma); and, probably, a herpes virus called KSHV (Kaposi's sarcoma and some B cell lymphomas). The ability of retroviruses to promote cancer is associated with the presence of oncogenes in these viruses. These oncogenes are very similar to proto-oncogenes in animals. Retroviruses have acquired the proto-oncogene from infected animal cells. An example of this is the normal cellular c-SIS proto-oncogene, which makes a cell growth factor. The viral form of this gene is an oncogene called v-SIS. Cells infected with the virus that has v-SIS overproduce the growth factor, leading to high levels of cell growth and possible tumor cells.

Viruses can also contribute to cancer by inserting their DNA into a chromosome in a host cell. Insertion of the virus DNA directly into a proto-oncogene may mutate the gene into an oncogene, resulting in a tumor cell. Insertion of the virus DNA near a gene in the chromosome that regulates cell growth and division can increase transcription of that gene, also resulting in a tumor cell. Using a different mechanism, human papillomavirus makes proteins that bind to two tumor suppressors, p53 protein and RB protein, transforming these cells into tumor cells. Remember that these viruses contribute to cancer, they do not by themselves cause it. Cancer, as we have seen, requires several events.

ENVIRONMENTAL FACTORS

Several environmental factors affect one's probability of acquiring cancer. These factors are considered carcinogenic agents when there is a consistent correlation between exposure to an agent and the occurrence of a specific type of cancer. Some of these carcinogenic agents include X-rays, UV light, viruses, tobacco products, pollutants, and many other chemicals. X-rays and other sources of radiation, such as radon, are carcinogens because they are potent mutagens. Marie Curie, who discovered radium, paving the way for radiation therapy for cancer, died of cancer herself as a result of radiation exposure in her research. Tobacco smoke contributes to as many as half of all cancer deaths in the U.S., including cancers of the lung, esophagus, bladder, and pancreas. UV light is associated with most skin cancers, including the deadliest form, melanoma. Many industrial chemicals are carcinogenic, including benzene, other organic solvents, and arsenic. Some cancers associated with environmental factors are preventable. Simply understanding the danger of carcinogens and avoiding them can usually minimize an individual's exposure to these agents.

The effect of environmental factors is not independent of cancer genes. Sunlight alters tumor suppressor genes in skin cells; cigarette smoke causes changes in lung cells, making them more sensitive to carcinogenic compounds in smoke. These factors probably act directly or indirectly on the genes that are already known to be involved in cancer. Individual genetic differences also affect the susceptibility of an individual to the carcinogenic affects of environmental agents. About ten percent of the population has an alteration in a gene, causing them to produce excessive amounts of an enzyme that breaks down hydrocarbons present in smoke and various air pollutants. The excess enzyme reacts with these chemicals, turning them into carcinogens. These individuals are about twenty-five times more likely to develop cancer from hydrocarbons in the air than others are.

EVOLUTIONARY AND ECOLOGICAL ASPECT OF CANCER

Development of cancer can be viewed as an evolutionary process. In the long-lived complex human body, comprised of trillions of cells, there is a need for continuous replenishment of certain kinds of cells such as skin, blood, and gut cells. DNA must be copied at each cell division and that means sometimes mistakes can be made during cell division. While mutations in all cells, except germ cells, will not be passed on to the next generation, some mutations can ignite a type of evolutionary arms race within our bodies that is the leading cause of death, namely cancer. Tumors arise through the process that involves the three key ingredients of evolution in nature. They are chance mutations, selection and time. The initial events in cancer formation are mutations that compromise the mechanisms that control how cells multiply and how they interact with their neighbors. Some combinations of mutations bestow selective advantages upon cells that enable them to proliferate unchecked. As these tumors grow, additional mutations occur that may give cells the ability to leave their original location and travel to, invade, and proliferate in other body tissues (metastasis). A critical event in tumor growth is that specific genes get mutated always in particular type of cancer. The identification of altered genes in cancer has lead to the discovery of new-targeted therapies using the antibody of the mutated genes. As a result, the cancer cells in a tumor become heterogeneous and some subpopulations of these cancer cells will, by chance, become drug resistant. Thus there is an ongoing competition between drug resistant and susceptible cancer cells in establishment of metastatic cancer in the body. Natural selection always prefers the most drug resistant phenotype in the cancer invasion to metastasis.

Increasing evidence indicates that tumor-stromal cell interactions have a crucial role in tumor initiation and progression. These interactions modify cellular compartments, leading to the co-evolution of tumor cells and their microenvironments. As mentioned above, tumorigenesis itself is an evolutionary process. In evolutionary terms, natural selection and species evolution are the result of complex interaction among living organisms and their environment. The tumorigenic process shares many similarities with the evolution of ecosystems; within tumors a continuous selection exists for tumor cells with the highest survival and reproductive (proliferative) advantage. Moreover, tumors persistently shape their microenvironments, thereby establishing an abnormal ecosystem. Similarly, during metastatic progression from a mix of tumor cells, those cells that can survive in the circulation and adapt to the new environment of a distant organ are the ones that will prevail and proliferate. Therapeutic interventions also place selective pressure on tumor cell populations and cells with acquired or pre-existing mutations that confer resistance will continue to proliferate. It is well established that gene expression changes occur in all cell types during breast cancer progression. A large fraction of genes abnormally expressed in tumors encode secreted proteins and receptors implying that alterations of paracrine and autocrine signaling occur as tumors evolve. The paracrine interactions between stromal and tumor epithelial cells promote the proliferation, invasiveness, tumorigenecity and metastatic potential of immortalized epithelial or cancer cells. For example, pancreatic tumor epithelial cells secrete factors (eg.Hh) that activate adjacent fibroblasts, thereby triggering their secretion of tumor growth promoting factors. Cancer- associated fibroblasts can originate from multiple sources. It is even suggested that the stroma is the crucial target of chemical carcinogen-induced mammary tumors and that mutations in epithelial cells are not sufficient for tumor initiation.

It may thus seem that evolutionary principles can be effectively applied to the origination and evolution of cancer in humans by way of chemical warfare to treat the disease and the subsequent ability to acquire resistance by the cancer cells for their own survival and spread to newer areas or organs. For this to happen, there occurs a co-evolution of the microenvironment with which the cancer cells themselves interact and grow to posterity. Hence, the evolution of cancer within us can be likened to the evolution of certain endoparasites such as malaria, but by keeping the enemy genes for cancer growth within us as well as producing more genomic and chromosomal alteration during the progression of cancer making the eradication of the disease more tedious and difficult. As we understand the molecular mechanisms underpinning the tumorigensis and well as their invasion to other organs, we try to make fresh chemotherapeutic approaches, but the cancer cells outwit our efforts by way of producing newer weaponry in the form of additional mutations and upsetting the chemotherapeutic programmes.

CANCER STEM CELLS

In recent years, stem cells have been the focus of a tremendous amount of biomedical research due to their potential for regenerative medicine. The discovery of cancer stem cells (CSCs) has stimulated even greater excitement among stem cell and cancer biologists. The idea that cancer stem cells could be the source for tumor cell formation came from the fact that both normal stem cells and cancer cells shared many features including their ability to proliferate indefinitely and remain undifferentiated. No doubt, the self-renewal properties of the CSCs are the real driving force behind tumor growth. These tumor propogating cells sustain tumor growth by maintaining a subpopulation of cells exhibiting the hallmark of stem cell properties of self-renewal and a capacity for differentiation.

It is well understood that the self-renewing multipotent stem cells exist in every organ in the body of mammals and most likely humans. These cells presumably are called on to proliferate in response to tissue injury and are involved in tissue repair. In addition, these stem cells derived from adult tissues are also capable of generating other tissue cell types in addition to their own. Since stem cells and cancer cells have a number of characteristics in common, the stem cells present in various tissues and organs can for the easy target for carcinogenic agents. It is already known that one of the drawbacks of embryonic stem cells as a source for tissue regeneration is that they can form teratomas, the tumors that are made up of a wide variety of cell types without any organized organ structure. Stem cells by their very nature are set to proliferate for several population doublings and thus have greater opportunity for carcinogenic mutations to accumulate than in most mature cell types. The ESC-like signature was detected in the CSC-enriched subpopulation of human breast cancers and it appears the ESC-like module is expressed by certain human tumors and that cancer stem cells may be partially responsible for the presence of this signature in bulk tumors. Indeed, many of the cell surface receptors expressed on stem cells are also found on cancer cells. Also, a ras-like pro oncogene, Eras, is expressed in mouse ES cells, which may give these cells tumor-like properties. Since normal stem cells and cancer cells share the same capacity for unlimited self-renewal, it is reasonable to assume that cancer cells acquire the machinery for cell proliferation that is expressed in normal tissue stem cells. However, cancer cells lose the feedback systems as to when to stop proliferating and when to start differentiating, unlike the stem cells. Expectedly, a number of signal transduction pathways and their regulatory mechanisms are shared by stem cells and cancer cells. They include the expression of genes involved in preventing apoptosis, e.g., bcl-2, and the developmentally regulated genes, Shh, Wnt, Notch and BMP. For example, Wnt and Notch pathways have been shown to contribute to the self-renewal of stem cells and progenitor cells in a variety of organs, including hematopoietic and nervous system. When dysregulated, these pathways can contribute to oncogenesis. Mutations of these pathways have been associated with a number of human tumors, including colon carcinoma and epidermal tumors for Wnt, medulloblastoma, basal cell carcinoma, and T-cell leukemias for Notch. Recent studies have indicated that genetic changes leading to aberrant Wnt signaling in either stem cells or progenitors have been implicated as early events in the development of leukaemia and other cancers. Other group of molecules implicated as regulators of self renewal of cancer initiating cells are the cytokines. Recent studies have identified the pleiotropic cytokine TGF-b to be involved in self-renewal of glioma initiating cancer cells through the induction of LIF and the JAK-STAT pathway

DISCOVERY AND ORIGIN OF CSCS

Applying the principles established from stem cell research, human CSCs are functionally defined by their enriched capacity to regenerate cancers using xenograft mouse models. Like normal stem cells, CSCs can reproduce themselves through the process of self-renewal, which can be studied in serial transplantation assays. Cancers derived from purified CSCs recapitulate the heterogeneous phenotypes of the parental cancer from which they were derived, reflecting the differentiation capacity of CSCs. These observations suggest that CSCs contain the complete genetic programs necessary to initiate and sustain tumor growth.

Table 26.2 lists the surface markers, used to describe CSCs from different cancers. Interestingly, cell surface markers for leukemia stem cells (CD34+, CD38-) are virtually identical to normal HSCs. Similarly, CD133, which is a marker for normal neural stem cells, has also been found to express in the brain tumor stem cells. Yet, CSCs have also their own distinguishing molecular profiles. As an example, the Bmi-1 proto-oncogene plays an important role in the regulation of self-renewal for both leukemia stem cells and HSCs. Bmi-1 is also required for neural stem cell self-renewal and is highly expressed in brain tumor CSCs. On the contrary, loss of the PTEN tumor suppressor functionally distinguishes leukemia stem cells from normal HSCs. Identification of such unique molecular features of CSCs that discriminate them from normal stem cells is pivotal for devising specific therapies that would spare normal stem cells. Thus, drugs more specifically targeted to the cancer stem cell population should result in more effective and durable responses. Determining the expression pattern of genetic and phenotypic markers at various stages of cancer formation is important to control cancer using such molecular methods. Expectedly, genes highly expressed at the invasive stage were already expressed in preinvasive stages, suggesting that the cancer stem cell population may be present early in tumor development. Thus, the expression of a subset of genes could be quantitatively correlated with the transition from preinvasive to invasive growth. To provide further evidence in this line it has been shown that hedgehog and Bmi-1 signalling pathways are activated in human breast cancer stem cells.

Organ	Normal stem cell markers	Cancer stem cell markers
Hematopoietic	(i) CD34+ CD38- Thyl- Lin- (ii) c-Kit+ Thyl low Scal+ Lin-	(i) CD34+ CD38- Thyl-Lin-
Breast	(ii) CD24^{med} CD49f^{hi} (ii) CD29^{hi} CD24+Lin-	(i) CD44+ CD24-/^{low} ESA+ Lin-
Brain	(i) CD133+ Lin-	(i) CD133+
Lung	(ii) Sca-1+CD34+Lin-	(ii) Sca-1+CD+Lin-
Skin	(ii) CD34+Cd71^{low}α6-integrin^{hi}	(i) CD20+
Prostate	(i) CD133+ a2bl^{hi}	(i) CD44+ α2β1^{hi} CD133+

Med = medium; hi = high

Table 26.2. List of cell surface markers for normal or malignant tissue stem cells reported in human (i) or mouse (ii) (Adapted from Li *et al.*, 2007)

EVIDENCE FOR CSC ORGIN FROM PROGENITOR CELLS

Several lines of evidence support the idea that a committed progenitor can be the cancer initiating cell as a result of oncogenic transformation. Co-expression of Bcl-2 and the BCR/ABL protein in committed myeloid progenitors is sufficient to drive leukemia development in mice. BCR/abl fusion protein is an oncogene fusion protein consisting of BCR and ABL. This protein is generally associated with chronic myeloid leukemia in humans. BCR/abl chromosome arises from a translocation in which one half of the long arm of chromosome 22 becomes attached to the end of the long arm of chromosome 9, creating dominant oncogene BRC/abl at the joint point. Similarly, researchin brain tumor development also indicate more committed neural progenitor cells are likely to be the targets of tumorigenic mutations.

Cell fusion has been shown to be one of the mechanisms for the apparent cellular plasticity associated with tissue stem cells. Theoretically, cell fusion between stem cells and mutated cells might lead to regaining of self-renewal capacity to allow further accumulation of transforming mutations. In fact, the fusigenic factor CD44 is used as a positive surface marker for CSCs in breast cancer implying that these cells may have the capacity to fuse with other cell types. If cell fusion is an originator of CSCs, it could easily explain the detection of both fusigenic proteins and aneuploid cells commonly associated with neoplastic malignancies.

CANCER STEM CELLS, TUMOR MIGRATION AND METASTASIS

By far the most detailed knowledge concerning normal stem cell migration comes from the hematopoietic system. Intriguingly, many of the factors known to govern hematopoietic stem cells (HSC) migration are also critical mediators of cancer metastasis. Thus, in response to stem cell mobilizing agents such as G-CSF and signaling through the laminar receptor, increased numbers of HSCs circulate out of the bone marrow. The laminin receptor is also now known to play a key role in cancer metastasis. Similarly, stromal cell derived factor and its receptor CXCR4 form a critical regulatory axis for HSC migration, engraftment and homing, and also function in the metastasis of breast, prostrate and other types of cancer. In addition to providing niches for HSCs, skeletal bones are also the most common sites for cancer metastasis. Thus elevated calcium sensing receptor in primary breast cancer samples correlates positively with bone metastasis. Therefore, bone specific factors such as the level of calcium ions may serve as chemo-attractants for guiding the migrating cancer cells into the bone.

Metastasis is a complex multi-step process. Tissue tropisms associated with cancer metastasis indicate that specific and distinct cellular mechanisms are involved in this process. Genetic mutations attained late in tumorigenesis are thought to provide a selective advantage for cells to metastasize. However, recent studies reveal that metastasis capacity is predetermined by genetic changes acquired at the initial stages of tumor development. Functional genomic studies made in recent times have begun to shed light on the cellular and molecular mechanisms of tissue specific metastasis.

A model for tissue- specific metastasis mediated by metastatic CSCs (mCSCs) suggests a transformation event in the normal stem or progenitor cells, followed by self-renewal leading to an expansion of the CSC pool. This pool of tumor-initiating cells has the capacity to expand into a fully heterogeneous primary tumor mass. Secretion of pre-metastasis niche- forming factors plays a critical role in determining the tissue tropism of the future metastatic lesion. Once the mCSCs begin to migrate through the blood, they are guided by homing and anchorage factors produced by the niche(not clear because the CSC would have left the niche on migration). After seeding, the local microenvironment in the niche helps determine if the mCSCs will either proliferate into a metastatic lesion directly, or will enter a quiescent period , which can be cut short by reactivation signals that promote expansion into a full-blown metastatic lesion.

An important outcome from this study is the identification of key steps of metastasis in providing potential targets for therapeutic intervention. Significantly, this model recommends that preventive treatment for metastasis should be started earlier than current practices. If mCSCs are highly mobile and able to generate metastases, then targeting the homing or seeding of these cells in the potential metastatic niches at the initial time of tumor presentation could significantly inhibit disease progression. In fact, blocking the homing factor CXCR4 has been shown to effectively prevent both primary tumor formation as well as metastasis in animal models. This treatment strategy has relevance to the fact that stem cells have the inherent property to pump drugs out with the use of the ABC family of drug transporters, thereby making the chemotherapies ineffective(Fig. 26.5).

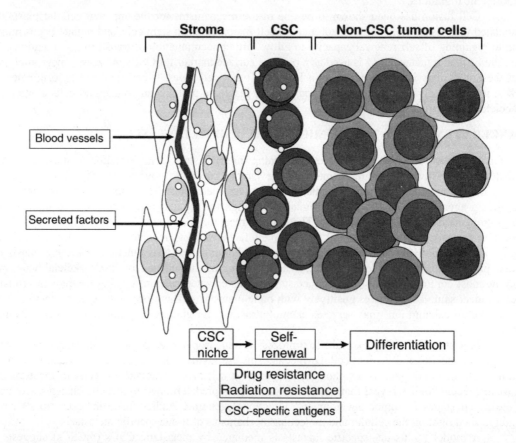

Fig 26.5 Model of cancer stem cells (CSCs) in solid tumours. CSCs are associated with the stromal components of tumor, including fibroblasts and/or blood vessels, which make up the CSC 'niche'. The niche cells secrete factors that support CSC self renewal. CSCs retain differentiation potential, giving rise to non-self renewing tumor cells that make up the bulk of the tumor. CSCs may be drug and/or radiation resistant, and may express CSC-specific antigens. Current research focuses on identification of the molecular mechanisms regulating these properties of CSCs highlighted, as they represent potential targets for therapy.(Redrawn from Ailles and Weissman, 2007)

Regulation of cell mobility and invasion is important for the dissemination of tumor cells from their primary location to lymph or blood vessels during metastasis. The epithelial to mesenchymal transition to cancer cells leads to increased cell mobility and tumor progression, and is known to involve transforming growth factor β (TGFβ).The transient and local factor activation of

TGFβ signalling in breast cancer cells swithces them from cohesive movement to single cell motility and promotes haematogenus metastasis.

Imaging of fluorescent reporter genes demonstrated that TGFβ is active primarly in single cells and that this correlates with the nuclear localization of Smad2 and Smad3, which are phosphylated by TGFβ and forms acomplex with Smad4 that accumulates in the nucleus. When cultured in the presense of TGFβ, tumor cells moved singly instead of growing in colonies. This was inhibited by knockdown of Smad4, suggesting a role for TGFβ mediated transcription in determining the mode mof migration. The genes upregulated due to cancer cell treatment with TGFβ have roles in the switch from collective to single cell motility, suggesting that TGFβ activates a programmed transcriptional mechanism to influence cell motility.The transient activation of TGFβ enables single cells to enter the blood and its subsequent inactivation permits growth at secondary site.

ROLE OF EPITHELIAL TO MESENCHYMAL TRANSITION IN CANCER INVASION AND METASTASIS-

Metastasis, being a multistep process involving the dissemination of cancer cells from the site of primary tumors, has been closely related to disruption of intercellular contacts and loss of adhesion molecules. The disruption of the cell-cell contact initiates metastasis through induction of an epithelial-to-mesenchymal transition (EMT), invasiveness, and anoikis resistance. Epithelial–mesenchymal transition is an important program during development by which epithelial cells acquire mesenchymal, fibroblast-like properties and is characterized by repression of E- cadherin, increased cell mobility and loss of cell adhesion. Induction of EMT plays a critical role in the initiation of tumor progression, malignant transformation, and metastasis in carcinomas of epithelial origin. Such an induction is brought about through several oncogenic pathways such as Src, Ras, integrin, Wnt/beta-catenin and Notch. Significantly, Ras-MAPK has been shown to activate two transcription factors Snail and Slug which are transcriptional repressors of E-cadherin. It has been demonstrated recently that cancer cells from solid tumors of epithelial origin can be induced to pass through the EMT by merely the repression of CDH1 gene, which encodes E-cadherin.

The role of EMT in the formation of cancer stem cells was first reported in 2008 by Mani et al., in mammary epithelial cells. They showed that epithelial cells that have undergone EMT not only appear mesenchymal but also behave like stem cells. To elucidate this, they induced EMT by the ectopic expression of Twist or Snail transcription factors both which regulates epithelial-mesenchymal transition and early embryonic morphogenesis and are capable of inducing EMTs in epithelial cells. They used non-tumorigenic, immortalized human mammary epithelial cells (HMLEs). The transition from the epithelial to mesenchymal characteristic was shown to be accompanied by the down-regulated expression of mRNAs encoding epithelial markers, such as E-cadherin, and upregulated mRNAs encoding mesenchymal markers, such as N-cadherin, vimentin, and fibronectin. Their results demonstrated that CSC is a state that can be generated from more differentiated tumor cells instead of solely via the transformation of local stem cells. Furthermore their finding also suggested that large numbers of stem-like cells can be generated from differentiated epithelial cells simply by inducing transient EMT. EMT may act as a central regulator of cancer metastasis by not only enabling cancer cells to disseminate but also by enforcing the initiation of tumor growth in the new location.

Recently, the EMT program was exploited to develop a high-throughput screening method to identify agents with specific toxicity for epithelial CSCs. The proportion of CSCs in breast cancer cell populations was enriched by inducing HMLER cells to pass through an EMT by short hairpin RNA (shRNA)-mediated inhibition of the E- cadherin encoding CDH1 gene. Another candidate molecule identified very recently to regulate the EM transition is a transcription factor KLF17 (also known as ZNF393) is a member of the Sp/KLF (specificity protein/Krüppel-like factor) zinc-finger protein family. KLF17 family member which are critical regulators of various cellular processes, including reprogramming of differentiated cells to stem cell, erythropoiesis and cell survival was shown to be a metastasis suppressor in human breast cancer.

CELL SIGNALING PATHWAYS IN CANCER STEM CELLS

CSC functions are regulated by major receptor mediated pathways. Behaviour of CSC are affected by signals from neighboring stromal, immune and non tumor cells presnt in their niche. Of these the major signaling programs are initiated from the outside by receptor tyrosine kinase, bone morphogenetic protein (BMP), hedgehog and Notch (refer Figure). Hedgehog inhibition inproves the efficacy of the standard chemotherapy by inhibiting CSC proliferation and promoting apoptosis. Activation of Notch signaling by over expression of Notch-1 and its ligand , Delta or Jagged has resulted in the increase in stem cell marker, nestin in gliomas and CSCs have displayed sensitivity to Notch inhibition. In glioma stem cells, a prodifferentiation BMP response mechanism is preserved. In these cells, BMP receptor is overexpressed and its binding with BMP inhibit the proliferation of these cells and induction into differentiation of CSCs into the astroglial and neuronal lineages.

IDENTIFICATION OF CANCER STEM CELLS AND THERAPEUTIC INTERVENTION

Strategies used to identify and isolate the CSCs from neoplasms rely on characteristics associated with normal stem cells, including the expression of particular cell surface markers (see above), the capacity to efflux fluorescent dyes by multidrug transporters and the ability to clonally expand into spheres under specific culture conditions. Cell surface markers can serve as either positive or negative criteria for a stem cell phenotype. The efflux capacity for the Hoechst 33342 fluorescent dye (that binds to DNA) is the most widely used strategy to identify cancer stem cells. This dye expulsion is mediated by multidrug-resistant membrane efflux pumps of the ATP-binding cassette transporter super family, in particular by ABCG2.

Common cell-surface markers used in the identification of CSCs of solid tumors are CD133 (or prominin-1), CD44, CD24, ALDH, ESA and epithelial-specific antigen (ESA; also known as epithelial cell adhesion molecule or Ep-CAM). For example, pancreatic cancer stem cells with either a CD44+ CD24+ ESA+ or a CD133+ phenotype have been identified from primary human pancreatic ductal adenocarcinoma. Isolated CD133+ cells formed tumors histologically indistinguishable from original tumors. However, metastatic potential of pancreatic ductal adenocarcinoma seems to be determined by a subpopulation of CD133+ cells that co expresses CXCR4. CXCR4 is the receptor of stromal cell-derived factor1 (SDF1), which is a potent chemokine produced by fibroblasts in the tumor microenvironment. The SDF1-CXCR4 signalling pathway has an important role in tumor invasion and metastasis. SDF1 is strongly expressed in target organs of metastasis (such as liver and lungs), lending support to the 'seed and soil' hypothesis of the metastatic process. Thus, continuous pharmacological inhibition of the CXCR4 receptor by AMD3100 abrogates the metastatic phenotype of pancreatic xenografts without affecting their tumorigenic potential.

Developmental signaling pathways that regulate self-renewal and cell fate in normal stem cells are also involved in (pancreatic) cancer stem cells and might serve as novel biomarkers and therapeutic targets. Two good examples are Notch and Sonic Hedgehog. Notch pathway components and Notch target genes are upregulated in invasive pancreatic cancer and cancer precursor lesions in mice and humans. A series of γ-secretase inhibitors are currently being investigated as potential agents to treat breast cancer with Notch activation. Similarly, Sonic Hedgehog (SHH) is a well known ligand that promotes the proliferation of adult stem cells from various tissues. Activation of the SHH pathway has been implicated in the development of various cancers. For example, aberrant SHH expression was observed in 70% of 20 resected human pancreatic ductal adenocarcinoma specimens, whereas no expression was found in normal pancreatic ductal cells. Thus, SHH pathway might serve as a promising target for new therapies. The SMO protein is a key signaling component of the SHH pathway and is specifically inhibited by cyclopamine, a natural steroidal alkaloid. This and other SMO inhibitors suppress the growth and metastasis of tumors that show activated Hedgehog signaling.

Recently, inhibition of the PI3 kinase pahway was found to sensitize putative CSCs to radiation induced apoptosis in the perivascular niche of medulloblastoma. Screens for agents that specifically kill epithelial CSCs are hampered by the rarity of these cells within the tumor cell populations and their selective instability in culture. Very recent study on the chemical screening for selective toxicity for breast CSCs has revealed that salinomycin, a potassium ionophore, is found to reduce the proportion of CSCs by 100 fold relative to paclitaxel, a commonly used breast cancer chemotherapeutic drug. In fact treatment of paclitaxel actually imposes a strong selection for CSCs survival and expansion. Furthermore, recent clinical observations have shown that after conventional chemotherapy, breast tumors have an increased proportion of cells with a CD 44 hi/ CD 24 lo marker profile and increased tumor sphere-forming ability. Another finding indicates the possibility that the elimination of the CSCs within a tumor may not result in its complete regression. Hence, it is preferable to develop combination therapies that apply agents with specific toxicity for CSCs together with agents that specifically target non-CSC populations within tumors. The refractoriness of CSCs to chemotherapy have been implicated to ABCG2, ABCG5 and multidrug resistant proteins (MDR1) transporters. Interestingly, the specific targeting of the ABCG5+ subset of tumor population with a monoclonal antibody inhibited tumour growth and melanoma progression.

CURRENT CHALLENGES AND FUTURE DIRECTIONS FOR CANCER THERAPY

In the words of Bruce Chabner and Thomas Roberts Jr., "Of the many challenges of medicine, none has had a more controversial beginning and none has experienced more hard-fought progress than the treatment and cure of cancer". Although the neoplastic process has been recognized for centuries, little was known about the biological mechanisms of transformation and tumor progression until the advent of molecular medicine in the latter half of the twentieth century. From the first experiments with nitrogen mustard 60 years ago to current attempts to develop drugs for specific cancer-related targets, researchers from multiple disciplines have joined together in the search for more effective cancer drugs. Over time, the development of anticancer therapies, based at first on empirical observations, has become increasingly dependent on an understanding of human tumor biology.

The major challenge(s) faced by clinicians during the cancer treatment was unanimously the "resistance" exhibited by the tumor to various therapeutic interventions. For example, the tumor that responded for chemotherapy initially demonstrated recurrence more often than not, surprisingly with an unexpected resistant phenotype. Similarly, the initial enthusiasm generated by radiation therapy could not be enjoyed for long (except for few isolated single nodule tumors) as it was slowly realized that the by-stander effect of radiation proved to be more lethal to the vital organs, eventually threatening the life of patients. It was realized that the therapeutic efficacy of a specific regimen relies on the stage of the tumor. For instance, if a primary liver tumor (not metastasized from other organs) was diagnosed at an early stage before it advanced to multiple nodular or invasive phenotypes, the chances of treatment for cure are high with a combination of surgical resection and chemotherapy. Unfortunately, this therapeutic option is limited by the lack of diagnostic tools for the early detection of liver cancer. With the result, majority of the patients when presented to the doctor are already in the advanced stage of liver cancer. With the advent of modern technology, for few cancers such as mammary or prostate, there are annual screenings that might help in the early detection of cancer.

Another challenging problem in treating the cancer cells is their biology itself. Tumor cells have some of the unusual phenotypes that make most of therapies less efficient. For instance, as majority of solid tumors rely on glycolysis for their energy requirement, it was opined that any energy blocker that can disrupt glycolysis could be an effective anticancer strategy. For example, 2-deoxyglucose (2-DOG), an analog of glucose impairs glycolysis by competitive inhibition. The underlying principle is that when 2-DOG enters tumor cell it would be phosphorylated to glucose-6-phosphate (like the normal glycolysis) but the lack of single oxygen (deoxy) makes it unsuitable for further metabolism, thus preventing energy production. Although this proved very fruitful in preclinical studies and in clinical trials (I, II), the results were discouraging, as the patients with hypoxic tumors did not respond to therapy. Nevertheless, tumor heterogeneity also poses a major hurdle in surpassing the therapeutic resistance. As majority of the tumors either coincidentally or strategically evade apoptotic network or disable the apoptotic pathways, it is probable that anticancer agents that promote apoptosis would be ineffective. In addition, he altered and abundant expression of MDR (multi drug resistance) genes resulting in the accumulation of MDTs (multi drug transporters) in tumor cell complicates the intracellular delivery and stability of any anticancer agent, as they would be pumped out by MDTs (for example, ABC transporters, p-glycoprotein etc) prior to their therapeutic function. Hence, tumor resistance that primarily depends upon the genetic make-up of a tumor cell (heterogeneity) has gained increased attention for the drug design and development.

Future directions of cancer cure primarily weighs in the philosophy of "personalized medicine" where every patient has to be thoroughly screened for the molecular phenotype in addition to the histopathology of the particular cancer, and would receive treatment accordingly. Unlike, conventional practices where one drug for one disease is administered to like-patients, personalized therapy would require considering variations in tumor biology and chemistry before designing a therapeutic strategy. Secondly, despite our current knowledge and understanding of tumor initiation, growth and development, there are key questions that still remain unanswered regarding their status as an effect or cause of cancer. For example, altered tumor metabolism is still argued intensely as the cause of cancer (as there is substantial evidence that proves targeting metabolism leads to tumor destruction). However, it is unclear how such an altered metabolism could be achieved without a prior oncogenic (transformation) cue or signal. Similarly, increasing body of evidence demonstrates the cellular microenvironment as the controller if not the initiator of tumorigenesis. Nevertheless, its role as the cause or effect of cancer requires further investigation. Thus addressing key biological questions related to tumor initiation would help in designing therapeutic strategies to target developed tumor.

27

MOLECULAR AND EVOLUTIONARY DEVELOPMENTAL BIOLOGY

ORIGIN OF THE NEW SCIENCE

Charles Darwin proposed the theory of evolution in his landmark book, *The Origin of Species*, in 1859. According to him; all life-forms change and develop over the course of time in response to changes and challenges in the natural environment. The ability or lack of ability to adapt to these changes is a key factor in the evolutionary process. Plants and animals that could not adapt to these changes became extinct; while those that were able to adapt survived, often in new forms. In the same manner, species that do not yet exist will someday develop from species living today. Darwin's theory of evolution through natural selection was based on solid scientific evidence that he amassed during his epoch-making voyage in H.M.S Beagle.

It is important to remember that Darwin explained the evolution of organisms on earth, but not the origin of life. However, the expedition results and the inferences he made from them led Darwin to the ultimate conclusion that all organisms on Earth had common ancestors and that all organisms on earth had descended with modification. As Darwin wrote, -- "and that from so simple a beginning endless forms most beautiful and most wonderful have been, and are being, evolved". With numerous studies, thereafter, supporting this conjecture of Darwin, his theory of *common descent* is convincingly confirmed.

Darwinian Theory clearly explained two aspects of evolution. One is the upward movement of a phyletic lineage, its gradual change from an ancestral to a derived condition. This is called anagenesis. The other aspect consists of the splitting of evolutionary lineages, to form new branches (clades) of the phylogenetic tree. This process of the origin of biodiversity is called cladogenesis.

Darwin himself realized that morphological changes must be the result of changes in development. In his time, however, both evolution and developmental biology remained as separate fields, but in the last two to three decades, there has been renewed interest in uniting these two fields, leading to the renaissance of evolutionary developmental biology, often abbreviated as Evo-Devo. This renewed interest is the outcome of our understanding of development at the molecular and genetic levels. When we combine this information with our understanding of the process of evolution at the same molecular and genetic levels, we get better insight into both processes. If evolution is the result of change in populations and species over time, development represents the change of an individual over time.

Evolution of morphology giving rise to diversity of live forms is a challenging subject for biologists. It is in this respect, Evo-Devo is going to offer revolutionary insights into the evolutionary process. It is increasingly understood that genes and the changes in their sequences are responsible for the evolution of morphological diversity. Evo-Devo provides a fundamental contribution to evolutionary theory by seeking to explore the mechanistic relationships between the processes of individual development and phenotypic change during evolution. Advances in techniques for gene cloning and visualization of gene activity in embryonic tissues facilitated the emergence of this new field. Comparison of developmental processes of different taxa at the molecular level permitted the discovery of extensive similarities in gene regulation among distantly related species with fundamentally different body plans.

Modern synthesis or "Neo-Darwinian Synthesis", a synthesis of Darwinism and genetics, emphasized genetic mutation as the source of new selectable variation and the subsequent natural selection. The last arrival in the evolutionary ideas is the "Developmental Synthesis" which maintains that evolution consists of inherited changes in the patterns of development and the mechanisms whereby such changes bring about adult variations. Evolution at this level is obviously the result of alteration in the use of regulatory genes (see below).

The famous quote of Dobzhansky in 1961 that "Nothing in biology makes sense except in the light of evolution" is henceforth modified by the evolutionary developmental biologist, Jernvall in 2006 who stated, "Nothing about variation makes sense except in the light of development". Long story short, evolution cannot be understood without understanding the evolution of development.

Recent advances in developmental genetics have cast new light on the evolutionary processes that led to speciation and natural selection. Basically, evolution is caused by heritable changes in the development of organisms. Certain modification in embryonic or larval development can create new phenotypes that can then be selected. Indeed, embryology has always been an integral component of the evidence for evolution and the principle of common descent. Recent genetic studies on development have uncovered the reconciliation between paleontological views of so-called macroevolution (evolution above the species level) and genetic views of microevolution, the variation detectable within the species. In essence, Evo-Devo attempts to explain the crucial mechanistic details of how genes link development to evolution. Accordingly, changes in the genes that control the development of organisms can bring about major evolutionary transitions.

DEVELOPMENTAL BASIS OF ANIMAL EVOLUTION

Evolution of animal form and complexity could result from the following factors:
1. The first one is the modularity of organization. The ground plan of bilateral animals involves repeated segments that can evolve independently.
2. The second factor is that most animals share a small but similar set of developmental genes that regulate the development of different modules. These genes produce the so-called regulatory proteins viz., transcription factors. Recent genomic studies have indicated that only a tiny fraction of our DNA, roughly about 1.5%, codes for 25000 proteins in our body. On the other hand, nearly 3% of our DNA, comprising about 100 million individual bits, is regulatory in nature. These regulatory DNA contains the instruction for bodybuilding, and hence evolutionary changes within this regulatory DNA lead to the diversity of forms.
3. Animal form evolves largely by altering the expression of functionally conserved proteins, and such changes largely occur through mutations in the *cis*-regulatory sequences of pleiotropic developmental regulatory loci and of the target genes within the vast network they control.
4. The promoters and enhancers in DNA, that regulate the transcription of protein-coding genes, promote evolution by causing existing genes to be expressed at new times and places.

5. In spite of genetic similarity among species, differences in the protein sequences are also causative factors for evolutionary changes.

DEVELOPMENTAL GENES AND EVOLUTION

The development of all metazoan organisms is controlled by a small number of developmental genes. These genes act hierarchically by switching other genes on or off. Many of the genes involved in regulating development in different species exhibit remarkable conservation of genetic pathways across evolution. As a result, different embryos of different animal forms use repeatedly the same kinds of strategies to achieve cellular specialization, tissue patterning and organogenesis.

Interestingly, two German scientists, Nusslein-Volhard and Eric Wieschans who received the Nobel Prize in medicine or physiology in 1995 for this discovery, first made the discovery of these genes in the fruit fly *Drosophila melanogaster*. It was later found out that these master genes that governed the formation and patterning of fruit fly body parts had homologues in higher vertebrates including humans. The comparison of developmental genes between species paved the way for the introduction of the new science of evolutionary developmental biology. On the contrary, an earlier view laid emphasis on population genetics, which explains that evolution is caused by changes in gene frequency between generations. Nevertheless, the new model of evolutionary synthesis integrates the developmental genetic approach to evolution with that of population genetics. In other words, this integrated scientific discipline endeavours to explain embryological evidence in favour of Darwinian evolutionary theory.

MECHANISM OF GENE REGULATION DURING DEVELOPMENT

Julian Huxley, one of the architects of 'Modern Synthesis' of evolutionary theory stated that a study of effects of genes during development is as essential for an understanding of evolution as are the study of mutation and that of evolution. It is therefore important to recognise which genes and what kinds of changes in their sequences are responsible for the evolution of morphological diversity among organisms. Consequently, there exists a causal link between gene mutation, development and the evolution of form.

A surprising discovery was that most animals used a small group of developmental genes in the construction of their diversified body plans. These genes came to be known as "toolkit genes", playing the same roles in development of all animal lineages. Initially, it was proposed that phenotypes could change by altering the timing and rate of protein synthesis. That means changing the gene regulation rather than protein expression by different genes could bring about alterations in protein synthesis. Concordantly, only changes in the regulatory apparatus of the toolkit genes were associated with morphological divergence. Because most animals share a conserved set of bodybuilding and-patterning genes, morphological diversity appears to evolve primarily through changes in the deployment of these genes during development. The patterning genes in particular encode mostly transcription factors and cell-signalling molecules in developmental gene activation. Furthermore, the complex expression patterns of developmentally regulated genes are typically controlled by numerous independent *cis*-regulatory elements (CREs). *cis*-regulatory elements have properties that make their evolutionary dynamics distinct from coding sequences. The modification of cis-regulatory element function can readily occur through point mutations that create or destroy transcription factor binding sites. Such regulatory evolution at the species level is sufficient to account for the large scale patterns of morphological evolution, including origin of humans. The modular organisation of the *cis*-regulatory regions of pattern regulating genes is highly complex with each CRE typically comprising of binding sites for multiple transcriptional factors. This condition enables controlling of gene expression within a special domain in a developing animal. Such modular organisation of CREs could pave way for changes in *cis*-regulatory sequences, controlling gene expression, leading to morphological evolution. Therefore, gene expression to bring about morphological evolution relies primarily on changes in the architecture of gene regulatory networks (transcription factors) and on functional changes within CREs. These evolutionary modifications of gene expressions (i.e., regulatory evolution) form the basis of morphological diversification among closely related species as well as profound anatomical divergence among groups at higher taxonomical levels.

It is now clear that regulatory DNA is the predominant source of the genetic diversity that underlies morphological variation and evolution. The factors contributing to the importance of *cis*-regulatory DNA in evolution include chiefly the modular organization of *cis*-regulatory systems. For example, individual CREs can act, and therefore evolve, independently of others; hence providing opportunity for the expansion and diversification of *cis*-regulatory systems in evolution. This evolvability of regulatory DNA sequence implies that it is a rich source of genetic and phenotypic variation. Furthermore, the combinatorial nature of transcriptional regulation, controlled by the diverse repertoire of transcription factors in animals, has important evolutionary ramifications. This dynamic evolutionary change in CREs suggests possible variations in the level, pattern, or timing of gene expression with consequential morphological evolution.

Evolutionary changes in the *cis*-regulatory regions of most toolkit loci are larger and more complex than those of the genes coding for cell-type-specific physiologically important proteins. In addition, regulatory sequences have different properties and are under different constraints than coding sequences. Hence, CRE divergence could occur in different ways. Mutations in existing CREs could create new binding sites for transcription factors, thus producing new regulatory linkages (i.e., between transcription factor and mutated CRE) and new gene expression patterns. New CREs can also arise by insertion of transposable elements (TE). It has been recently shown that there are thousands of transposable elements insertion near developmentally regulated human genes and that these TE-derived sequences are under strong purifying selection. Purifying selection, also called stabilizing selection, is the most common mechanism of natural selection in which genetic diversity decreases as the population stabilizes on a particular trait value. In other words, this selection acts to keep a character constant in a population.

There can be loss of transcriptional inputs in existing CREs via mutational loss of transcription factor binding sites in CREs. In addition, there can be remodelling of CREs by way of changing the number, affinity, or topology of transcription factor binding sites within a CRE. These changes can alter the output of a CRE in such a way that the pattern or level of gene expression is altered. Concordantly, CRE evolution is an essential ingredient of genetic regulatory network evolution, and hence is the predominant mechanism underlying the evolution of development and form. Furthermore, CRE regulatory evolution also creates novelty. The exploration of new morphologies is facilitated by new combinations of gene expression that can arise without changes in protein function. Thus regulatory DNA is a rich and continuous source of potential genetic, developmental, and phenotypic variation, and thus evolutionary change. In other words, the link between DNA sequence evolution and phenotypic diversity often involves *cis*- regulatory elements acting as units of evolutionary change.

CRE CONTROL OF WING PIGMENTATION IN INSECTS

The best example for evolution through developmental regulatory genes activity is found in the diversified pattern of wing pigmentation in cyclorrhaphan dipteran flies. Fruit flies display all sorts of conspicuous patterns of black pigmentation on thewir head, thorax, abdomen, and wings. These patterns are regulated by a variety of well-conserved signalling pathways and transcriptional factors that control the special expression of the enzymes that promote or inhibit the formation of the pigment melanin.

Wing pigmentation patterns in dipteran flies result from the local conversion of precursor metabolites into pigment deposits by enzymes. Therefore, the expression patterns of the genes encoding these enzymes change the pigmentation patterns among different fly species. Accordingly, pigmentation patterns could evolve by changing the activity or spatial deployment of transcription factors that regulate pigmentation genes and/or by changes in the CREs of pigmentation genes themselves. As an instance, the evolutionary divergence in Yellow expression results from functional changes in a CRE controlling *yellow* (*y*) gene (gene responsible for black pigment) in the developing wing. In the unspotted species, this CRE drives a uniform expression pattern throughout the wing (Fig 27.1). In the spotted species, the regulatory activity of this element has changed to also drive high levels of yellow in the spot area, by co-opting another CRE (*spot* CRE) to generate a novel pattern. In summary, it may be said that wing pigmentation patterns can arise from the evolution of regulatory connections among pigment gene CREs and different combinations of transcription factors.

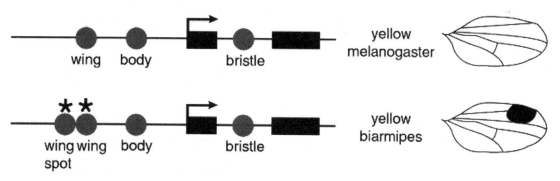

Fig 27.1 Expression of the yellow pigmentation gene of Drosophila is controlled by several different cis-regulatory elements (circles). Differences in the activity of selected elements (wing and wing spot) underlie differences in pigment patterns between species. Asterisks denote gene duplication followed by mutations in the cis-regulatory region.
(Redrawn from Carroll, 2005)

STUDY OF HOMOLOGY

In discussing the developmental mechanisms of evolutionary change, a study of homology is pertinent, as it is one of the central concepts in developmental evolution. Embryologists dealt with homology at morphological level, but it is now viewed at molecular and genetic level. According to Darwin's concept, similarities of structure (homology) were due to common ancestry, while the modifications in the evolved structures were due to natural selection, by way of adaptation to the environmental circumstances. Molecular biologists can now discuss an even deeper homology, the homology of genes. The commonality of certain developmental genes found among animals ranging from lower metazoans to the highly evolved humans suggests molecular evolution of body plan and its complexity from the urbilaterian ancestor, which has given origin to protostomes and deuterostomes.

With the advent of sequencing technology for proteins and genes, the available sequences of molecules and their degree of resemblances were used to trace species relationships in the constructed phylogenetic tree. Because genes are inherited, the sequences of genes and the proteins they encode are passed to descendants of species. Similarly, changes that occur in one species are passed onto its descendants throughout life's long genealogy. Indeed, the degree of similarity in DNA and also proteins is an index of the relatedness of the species. Therefore, the study of homology in gene and protein sequences is highly instructive in not only understanding relationship between species, but also tracing the origination and evolution of life forms themselves. There are now many sophisticated mathematical and statistical formulas to analyse the sequences and decipher and determine the kinship among the species.

HOX GENES AND BODY AXIS FORMATION

Developmental biologists have long recognised many profound differences in the basic processes of patterning and morphogenesis occurring in embryos of different phyla. With the discovery of developmental genes, it became apparent that the basic same sets of genes (toolkit genes) are found to regulate development throughout the animal kingdom. Surprisingly, these developmental regulatory genes exhibited great homology among organisms with different body plans. Among many such developmental genes, the most extensively investigated is the *Hox* genes, which are known to play an ancient and well-conserved role in anterior-posterior patterning during animal development. Recent studies have indicated that changes in the function and expression of these genes can lead to evolutionary changes in morphology at both macroevolutionary and microevolutionary scales. *Hox* genes also exhibit remarkable phylogenetic distribution among animal taxa, including basal metazoan groups. Further, these regulatory genes are expressed in roughly similar patterns during development but produce very different phenotypes, suggesting extensive "rewiring" of developmental gene networks during the course of evolution.

Expectedly, *Hox* genes were identified first in insects. Developmental pathway of undifferentiated cells in the *Drosophila* embryo is specified by master regulatory genes, known as homeotic selector genes. Their expression is required throughout development. These genes occur as two clusters in the third chromosome of the fruit fly. One cluster, the *Bithorax* Complex, contained three genes, which influenced the posterior half of the fly; the other, the *Antennapedia* Complex, contained five genes that affected the front half of the fly (for hierarchical activation of these genes, see chapter 12). The relative order of the genes in these two clusters corresponded to the relative order of the body parts they affected. A surprising finding about these regulatory gene complexes (bithorx and antennapedia) is that any mutation in these genes leads to homeotic transformation. Mutations in *D. melanogaster* bithorx gene caused the transformation of the third thoracic segment into the second thoracic segment, resulting in a fly with two-second thoracic segments and no third thoracic segment. In the normal flies, the second thoracic segment possesses the fly's single pair of wings, whereas the third thoracic segment contains just a small pair of halteres. Thus, the mutant fly with two-second thoracic segments possessed two pairs of wings instead of the normal single pair. As shown in Fig.27.2, homeotic mutation occurring in *Antennapedia* loss of function alleles of *D. melanogaster* causes transformation of antennae into enormous, nearly perfect legs.Not all developmental regulatory genes produce homeotic phenotypes when they are mutated. Many other mutations are known to affect developmental processes in such a way that they either delete specific organs or cell types, alter body proportions, change the timing of reproduction or fecundity, modify color patterns or other integumentary structures, and create a host of other specific phenotypes. The equivalent of *Drosophila* homeotic selector genes was found surprisingly in all known animal groups and is termed as *HOX* genes. *HOX* genes are a particular family of homeobox-containing genes and their expression provides the basis for anterior-posterior axis specification throughout the animal kingdom.

The homeobox (*HOX*) genes encode proteins that bind to the DNA of other genes, triggering a cascade of processes that ultimately yield eyes, limbs, hearts and other complex structures. Over the 60 amino acids of the homeodomain (protein chain derived from the homeobox domain of the homeotic gene), some mice and frog proteins were identical to the fly sequences up to 59 out of 60 positions. Thus *HOX* genes are so important in evolution of metazoan body plan that their sequences have been preserved throughout the span of animal evolution.

Fig. 27.2: A scanning electron microscope image of a Drosophila flies with the Antennapedia mutation. This mutation causes the fly to grow legs where it should grow antenna. (Courtesy: http://www.learner.org)

Some of these developmental genes encode proteins with regulatory functions; some are transcription factors, proteins that regulate the expression of other genes; others are involved in communication among cells (signals, receptors, and the machinery that transduces a perceived signal); and some regulate key cellular processes, such as mitosis and cell death. Communication among cells endows them with distinct fates, which in turn activates different batteries of genes that mediate the processes involved in growth, morphogenesis and differentiation. These studies nevertheless, demonstrated the importance of mutations in developmental genes contributing enormously to phenotypic evolution.

266 Molecular Developmental Biology

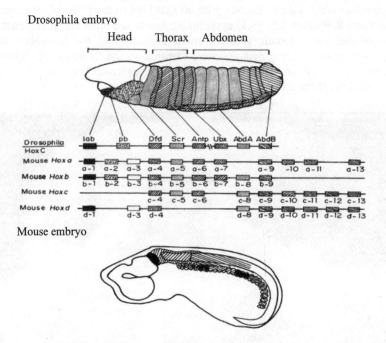

Fig. 27.3 Comparison of *HOX* gene clusters showing the evolutionary conservation of homeotic gene organization and spatiotemporal expression patterns in the *Drosophila* and mouse embryos. In *Drosophila*, two gene clusters namely, Antp-C and BX-C form head, thorax and abdomen. The corresponding *HOX* genes in the mouse are present in four clusters (*HOX*-a, *HOX*-b, *HOX*-c, *HOX*-d). Matching numbers or shades in column indicate strong structural similarities across the two species. (Redrawn from Sean Carroll 2005).

Understandably, variation of morphological form in the animal kingdom is underlain by a common set of genetic instruction. Interestingly, not only are the *HOX* genes themselves homologous, but they are found in the same order on their respective chromosomes. The expression patterns are also remarkably similar between the *HOX* genes of different phyla: the genes at the 3' end are expressed anteriorly, while those at the 5' end are expressed more posteriorly. The *HOX* genes, laid end-to-end in the chromosome express the first gene at the head of the animal in every species so far investigated. So the co-linear expression of the *HOX* genes follows a temporal sequence, and the switching on of each *HOX* gene somehow switches on the next one in line. Fig. 27.3 indicates the similarity in the positioning of the *HOX* genes in their respective chromosomes of both fruit fly and mouse. Interestingly, the ancestral animals seem to have grown more complicated bodies by lengthening and developing the rear end, and not the head end. So the *HOX* genes replay an evolutionary sequence, supporting the famous theory of Ernst Haeckel that 'ontogeny recapitulates phylogeny'.

Furthermore, the human *HOXB4* gene could mimic the function of its *Drosophila* homologue *deformed* (*dfd*), when introduced into *dfd*-deficient *Drosophila* embryos. Another example in this regard is the *Pax-6 HOX* gene, which triggers eye formation. Remarkably, *Pax-6* helps to organize compound eyes in fruit flies and camera eyes in both squid and vertebrates, in spite of the fact that these structures were once thought to have evolved independently. Thus, *Pax-6 HOX* gene from mice has the ability to induce in the fruit fly *Drosophila* the formation of fly eyes all over the body, even on the wings. Another *HOX* gene, *tinman*, induces heart formation in both insects and vertebrates and Distal-less controls the development of fly legs, fish fins and the tube feet of sea urchins. Yet another example of gene homology between fruit fly and mammals is the head forming genes. Two pairs of genes in mice namely, *Otx* (1and 2) and *Emx* (1 and 2) are responsible for building an 'encephalised' front end in the embryo. Surprisingly, these two genes have been found to have equivalents in the development of the head end of the fruit fly. Using techniques such as genetic rescue, it has been shown that human *HOX* genes such as *Otx* and *Emx* genes can rescue their fly equivalents.

Evidently, *HOX* gene expression pattern defines the development of all animals, but variation in *HOX* expression pattern might lead to evolutionary change. This is in contradistinction to the earlier premise that vertebrates invented a whole set of new genes for body planning and construction. To summarize, there exists incredible conservatism of embryological genetics between organisms/ species of disparate appearance and phylogenetic unrelatedness. The fly and humans may look different, but their embryos are similar. All these evidence favour a funny exclamation that human beings are nothing but glorified flies, embryologically speaking.

HOX GENES AND THE EVOLUTION OF TETRAPOD LIMB

Developmental genetic studies have revealed that the vertebrate appendages are derived from a common ancestral design. Tetrapod limbs of terrestrial vertebrates evolved from paired pectoral and pelvic fins of fish in the Devonian period. The principal difference between the paired fins of these fish and the limbs of full-fledged tetrapods is the presence in the latter of hands, feet, and digits. Homologues of the upper arm/ thigh (stylopod) and lower arm / calf (zeugopod) are present in primitive fish fins, but only late Devonian vertebrates possess the third major limb element, called autopod. Therefore, the origin of autopod has direct relationship with the transition from fins with two limb elements to limbs with three limb elements, as found in the fish fossil specimens of Sauripteris and Acanthostega. Both these species possessed autopods, but the 'fingers' were many in numbers. Further evolution of autopod structures involved both reduction in number and specialization of the digits in the land tetrapods. Later modifications in the tetrapod limbs were made under various ecological demands.

In addition to specifying anterior-posterior axis during animal development, *HOX* genes also possess genetic instructions to form limbs. Particularly, *HOX* gene expression has been correlated to the evolution of tetrapod limb from the fish fin. These limbs are created by the interaction of ectoderm and mesoderm. Interestingly, genetic instructions carried in the *HOX* genes to form these two distinctly different types of limbs are extremely similar. In tetrapod, all three phases of limb development (see Fig. 27.3) involve the deployment of two particular sets of *HOX* genes, resident in two of the four *HOX* clusters. Between each phase, the spatial patterns of *HOX* expression change and correlate with the specification of each limb element. Evolution of the third phase of *HOX* expression in the autopod is controlled by separate regulatory DNA, which activates the *HOX* genes in a new distal part of the embryonic limb.

As the tetrapod limb grows outward, the pattern of *HOX* gene expression changes. When the stylopod is forming, *HOXd-9* and *HOXd-10* are expressed in the progress zone mesenchyme (see vertebrate limb development). When the zeugopod bones are being formed, the pattern shifts remarkably, displaying a nested sequence of *HOXd* gene expression (Fig. 28.2). The posterior region expresses all the *HOXd* genes from *HOXd-9* to *HOXd-13*, while only *HOXd-9* is expressed anteriorly. In the third phase of limb development, *HOXd-9* is no longer expressed. But *HOXd-13* is expressed in the anterior tip of the limb bud, but there is complete inversion of *HOX* gene expression during the autopod formation (Fig. 27.4). Such inversions in *HOX* gene expression are due to the acquisition of new regulatory DNA sequences that activate the genes to produce new structures, as found in the autopods.

268   Molecular Developmental Biology

In addition to the above-described changes in the *HOX* gene expression, there are other developmental changes involving other genes in the shaping of the autopod. Particularly, bone-promoting BMP family and joint- making GDF family genes play pivotal roles in shaping and patterning the autopod.

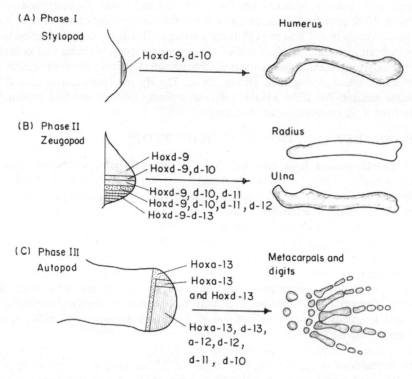

Fig. 27.4 Changes in the *HOX* gene expression during the formation of the tetrapod limb. (A) During the formation of stylopod, *HOX*d-9 and *HOX*d-10 are expressed in the newly formed limb bud. (B) During Zeugopod formation, there is a nested expression of *HOX*d genes such that *HOX*d-9 through *HOX*d-13 are expressed in the posterior of the limb bud, while only *HOX*d-9 is expressed both anteriorly and posteriorly. (C) Inversion of *HOX* gene expression during autopod formation. *HOX*d-13 and *HOX*a-13 are expressed anteriorly and posteriorly, while *HOX*d-10 through *HOX*d-1 through *HOX*d-12 and *HOX*a-12 are expressed posteriorly. (Redrawn from Gilbert, 2000).

Differences exist in *HOXd-11* and *HOXd-13* expression in fish and tetrapod embryonic appendages. Whereas *HOXd-11* expression is distal to *HOXd-13* expression in fin of a fish, in the tetrapod limbs, *HOXd-13* expression becomes distal to *HOXd-11* expression. Furthermore, expression of 5' genes of the *HOXd* group may be crucial in the change from fin to limb. Tetrapods and fishes share the first two phases of the *HOX* expression pattern in their appendages. Thus, both groups form stylopods and zeugopods. Phase III pattern of *HOX* gene expression is however unique to tetrapods and not found in the fishes.

This change in *HOX* gene expression is mediated by a single enhancer element that is not found in fishes. Thus, the foot and hand in tetrapods appear to be new structures in evolution, and they appear to have been formed by the repositioning of *HOXD* gene expression during fin development into tetrapod limb. In a significant discovery of the fossil, called Acanthostega, which is half fish and half tetrapod, it was found to have a tetrapod limb with eight digit hands on the end of them. Understandably, the hand we all possess developed in a peculiar way from the fish's fin: by the development of a forward curving arch of bones in the wrist from which digits were flung off towards the rear side. There is now corresponding genetic evidence to support this. The *HOX* genes set up a gradient of expression curving towards the front of the growing limb, to divide it into separate arm and wrist bones, and then they suddenly set up a reverse gradient on the outside of the last bones to throw off the five digits.

In addition to *HOX* genes, there were many other genes involved in the shaping of the autopod. They include members of the bone-promoting BMP family and the joint-making GDF family.

HOMOLOGY IN *WNT* GENE FAMILY

The homology does not just end up with *HOX* genes; it extends to many other developmental genes also. Similar to developmental genes, whose products control transcription of zygotic genes during embryogenesis, signal molecules involved in cellular differentiation also show homology among animal forms in an evolutionary perspective. Like developmental genes, most of the signalling pathways are also highly conserved. Cells communicate with one another by sending signals in the form of proteins that are exported and travel away from their origin. These proteins then bind to receptors on other cells, where they trigger a cascade of events including changes in cell shape, migration, cell multiplication, and the activation or repression of genes. As tissues grow, signals between populations of cells induce local pattern formation within the developing structures. As an example, the *Wnt* gene family encodes secreted signalling molecules that control cell fate in embryonic development. The expression of this family of signal proteins, across animal phyla in bilateria reflects different stages in the metazoan evolution.

Recent discovery of *Wnt* signalling molecules in the diploblastic, radially symmetric cnidarians has thrown open the possibility that these molecules would have originated as the first signalling system in the control of axis specification. In *Hydra*, not only the *Wnt* proteins, but also the member molecules in the *Wnt* signalling pathway have been characterized. Essentially, *Wnt* expression predominates in the *Hydra* head organizer, situated in the upper part of the hypostome. Upregulation of *Wnt* expression also occurs during bud formation and head regeneration. *Wnt* expression is however meager towards the basal disc. Thus, *Wnt* signalling may be involved in axis formation (oral to aboral) in *Hydra*. When *Hydra* β-catenin (Hyβ-Cat) (an important member in the *Wnt* signaling pathway) mRNA is ectopically expressed in the ventral blastomeres of 8-celled *Xenopus* embryos, complete secondary body axis containing the most anterior structures was formed. Evidently, Hyβ-Cat has the ability to regulate transcription of axis-inducing downstream target genes in *Xenopus* embryos. Thus, a *Wnt* signalling center, involved in axis formation is important in the evolution of body plan of the earliest multicellular animals such as the common ancestors of cnidarians and bilateral metazoans. Interestingly, although *HOX* genes have taken over the function of axis determination in higher metazoans and vertebrates, *Wnt* signalling cascades still act in the establishment of dorsal-ventral polarity and axis formation in *Drosophila* and vertebrates.

Recently, *Wnt* family genes such as *Wnt-1,-6,-10,-9* and *-3* have also been found in a sea anemone, suggesting that both cnidarians and bilaterians originated from a common ancestor. The co-expression of *Wnt* genes with *HOX* genes along the oral-aboral axis of the planula larva of the sea anemone *Nematostella vectensis* also suggests its synergistic function in axial differentiation in cnidarians. It is important to reiterate here that *Hox* genes are the central genomic components for metazoan patterning and development.

A search down under the cnidarians, such as in sponges have revealed the expression of not only the *Wnt*, but also its activated receptor, namely, frizzled protein. In addition, sponges make a variety of metazoan-like transcription factors that control gene expression during development. This

270   Molecular Developmental Biology

indicates that the urmetazoan, the ancestor of all multicellular organisms would have possessed the entire genetic toolkit to direct a body plan containing multiple cell types.

Looking even further back in time, the single-celled ancestors of the ermetazoan, have possessed these developmental genes. The evidence for this contention is provided by the fact that choanoflagellates, the modern-day descendants of such single-celled ancestors, possess many of the toolkit genes needed for multicellular living. Although coanoflagellates did not possess any *Hox* gene homologues, they contain many cell adhesion and signalling proteins domains that are otherwise restricted to metazoans. Cell organisation leading to pattern formation might have developed only after the origin of muticellular metazoans, but the presence of adhesive proteins only indicated their precursor-functional roles of cell-cell adhesion and cell signalling that constituted the main ingredients of multicellularity and metazoan evolution. Evidently, the molecular machinery for multicellularity was on site before the transition to multicellularity took place.

Taken together, it is suggestive that the common metazoan ancestor exhibited ermetazoan-like developmental patterning. Furthermore, most major animal lineages would have arisen from a sponge-like ancestor in response to environmental and ecological changes during the Cambrian by undergoing rampant morphological radiation.

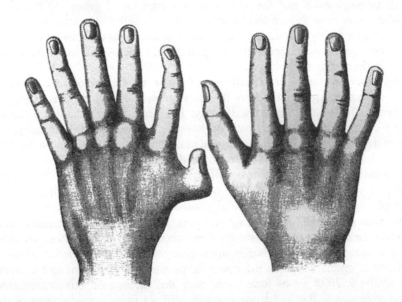

Fig. 27.5. Showing Polydactyly of a human hand compared with normal (Redrawn from S.B.Carroll 2005)

HEDGEHOG GENE FAMILY

Another classical example is the *hedgehog* gene discovered in *Drosophila melanogaster* by Nusslein-Volhard and Wieschaus. This gene is so named that its mutant larva is covered with fine hairs or denticles, resembling the fur pattern of a *hedgehog*. Its importance in development came to limelight when its counterparts were discovered in vertebrates. The vertebrates have three *hedgehog* genes called *Sonic hedgehog, Desert hedgehog* and *Indian hedgehog. Sonic hedgehog* (*Shh*) is expressed in a developing chicken limb in the posterior edge of the limb bud. This region corresponds to the zone of polarizing activity (ZPA). It is now known that the activity of ZPA is due entirely to *Sonic hedgehog* expression. Ectopic expression of *Shh* in the anterior edge of the limb bud resulted in the formation of a mirror image duplication of normal digits. Interestingly, a type of polydactyly in humans is also the result of a mutation that affects *Sonic hedgehog* expression during human limb development (Fig. 27.5).

Sonic hedgehog signalling is important in vertebrate development from its activity along the ventral midline of the developing embryo. Signalling from the floor plate cells is critical for the patterning of overlying tissues and their subdivision into the left and right parts of the eye field and brain hemisphere. Mutation in the *Shh* pathway is also illustrated in new born sheep, afflicted with cyclopia, a lethal malformation in which they bear a single central eye, lack most nasal and jaw structures, and have incompletely developed brain hemispheres. This abnormal condition is also called holoprosencephaly, meaning single forebrain, and the key defects are that the forebrain and eyes fail to become separated into two symmetrical structures. In the sheep, cyclopia is caused by the ingestion of a plant, the lily *Veratrum californicum* by a gestating mother sheep. The chemical factor, namely, cyclopamine found in this lily plant is a known teratogen. Cyclopamine is an inhibitor of the *Sonic hedgehog*-signalling pathway in mammals. It blocks part of a receptor so that cells that should respond to the *Sonic hedgehog* protein cannot do so. In humans, fetal alcohol syndrome is a manifestation of alcohol toxicity at critical stages of human gestation and can produce holoprosencephaly. Mutations that abolish *Shh* gene activity or that of other components of the pathway also cause cyclopia.

EVOLUTION AND SPREAD OF *HOX* GENES

In the majority of invertebrates, there is only one cluster of *HOX* genes. Interestingly, prochordates, such as the cephalochordates, believed to be nearer the hypothetical common ancestor of all chordates, have only one cluster of *HOX* genes. During evolution from this basic prochordate-like precursor stock, cluster duplication has occurred several times on the line of jawless fish, cartilagenous fish (sharks), and again lampreys (see Fig. 27.6).

This has finally given rise to the present condition in the highly evolved vertebrates, mammals, having four clusters of *HOX* genes to build up the complex body form. Thus, mice and humans have four *HOX* clusters containing 39 genes in all. Evidently, an increase in *HOX* cluster number has occurred by gene duplication all through the evolution of vertebrates in a chequered manner. Possibly, the first duplication of *HOX* gene cluster would have happened sometime after the split between the vertebrate and prochordate line. This is further evidenced by the fact that all other deuterostomes such as tunicates and echinoderms possess single *HOX* clusters. Presumably all vertebrates, such as all mammals, birds and certain fish, including the primitive coelacanth, have four *HOX* clusters. In other words, there were four *HOX* clusters in the common ancestor for all of these jawed vertebrates.

Recent studies on the occurrence of *HOX* genes in the diploblastic cnidarians and other lower bilaterians have indicated their axializing properties in these lower metazoans. Based on the chronological appearance of these genes during animal evolution, the increase in number and the complexity of of the *HOX* gene clusters has also happened parallely (Fig. 27.5). To start with, *HOX* genes appeared both in the Cnidarian- bilaterian ancestors as a single cluster. Obviously, they performed the basic function in axialization. The same trend was continued in other invertebrates (including insects) to effectively control the anterior-posterior axis determination. Echinoderms and the prochordates also possessed a single cluster of *HOX* genes. Apparently, only with the emergence of true chordates, *HOX* cluster duplication started.

A feature of interest is that, just like *HOX* gene duplication, there occurred duplication of several other developmental genes during vertebrate evolution, supporting the idea that the evolution

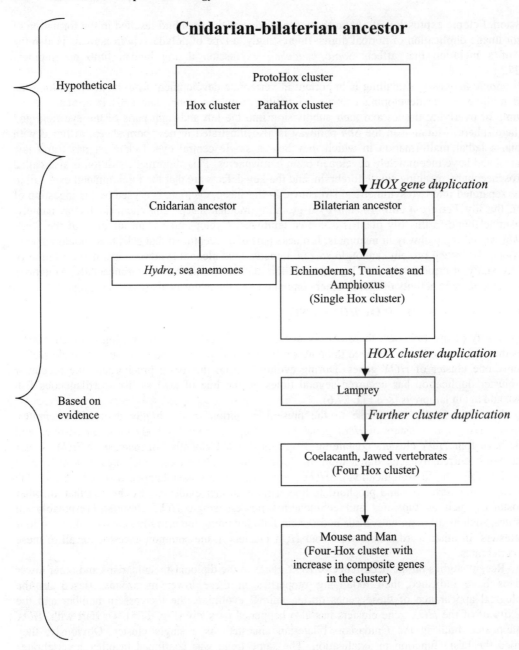

Fig. 27.6 Showing the evolutionary origin of *HOX* genes (see text for explanation).

of more genes played a role in the evolution of complex body design. This increase in the number of genes is reflected in the increased cell types as found in higher animal groups. Mammals and other higher vertebrates have many more cell types than cephalochordates, which lack the cells that could give rise to cartilage, bones, head and certain sensory structures. Thus, greater cell type and tissue complexity is achieved by using more number of genes, which generated more combinations of developmental instructions.

Yet another illustrative example of *HOX* gene-specific control of common body development in vertebrates is the positioning of neck vertebrae, at the cervical / thoracic region along the main body axis of the embryo. The forward boundary of expression of the *HOX* gene, *HOXc8* falls at the cervical / thoracic boundary in mice, chicken and geese, despite differences in the number of thoracic vertebrae in these animals. However, the relative position of *HOXc8* expression has shifted among these animals.

As a result, the relative length of the neck varies among these animals. In the snake, the cervical / thoracic boundary is lacking and the *HOXc8* expression extends way up into the head. Hence, all of these vertebrae bear ribs, which indicate that they are of thoracic type (Fig.27.7). This condition results in the loss of neck by extending their thorax via shifts in *HOX* zones. Evolutionary shifts in *HOX* zones also arise through changes in the DNA sequences of *HOX* gene regulatory region, namely promoters and enhancers. The important inference from the above account is that the developmental genes such as *HOX* genes control large-scale change in animal design during evolution and adaptation. The foregoing discussion on the role of developmental genes in the evolutionary origin of animal form strengthen the fact that animal body plan is established in the embryo and determined by the developmental program.

HOX GENES AND BODY SEGMENTATION IN ARTHROPODS

Cambrian explosion of diversity of forms is occasioned with the evolution of arthropods. The joint-legged arthropods have been supposed to evolve from the lobopodians with simple unjointed appendages. A dominant feature in the Cambrian arthropods is the similar appearance of many segments and their associated appendages. The repetitive organization of body segments is also found in lobopodians, which display large numbers of relatively few different types of body parts. A series of innovations, such as segmentation, a hard exoskeleton, and a biramous limb took place in the lobopodians to be evolved into the basic features of all arthropods. In all modern arthropods, *HOX* genes specify different body segment types and the associated appendages. Hence in the early insect and arthropod ancestors, there existed a smaller set of *HOX* genes that specified the smaller number of distinct segment types. A living evidence to illustrate this concept is that the onychophorans, which are considered to be the connecting link between the arthropods and their precursors, which still walk on lobopods like their Cambrian ancestors. More interestingly, onychophores possess all of the *HOX* genes known from insects and other arthropods. Thus, all of those Cambrian lobopodians and the arthropods have the same large set of ten *HOX* genes and the bodies of all of the later arthropods such as spiders, myriapods, insects, and crustaceans were sculpted by the same set of *HOX* genes.

Arthropod evolution not only involved the increasing number of segments, but also the limb type diversity. Furthermore, appendages on the head, trunk, and tail of arthropods have become specialized for feeding, locomotion, respiration, burrowing, sensation, copulation, brooding young, and defense. In fact the adaptational success of arthropods mainly depends on the increased specialization of the limb types. The genetic control of appendage types is best known in fruit fly embryos, where the *HOX* proteins control all the formation of limbs in all segmental body compartments such as head thorax and abdomen. Again, the great diversity of appendage types and functions has been achieved by deploying different *HOX* genes in different zones along the main body axis. The increase in the number of different appendage and segment types in arthropod evolution is the product of generating a greater number of unique zones in the embryo in which specific individual or combinations of *HOX* genes are expressed. The insect embryos contain many unique individual or combined *HOX* zones to build up different body parts along with their appendages. In addition, shifting *HOX* zones have sculptured the prominent differences along the main body axes of living arthropod groups such as arachnids, myriapods, insects and crustaceans.

CONSERVED ROLE OF PAX6 IN EYE FORMATION

Vertebrate eye development is dependent upon coordinated inductive interactions between presumptive neural retina and the surface ectoderm (refer section on organogenesis for details). The developmental genetic pathway for eye development in all animals is now known to be specified by the paired-class transcription factor Pax6. An interesting fact is that homologues of Pax6 gene is expressed in a variety of animal forms such as insects, all vertebrates, planarians, squids, rag worms, irrespective of the type of eyes found in them. For example, vertebrates have camera type eyes with a single lens, whereas all arthropods have compound eyes in which many independent unit eyes gather visual information. Molluscs such as octopus and squids have camera type eyes, with other molluscan species possessing other types of eyes. However, they all require *Pax6* eye-building gene for their normal formation. In accordance with their functional conservancy, the primary structure of this transcription factor shows great similarity among diverse animal species. Fig.27.7.shows a portion of the amino acid sequence similarity in the fruit fly, mouse and humans. Expectedly, mouse and human sequences are identical and only four amino acid residues show alterations in the fly sequence. The ubiquitous presence of this toolkit developmental gene is an indication that it was present in their common ancestors also.

Fruit fly LQRNRTSFT**ND**QI**DS**LEKEFERTHYPDVFARERLA**G**KI**G**LPEARIQVWFSNRRAKWRREE

Mouse   LQRNRTSFTQEQIEALEKEFERTHYPDVFARERLAAKIDLPEARIQVWFSNRRAKWRREE

Human   LQRNRTSFTQEQIEALEKEFERTHYPDVFARERLAAKIDLPEARIQVWFSNRRAKWRREE

Fig. 27.7 The Pax-6 eye-building gene. A portion of the amino acid sequence of the fruit fly, mouse, and human proteins is shown. Note the great similarity between the fly and mammal proteins, and that mouse and human sequences are identical. The different amino acid residue in the fruit fly sequence is in bold and underlined

Mutations in the *Pax6* gene cause the *ANIRIDA* syndrome in humans, the *Small eye* phenotype in mouse and the eyeless phenotypes in flies. In vertebrates, Pax6 is expressed in presumptive eye tissues with a pattern implying a role in early development of both lens and retina. Evolutionary similarity of Pax6 expression in both vertebrates and the invertebrate insects has been elegantly illustrated in several misexpression studies using this gene product. Despite the enormous morphological differences in insects and vertebrate eyes, the mouse *Pax6* specifies insect eye development when placed in a fly embryo. Similarly, the Swiss scientists from Walter Gehring's lab created transgenic flies, which could express the eyeless gene in various places in the developing fly. By ectopic expression of eyeless gene, they were able to produce flies with eyes on their antennae, legs, wings and various other places (Fig..27.8). So eyeless looks like a control switch gene for making eyes.

EVOLUTIONARY CONTROL OF EYE DEVELOPMENT

In spite of the suggested independent evolution of eye in various animal taxa over 550 million years, the control switch for eye development has been fairly conserved. An alternative explanation is that a primitive eye, present in the common ancestor of all bilaterian animals,was patterned by Pax 6. During the course of evolution, modifications in eye structure occurred that led to the diversity of eye types we see in extant animals. Yet another alternative theory is that Pax 6 could have been recruited independently on several occasions to serve as transcriptional factor guiding eye development.

In view of the occurrence and expression of *Pax 6* or its orthologs in the eyes of almost all bilaterian phyla ranging from lobophorates to vertebrates, It is suggestive that the common bilaterian ancestor such as the hypothetical Urbilateria may have deployed an ancestral Pax 6 gene in the photoreceptor cells during the development of a primitive eye or light-sensing organ. In support of this idea, it has been found that among the many transcriptional targets of *Pax 6* are the rhodopsin genes (a light sensing molecule with the protein moiety of opsin found in photoreceptor cells). As this primitive eye independently evolved in different animal lineages into many morphologically complex eye types, *Pax 6* continued to serve as a key regulator of eye development and acquired additional transcriptional targets, all of which playing a role in eye development to the present day condition.

Significantly, a related *Pax* gene (with a *Pax 6*-like homeodomain) is also expressed in the photoreceptor cells of a cnidarian, providing more evidence that the functional similarities between the mouse *Pax6* and *Drosophila eyeless* genes in eye development may be inherited from a common ancestor. The remarkable conservation of *Pax 6* expression and the *Pax 6* –regulated circuit suggests that all bilaterian eyes share a common developmental genetic circuit and that this circuit was present in the bilaterian ancestor.

If the fly *Pax6* gene is expressed in frog skin the resultant fly Pax6 protein initiates eye development, causing the otherwise epidermal cells to become an amphibian retina. The differentiated ectopic eyes in *Xenopus laevis* contains mature lens fibre cells, ganglion cells, Muller cells, photoreceptors and retinal pigment epithelial cells in a spatial arrangement similar to that of endogenous eyes. Obviously, the lens, retina and retinal pigment epithelium arose as a consequence of the cell-autonomous functions of Pax6. The misexpressed *Pax6* causes the ectopic expression of a number of genes including *Rx, Otx2, Six3* and endogenous *Pax6*, which are implicated in eye development in a sequential manner. Further, the formations of ectopic and endogenous eyes could be suppressed by co-expression of a dominant-negative form of *Pax6*. Taken together, the ability of *Pax6* to direct eye formation in flies and *Xenopus* underscores the conservation in the hierarchy and regulation of the genetic machinery controlling eye development. Again, it provides further support for the monophyletic evolution of light sensing organs in both invertebrate and vertebrate animals with different types of eyes. Additionally, the eye formation also demonstrates how evolution works with common genetic tools to build complex organs.

In addition to being the main architect in the eye formation, *Pax6* is also involved in building part of the brain and nose in mammals. In *Drosophila*, *Pax6* homologues eyeless (*ey*) and twin of eyeless (*toy*) are expressed both in the eyes and central nervous system (CNS). In the brain, the *ey* plays important roles in axonal development of the mushroom bodies, centres for associative learning and memory. The analysis of *cis*-regulatory regions of *ey* that promote gene expression in various parts of the developing CNS has indicated that *ey* is independently regulated in brain development, whereas in the eye primordial, its expression requires both *ey* and *toy*. Thus, the mechanisms responsible for the complex expression patterns of the *Pax6* genes in controlling eye formation and brain development of *Drosophila* differ significantly.

276   Molecular Developmental Biology

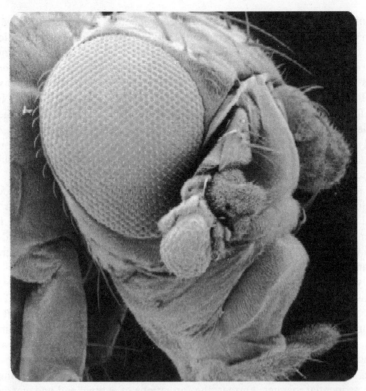

Fig 27.8: A scanning electron microscope image of the head of D.melanogaster. Targeted expression of the eyeless gene induced the formation of the eye facets on the antenna (To the lower – right of the eye), which are very similar to the facets of the normal eye (Courtesy: http://www.learner.org).

DEVELOPMENTAL BASIS OF MORPHOLOGICAL VARIATION IN THE BEAKS OF DARWIN'S FINCHES

The potential role of regulatory genes in mediating dramatic phenotypic transformation is well demonstrated in the beak morphology of Darwin's finches in the Galapagos Island. The beak morphology also provided an example for heterometry, a type of macroevolutionary change, brought about by a change in the amounts of a gene product during morphogenesis. Finches also helped Darwin to frame his evolutionary theory of descent with modification, in addition to being the most celebrated illustrations of adaptive radiation. The impressive array of specialization in beak morphology and function in the finches is in accordance with the diverse feeding niches that different species have come to occupy. Fig. 27.9 shows the examples of Darwin's finches with different beak shapes adapted to different feeding stuffs, big strong beaks for hard nuts and the like, fine beaks for small seeds and the like. Beak morphology in finches has thus evolved by means of natural selection in precise correspondence to changing ecological conditions, including food availability and interspecific competition.

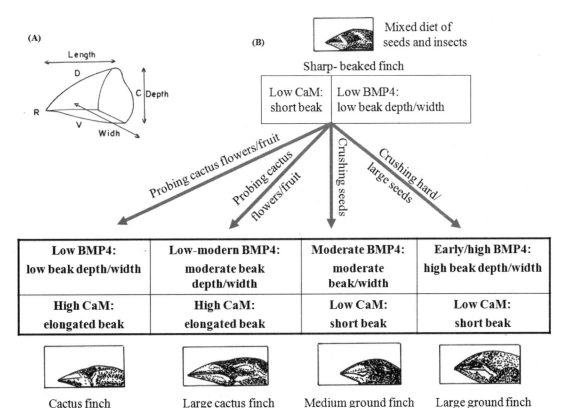

Fig 27.9 BMP and CaM – dependent signaling regulates growth along different axes, facilitating the evolution of distinct beak morphologies in Darwin's finches. (A) Developing avian beak is a three – dimensional structure that can change along any of the growth axes. (B) A beak of the sharp - beaked finch reflected a basal morphology for Geozpiza. The model for BMP4 and CaM involvement explains development of both elongated and deep/wide beaks of more derived species. Abbreviations: C, caudal; D, dorsal; R, rostral; V, ventral (After Abzhanov et al. 2006).

The external differences in beak morphology reflect differences in their respective craniofacial skeleton. A recent study compared the beak development in six species of Darwin's finches belonging to the monophyletic ground finch genus *Geospiza*. Species-specific differences in the morphological shape of the peak prominence appear during embryonic stage 26. Among the many growth factors that are known to be expressed during avian craniofacial development, *Bmp2* and *Bmp7* correlated with the size of the beak but not its shape. In contrast, a striking correlation between beak morphology and the expression of Bmp4 was observed. The species with deeper, broader beaks relative to their length express Bmp4 in the mesenchyme of their beak prominences at high levels and at earlier stages (a heterochronic shift) than species with relatively narrow and shallow beak morphologies (Fig.27.9). Moreover, the differences in Bmp4 expression are coincident with the appearance of species-specific differences in beak morphology. However, this specific Bmp4 expression was apparent only in the upper beak, whereas the expression in the lower beak remained constant, in spite of the fact that lower beak morphology varies in concert with that of the upper beak. It is important to note here that Bmp4 has been shown to be important in the production of skeletogenic cranial neural crest cells and are capable of affecting patterning, growth and chondrogenesis in derivatives of the mandibular and maxillary prominences. These findings suggest that Bmp4 may have a proliferating effect on skeletal progenitors in the upper jaw. Furthermore, it has been speculated that differences in the *cis*-regulatory elements of Bmp4 may underlie the distinct pattern. Alternatively, they could also be explained by differences in the timing or amounts of upstream inductive factors or differences in the transduction of such signals like *Shh* and *Fgf8*. Artificial increase in the Bmp4 levels in the beak mesenchyme also alters beak morphology in the same direction as is seen in the larger ground finches. Evidently, variation in *Bmp4* regulation is one of the principal molecular variables that provided the quantitative morphological variation acted on by natural selection in the evolution of the beaks of the Darwin's finch species.

Yet another developmental pathway responsible for evolutionary morphological changes in the finch beak involves calmodulin (CaM), a molecule that mediates Ca++ signaling. A study showed that CaM expression levels are higher in the long and pointed beaks of cactus finches than in more robust beaks type of other species (see Fig.27.9). When this upregulation of CaM- dependent pathway is artificially replicated in the chick frontonasal prominence, it caused an elongation of the upper beak, very similar to that in cactus finches, suggesting a role for CaM-dependent pathway in the developmental regulation of craniofacial skeletal structures. More importantly, these two examples emphasize the genetic basis for evolutionary changes in phenotype formation during embryonic development.

EVOLUTIONARY PELVIC REDUCTION IN THREE-SPINE STICKLEBACKS

It is well understood that DNA regions that form the enhancers of developmental genes are modular. Hence, modularity of development is determined by the modularity of the gene enhancers. Each enhancer element allows the gene to be expressed in a different tissue. If a particular gene loses or gains a modular enhancer element, the organism containing that particular allele will express that gene in different places or at different times than those organisms retaining the original allele. Such mutability can result in the development of different morphotypes. The evolutionary importance of such enhancer modularity is demonstrated in the pelvic reduced three-spine stickleback fish. Stickleback species inhabiting the sea have a pelvic spine that serves as protection against predation, whereas the freshwater species that evolved from marine sticklebacks do not have it. The gene responsible for the reduction of the pelvis skeleton (and hence pelvis spine) in the freshwater species has been identified as a toolkit gene called *Pitx1*, a homeobox-containing transcriptional factor that is critical for hind limb identity and growth. Since pelvic fin is a homologue of mammalian hind limb, the same *Pitx1* gene is responsible for the hind limb formation in the mouse. *Pitx1* is also expressed in the precursors of the thymus, nose and sensory neurons. Interestingly, in marine (with pelvic spike) and freshwater (without pelvic spike) sticklebacks, amino acid sequence and splicing of the *Pitx1* transcript is identical. Since no mutation was found in the coding region of *Pitx1* in the pelvic reduced freshwater population, no alteration has occurred in the amino acid sequence of the gene product. However, an examination of the spatial patterns of *Pitx1* expression during normal development of both marine and freshwater larvae revealed the difference. In the marine stickleback species, *Pitx1* was clearly expressed in bilateral patches of mesenchymal cells at the site where the pelvic fin bud normally develops. *Pitx1* expression was also detected in developing thymus, olfactory pits, sensory neuromasts on the head and the ventral portion of the developing caudal fin. On the other hand, no *Pitx1* expression was detectable in the prospective pelvic region of the freshwater larvae, although the gene expression was visualised in other parts (except caudal fin) mentioned above for the marine species. Evidently, freshwater fish showed an altered pattern of *Pitx1* expression in some tissues but not others; consistent with a regulatory mutation that disrupts expression in both the prospective pelvic region and caudal tail.

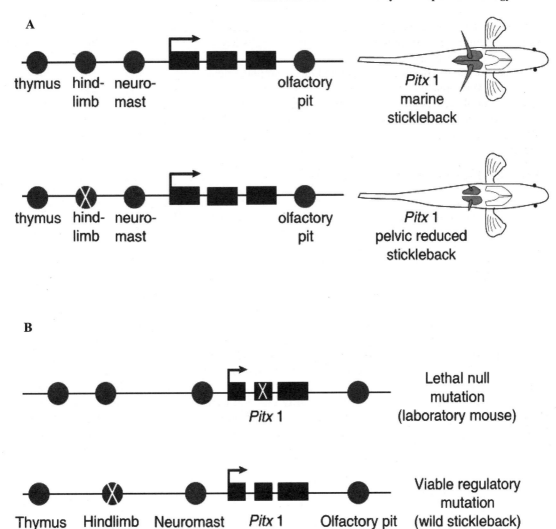

Fig.27.10. The Modular Architecture of the cis-Regulatory Regions of Pleiotropic Genes Enables the Independent Evolution of Gene Expression in Different Body Parts. (A)Shows that the expression of the Pitx1 gene of the three-spined stickleback is controlled by multiple cis-regulatory elements (circles). In pelvic-reduced stickleback fish, Pitx1 expression is absent from the pelvic region. This is proposed to occur through the selective loss of activity of the hind limb regulatory element (cross through circle) (B) Null mutations in Pitx1-linked in mice disrupt multiple functions and lead to neonatal lethality. In contrast, regulatory mutations in modular cis-acting regulatory elements of sticklebacks could produce major morphological changes in particular body regions without disrupting functions in other tissues (Figure based on [Shapiro et.al 2004.].)

From an evolutionary perspective, the altered pattern of *Pitx1* expression seen during normal development suggests that *cis*-acting regulatory mutations in *Pitx1* are a major cause of pelvic reduction in this rapidly evolving system (Fig.27.10). Obviously, regulatory mutations in key developmental control genes may provide a general mechanism to selectively alter expression in specific structures and yet preserve expression at other sites required for viability. In addition, this differential gene expression study suggests that regulatory mutations in key developmental control genes may also be responsible for major and rapid morphological changes in limb and fin structures so important in vertebrate evolution. This study clearly demonstrates that duplication of regulatory genes and their control regions must have contributed greatly to the evolution of vertebrates. In other words, evolutionary changes in "regulatory systems" were responsible for the evolution of anatomy and that divergence and specialization result from mutations altering "regulatory circuits" rather than chemical structures.

EVO- DEVO AND HUMAN EVOLUTION

In the origin and evolution of life on earth, the most important phenomenon is the evolution of form as found in fossil record and in the diversity of living species. Darwin's hypothesis of common descent with modification and adaptation by natural selection is the crux of evolutionary theory. He reasoned that organs of extreme perfection and complication such as the eye evolved by incremental transitions through many intermediate states. Evolutionary developmental biology emphasizes that increments of change in developmental process are assimilated over greater time periods to produce increasingly diverse forms. It is now well understood that developmental genes, playing pivotal role in deciding the shape and form of body, also undergo evolutionary changes to give rise to species diversity.

As we have seen in the preceding section, expression of several developmental genes occur in the embryos in a strictly temporal and spatial pattern, bringing about the body axis formation and orderly appearance of organ systems to give bilaterality and complexity to the adult organism. Unfortunately, such gene expression studies on living human embryos could not be carried out due to ethical reasons. Nevertheless, successful sequencing of human genome as well as of chimpanzee genome recently, has uncovered genetic similarity between these two primates, which have evolved from a common ancestor. Comparing the genetic code of humans and chimps will allow us to comb through each gene or regulatory region to find single changes that might have made a difference in evolution. Although the two genomes are very similar, there are about 35 million single-nucleotide changes, 5 million insertions / deletions events and many chromosomal rearrangements that have accumulated since the human and chimpanzee species diverged from our common ancestor. The weight of genetic evidence on the evolution of primates, great apes and humans is however due to changes in the control of genes rather than in the proteins the genes encode.

To highlight the genetic differences at developmental level between apes and humans, two genes implicated with human may be discussed here. They are: 1) the human gene encoding a protein called myosin heavy chain 16 (MYH 16) and 2) *FOXP2*, the so-called gene for speech and language. An important human trait that distinguishes us from apes is the reduced size of our jaw muscles. MYH 16 is a specialized myosin that form an integral part of temporalis muscle (jaw muscle) of apes. The human *MYH 16* gene is also expressed in the human temporalis muscle. But in humans, a deletion of two bases disrupts the code of the gene and is associated with the reduction of two muscles involved in chewing, which is massive in apes. This is an instance of fossilization of a human muscle gene, causing evolutionary change brought about by natural selection. The loss of function of MYH 16 protein led to the reduction of the temporalis muscle with consequent reduction in jaw muscles in hominid evolution. This reduction in the human jaw musculature, and the force imposed on the mandible, would reduce the stress on bones in the skull. This again resulted in the thinning of braincase making way for the expansion of brain size. Another consequence of reduction in jaw musculature is the eventual evolution of finer control of the mandible, which is required for speech.

Another important example of protein evolution in the context of the ascent of human being from other primates is the evolution of a gene affecting speech. Recent discoveries have uncovered the presence of a gene in humans whose mutation has caused impairment in speech and language. This is called *FOXP2*. Like other developmental gene products, the FOXP2 is a transcriptional factor that binds to DNA and regulates the expression of other genes. A one amino acid mutation in *FOXP2* causes speech impairment, but not complete loss. Recent studies have indicated that individuals with disruption of *FOXP2* have multiple difficulties with both expressive and receptive aspects of language and grammar. A predominant feature of the phenotype of affected individuals is an impairment of selection and sequencing of fine orofacial movements, an ability that is typical of humans and not present in the great apes. Although FOXP2 protein is identified in primates, rodents and birds, the human FOXP2 protein differs at only 4 out of 716 positions from that of mouse, at 3 positions from the FOXP2 of the orangutan, and at just 2 positions from those of the gorilla and chimpanzee. This is a good indication that for some period of the past 200,000 years, during human evolution, mutations in the *FOXP2* gene were favoured and spread throughout *Homo sapiens*, making us different from our present day ancestral relatives by the possession of speech and language.

PROTEIN MODIFICATION AND EVOLUTION

Mice and humans have nearly identical sets of about 25000 genes. Chimps and humans are almost 99% identical at the DNA level. However, even a 1% difference in DNA sequence implies a substantial difference in protein sequence. Consequently humans and chimps have different amino acid sequences in at least 55% of their proteins, a figure that rises to 95% for humans and mice. Such a protein- sequence evolution may explain the enormous phenotypic differences that exist between man and mice.

Natural selection is commonly thought to operate mainly at the protein level. Protein changes occur due to nucleotide changes in protein- coding regions of the gene. There are two types of changes; one is 'synonymous changes' (which do not cause any change in amino acids) and another 'non-synonymous changes' (which do cause amino acid changes). Synonymous DNA substitutions do not alter the amino acid sequence because they occur at degenerate sites in the codon (such as a CGT to CGG change, as both codons encode arginine).

Gene duplication is by far the most common method by which new protein structure could evolve. Extra copies of a gene can arise by unequal crossing over or by reverse transcription, allowing one copy to retain its function while the other assumes a new function. This process is considered to be a major force in evolution. Clearly, the evolution of new protein function after gene duplication has an important role in the evolution of the diversity of living organisms. Thus, a significant number of genes in humans (39%) are members of families derived from repeated duplications and diversification of ancestral genes. A few such families include the globins (such as myoglobin and the various haemoglobins); immunoglobulins; opsins; and olfactory receptors.

Ohno proposes a general hypothesis to explain how new protein functions arise due to gene duplication. According to his model, one copy of a duplicated gene is completely redundant and is thus freed from functional constraint. This redundant copy can accumulate mutation at random. Occasionally, by sheer chance, these randomly accumulated mutations will give rise to a protein with a new useful role. Furthermore, duplication of a generalized, multi-functional ancestral gene can also permit the daughter genes to specialize- a process termed 'subfunctionalism'. An alternative view to this hypothesis is that new protein functions do not often arise through chance accumulation of random mutations, but rather occur as a result of natural selection acting to favour specific changes at the amino acid level.

In addition to gene duplication, there are other ways in which proteins have evolved adaptively. These include gene conversion, recruitment of genes to new functions (antifreeze glycoproteins that allow fish to live in frigid waters), exon shuffling and the addition of transposable elements to coding sequences. The relative importance of gene and protein changes in creating biological diversity is unknown, but both must have evolved in tandem, as different members of gene families are often expressed in different tissues or at different times.

REGULATORY RNAs

Recent advances in molecular research have revealed a novel function of RNA. The central dogma of biology is that DNA makes messenger RNA, messenger RNA makes proteins, and proteins do everything else that needs to be done in a living cell. With protein-coding genes, a copy of the gene's DNA is transcribed into RNA, and this mRNA, after processing, arrives at the cell's protein- making units where its information is used to manufacture a specific kind of protein molecule. The new regulatory role of RNA emerged with the discovery of a class of short RNA molecules known as silencing RNAs and a second class called micro-RNAs. Both types of RNAs are generated from short genes, or pieces of DNA in the genome. In the case of regulatory RNA, the RNA transcript of a gene is processed by certain enzymes known as Dicer, Slicer, and Argonaute to produce short fragments of RNA, some 20 or so units in length. These short fragments of micro- RNAs find a match to messenger RNAs in the cell and curb the activity of the mRNAs. Similarly, the silencing RNAs destroy their mRNAs targets. Such interference of these regulatory RNAs in the normal functioning of the cell's mRNAs imposes an evolutionary constraint on the cells mRNAs. Because regulatory RNAs can influence so many mRNAs at a time, they are likely to play a major role in the origin of new proteins during evolution.

In conclusion, it may be said that morphological variation and divergence are associated with genes that regulate development. Concurrent evolution of genetic regulatory systems together with spaceo-temporal deployment of regulatory genes provide the molecular means by which morphological variation and divergence from the common ancestor could have come to effect. Other developmental parameters such as those genes that encode transcription factors and molecules that take part in signal- transduction pathways also constitute important factors in bringing about developmental changes during ontogeny, leading to phenotypic variation and evolution of new species.

To put it alternatively, it may be said that phenotypic variation, as occurring at molecular level, could underlie variation (mutation) in genes, insertions or substitutions in gene regulatory regions and non-coding regions as well as protein level differences. It has indeed been anticipated that changes in gene regulation are a more important force than coding- sequence evolution in the morphological and behavioural evolution of hominus. Thus regulatory sequences are central to changes in gene expression and morphology. Both intraspecific variations and interspecific divergence in gene expression are probably due to substitutions in non-coding regions that influence transcript or protein abundance through transcriptional or post-transcriptional mechanisms. In conformance with the above conclusions, Sean S. Carroll, the noted evolutionary developmental biologist has to say this; genetic theory of morphological evolution could be condensed into two statements: (1) forms evolves largely by altering the expression of functionally conserved proteins; and (2) such changes largely occur through mutations in the *cis*-regulatory regions of mosaically pleiotropic developmental regulatory genes and of target genes within the vast regulatory networks they control. Thus, Evo-Devo offers another evolutionary thought that links changes in genes to alterations in development and the evolution of form.

A CONCLUSION TO GENETIC THEORY OF MORPHOLOGICAL EVOLUTION

With the recognition that genetic and regulatory networks form the essence of animal development, an understanding towards the forces that drive changes on gene regulation, regulatory networks and regulatory sequences as sources of morphological variation and diversity is imperative. Sean B. Carroll (2008) has elegantly summarised the genetic theory of morphological evolution in the following way:

1. Most proteins regulating development participate in multiple, independent developmental processes and the formation and patterning of morphologically disparate body structures.
2. Morphologically disparate and long-diverged animal taxa share similar toolkits of bodybuilding and body-patterning genes.
3. Many animal toolkit proteins, despite over 1 billion years of independent evolution in different lineages, often exhibit functionally equivalent activities *in vivo* when substituted for one another. These observations indicate that the biochemical properties and their interactions with receptors, cofactors, etc. have diverged little over vast expanses of time.
4. The formation and differentiation of many structures such as eyes, limbs, and hearts- so morphologically divergent among different phyla that they were long thought to have evolved completely independently- are governed by similar sets of genes and some deeply conserved genetic regulatory circuits.
5. Duplications within several prominent toolkit gene families have been surprisingly rare in the course of animal diversification relative to duplications of other gene families.
6. Changes in the spatial regulation of toolkit genes and the genes they regulate are associated with morphological divergence.
7. Large, complex, and modular *cis*-regulatory regions are a distinctive feature of pleiotropic toolkit loci, compared to those of loci encoding cell-type-specific proteins engaged in the chemistry of physiological processes.
8. Transcription factor-regulated gene network, linking a transcription factor and a CRE, regulates scores to hundreds of individual target genes.

28

ECOLOGICAL DEVELOPMENTAL BIOLOGY

Ecological developmental biology refers to embryonic development that studies the interactions between a developing organism and its environment. This study also concerns the developmental plasticity of animals by integrating signals from the environment into their normal developmental programs. The expression of many genes is sensitive to environmental conditions, and this sensitivity provides a general means of transducing external cues into phenotypic responses. In short, ecological developmental biology focuses on the interactions and interrelationships between genes and the environment. We have seen in the previous chapters how cell-cell interaction within the developing embryo is achieved by taking molecular signals from each other, leading to the formation of different organs and ultimately the whole body. This chapter deals with molecular signals that come from sources outside the organisms.

Animals and all other living organisms do not live in isolation, but always live in association with both abiotic and biotic factors around them. This gives opportunities for them to interact with the environment both for their living and development. Development in particular could be immensely influenced by external factors such as diet, temperature, maternal conditions and social interactions prevailing in the given environment. In many instances, this occurs through the altering of gene expression by direct induction or through neuroendocrine influence. Alteration of gene expression due to environmental interaction could also be mediated by other methods such as DNA methylation and by transcription factor activation. Therefore, environment is considered to play a role in the generation of phenotypes, in addition to its well established role in the natural selection of which phenotypes will survive and reproduce. In a similar way, an organism may respond to the presence of other organisms in the environment by altering its development in specific ways to produce a phenotype, best suited to survive and thrive under the particular environment. Such a developmental influence is termed as developmental symbiosis, wherein a host responds to cues from its symbionts and outsources development cues. Best examples for developmental symbiosis come from the relationships between animals and the microbes that live inside them.

ENVIRONMENTAL INFLUENCE ON NORMAL DEVELOPMENT

POLYPHENISM

Environment contains signals that will enable a developing organism to produce a phenotype that will increase its fitness in that particular environment. In most organisms, a genotype can produce many different phenotypes. The exact phenotype that is expressed depends on the environment in which the organism develops. Such a developmental characteristic is called phenotypic plasticity (responsiveness to the external environment) and the development and evolution of such adaptive phenotypes in an environmental variable is termed as polyphenism. An illustrative example of polyphenism is found in the migratory locust, *Schistocerca gregaria*. These plant-eating grasshoppers exist in two mutually exclusive forms. One is short-winged, uniformly colored with green pigmentation, and solitary. The other is long-winged, brightly colored, and gregarious. The striking phenotypic differences between these two morphs are determined by the environmental cues that determine which morphology the young locust will develop. The major environmental stimulus is the population density, as measured by the rubbing of the legs. When locust nymphs get crowded enough that certain neuron in the hind femur is stimulated by other nymphs, their development pattern changes, and after the next molt, they emerge with long, brightly colored wings, leg development suitable for migration and migratory behaviors. The gregarizing agent in *S. gregaria* is an aqueous substance found in the foam surrounding the egg pod of the greagarious morph only. This chemical agent, which appears to be a modified form of the neurotransmitter L-dopa, is thought to be synthesized in the accessory glands of the female reproductive tract and to act during the time of egg laying. If the foam is transferred from egg masses laid by gregarious females to egg masses produced by solitary females, the solitary eggs turn into gregarious locusts. This confirms the transgenerational polyphenism produced by triggering of high density in the population of locusts.

Different phenotypes thus produced by the environment are called morphs or ecomorphs. Genetically identical animals can have different morphs depending on the season, their larval diet, or other signals present in the environment (as in *S. gregaria*). Incidentally, phenotypes produced by different genetic alleles are called by the name polymorphism. Polymorphism is the result of genetic differences, whereas morphs are the result of environmental signals.

Another example of polyphenism, this time brought about by season, is found in butterflies. In North America, the pigmentation of many butterfly species follows a seasonal pattern. For example, in the polyphonic *Pontia* butterfly found throughout Northern Hemisphere, the pupae that eclose during the long days of summer have lighter hind wind pigmentation, whereas in the shorter cooler days of spring the eclosion results in the spring morph having a more highly pigmented ventral hind wing. Pigmentation of wings has a functional advantage during the cooler months. The darker pigments absorb sunlight more efficiently than the lighter ones in summer, raising the body temperature more rapidly.

ENVIRONMENTAL CUES AND DEVELOPMENT OF NEW PHENOTYPES IN TADPOLES

Environmental factors such as the presence of predators can also generate a new phenotype that is suited for that particular environment. The best example is found in the tadpoles of the gray tree frog *Hyla chrysoscelis*. These tadpoles living in a freshwater pond has the ability to sense soluable biochemicals in the water, given off in the saliva or urine of its major predator, such as the dragonfly larvae. In the presence of these insect larvae that feed on them, these tadpoles develop bright red tails that give a "warning colouration" thereby deflecting the predator's attention, and a set of trunk muscles that enable them to make swift turns to escape being eaten.

HORMONAL REGULATION OF TEMPERATURE-DEPENDENT POLYPHENISM

In insects, hormones can change gene expression by transducing sensory information from the environment into the body, thereby bringing about different phenotypes by changing gene expression. The expression of such developmental genes requires an environmental signal; otherwise, the genes will remain inactive. That is to say, the environmental stimulus alters the endocrine mechanism of metamorphosis by altering either the pattern of hormone secretion or the pattern of hormone sensitivity in different tissues. Such changes in the patterns of endocrine interactions result in the execution of alternative developmental pathways. This enables substantial localized changes in morphology that remain well integrated into the structure and function of the organism. The European map butterfly, *Araschnia levana*, is an excellent example to demonstrate environmental determination of phenotype. The butterflies emerging from their pupal cases in the spring are different from the butterflies eclosing in the summer. The spring morph is bright orange with black spots, while the summer form is mostly black with a white band. These butterflies are the same species, but the temperature conditions during their larval development resulted in different phenotypes. The temperature and daylight signals produce the phenotype by regulating the amount of hormone ecdysone during the larval stage of development. In the short photoperiod (below critical daylength), there is no pulse of 20-hydroxyecdysone (20E) during early pupation, and the spring form of the butterfly is generated. When these spring butterflies mate, the larvae experience a long photoperiod and generate the summer pigmentation. By injecting ecdysone into the larvae, the spring form can be converted to a summer form.

Another example of temperature- dependent polyphenism, mediated through hormone action is found in another butterfly species *Bicyclus anynana*. In these butterflies, the presence of eyespots is controlled by hormonal regulation of the *Distal-less* gene. Two environmentally induced phenotypes have been obtained in this butterfly species. The dry season morph is sluggish with cryptic coloration resembling the dead brown leaves of its habitat. The wet season morph is an active flier and has ventral hindwing eyespots that deflect attacks from predatory birds and reptiles. The determining factor in this seasonal polymorphism appears to be the temperature during pupation. The low temperatures produce the dry-season morphs and the high temperatures produce the wet –season morphs. The high temperature allows the accumulation of 20E, which is able to sustain *Distal-less* expression in the pupal imaginal disc. The region of *Distal-less* expression becomes the focus of each eyespot. In the cooler weather, 20E is not formed, and hence, *Distal-less* expression in the imaginal disc is not sustained, and the eyespot fails to form. Evidently, environmental temperature is sensed and converted into an endocrine signal that regulates gene expression of eyespot formation. An interesting aspect of hormonal regulation of development is that development comes under the control of the central nervous system. The developmental hormones (Ecdysone, for instance) are directly regulated by neurosecretory factors. The central nervous system can integrate information about the animal's internal and external environments and use this information to regulate the secretion of hormones.

Insects provide plenty of examples in which hormones have been correlated to the expression of particular genes. In the honeybee larvae, a protein-rich diet, called royal jelly, formed under high titers of juvenile hormones causes formation of fertile queens. Without this enriched diet, the larvae become sterile workers. In this case, Juvenile hormone alters gene methylation, thereby causing the fertile phenotype to develop.

TEMPERATURE AND SEX DETERMINATION IN REPTILES

The classical example of environmental temperature controlling the sex determination is found in many reptiles. In turtles and crocodilians, there are many species in which the temperature at which an embryo develops determines whether an individual is male or female. Such temperature-dependent sex determination in reptiles is mediated by the ezyme activity of aromatase. Aromatase is an enzyme that converts androgens into estrogens such as estradiol. In the turtle, *Emys,* high temperatures during the sex determining period cause high aromatase levels, leading to the production of estrogen and ovary formation. When the egg incubation temperatures are low, the aromatase activity is not promoted thereby preventing ovary from forming. This condition also favours testis formation (Fig.28.1). Under

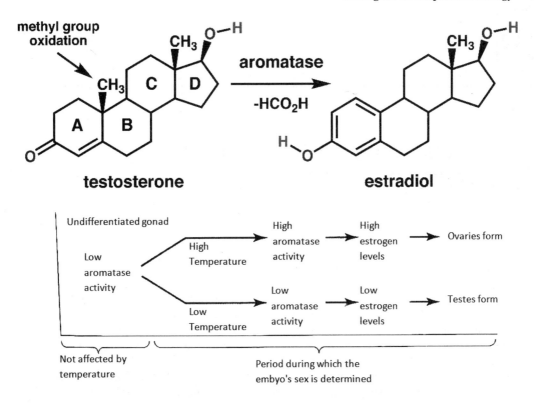

Fig 28.1 Showing the conversion of testosterone into estradiol to effect temprature dependent sex determination in reptiles.

laboratory conditions, when aromatase inhibitors are used to inhibit estrogen synthesis in the turtle eggs, male offsprings are produced irrespective of temperature. Obviously, hormones play direct role in the temperature-sensitive sex determination in reptiles.. Aromatase also appears to be involved in the temperature –dependent sex differentiation of lizards and salamanders. Possibly, the expression of aromatase is activated differently in different species. In some reptiles, the aromatase itself may be temperature-sensitive. In other species, the expression of the aromatase gene may be differentially activated at higher temperatures. In the turtle, *Trachemys scripta*, two genes namely, *Sox9* and *Dmrt1* which are involved in mammalian testis determination, are also active very early during the period of sex determination of this turtle and may act to turn on the aromatase gene.

29

TERATOGENESIS AND MEDICAL EMBRYOLOGY

We have seen in the last chapter how normal development can be influenced by environmental factors and the environmental signals could induce new phenotypes for adaptation to that particular environment. Environment could also provide signals that could disrupt normal development in many instances. In humans, these alterations in embryonic development could result in anomalies may include missing limbs, missing or extra digits, cleft palate, eyes lacking certain parts, hearts that lack valves, and so forth. In medical terms, these birth defects are referred to as congenital anomalies. When genetic events ssuch as gene mutations, chromosomal aneuploidies, and translocations are the underlying mechanisms of such abnormalities, they are called malformations. If the abnormalities are the direct effect of exogenous agents such as chemicals, viruses, radiation and excessive heat, then the effect is called disruptions. The environmental agents that cause genetic damage, bringing about malformations are called mutagens. On the other hand, the exogenous factors responsible for disruptions are known as teratogens, and the study of this abnormal development by disruption, caused by environmental agents, is termed as teratology. Teratogenesis not only helps the physicians to counsel prospective patients but also helps them to understand how the human body is normally formed. The discovery of genes responsible for inherited malformations is helpful in the identification of steps in development being disrupted by teratogens.

The teratogenic effect of an agent depends on the genotype of the embryo and the mother, as well as the ways their genotypes interact with adverse environmental factors. Hence, genetic differences can predispose individuals to being affected by teratogens. There are critical periods of development (windows) when embryos are susceptible to being disrupted by teratogenic agents. Teratogenic agents act at the level of genes, cells, and tissues in developing organisms to disrupt normal sequences of development. In humans, disrupted development could bring about death, malformation, growth retardation, and functional defects.

While some of the teratogens are naturally occurring, others are the products of industrial manufacture. Other compounds that affect the embryonic and fetal development come from drugs that are used to treat pregnant women. Medical embryology primarily deals with the study of teratogenicity and the congenital malformation in humans. Some of the drugs and environmental chemicals and the mechanism by which normal development is impaired is detailed below.

THALIDOMIDE AND PHOCOMELIA

Thalidomide is a drug given to pregnant women as an effective mild sedative and remedy for morning sickness. This caused a rare syndrome of congenital anomalies such as phocomelia, a condition in which proximal limb truncation occurs or amelia, referring to absence of limbs. An estimate indicates that more than 7000 affected infants were born to women who took thalidomide all over the world. This drug also caused other deformities such as heart defects, absence of the external ears, reduced size of the eye (Microphthalmia), and malformed intestines. Thalidomide exposure to certain nonhuman primates in vitro also induces limb malformations identical to those seen in humans. The drug was withdrawn from the market in the year1961.

Thalidomide induces oxidative stress through the formation of free radical- initiated reactive oxygen species (ROS) in limb buds. This causes enhanced BMP signaling, which leads to the hyper expression of the Wnt antagonist Dkk1 and subsequent suppression of Wnt. As a result, there is a diminution in β-catenin activity, ultimately leading to increased programmed cell death (PCD), which is responsible for the limb and eye defects during embryonic development. Under normal conditions, both BMP and Wnt proteins are secreted by cells to regulate the cell division and differentiation of the neighbouring cells The down-regulation of Wnt/ β-catenin signaling is important for the teratogenic activity of the drug thalidomide. The same mechanism of action for this drug has been implicated for fibroblast isolated from whole embryos and for eye development. Similarly, in chicken embryos, thalidomide induces apoptosis in distal mesenchyme cells of the limb bud as well as in the apical ectodermal ridge (AER). Consequently, proximo-distal outgrowth is disturbed resulting in truncated limbs lacking both proximal and distal structures. Another mechanism of action of thalidomide is by disrupting development through inhibition of blood vessel formation in the area.

There is a definitive window of susceptibility during which thalidomide caused abnormalities. Thalidomide disrupts different structures at different times of human development. Human development is divided into two periods, the embryonic period (first two months of gestation) and the fetal period (the remaining time in utero). During embryonic period, most of the organ system form; the fetal period is the one of growth and modeling. Thus, thalidomide is found to be teratogenic only during days 34-50 after the last menstruation. From 34-38 days, no limb abnormalities are seen. But thalidomide can cause the absence or deficiency of ear components during this period. Malformation of the limbs occurs mainly during 38 to 50 days in a progressive manner of thalidomide exposure. Prior to week 3, exposure to teratogens does not usually produce any abnormality, because the teratogen encountered at this time period kills all the embryonic cells resulting in embryo's death. The teratogen during this exposure period may alternatively damage only a few cells, allowing the embryo to recover, as the embryonic cells at this stage remain totipotent to give rise to all different cell types. Weeks 3-8 is vulnerable to teratogen exposure, because this is the period when cell types are differentiating and most organs are formed. The nervous system is forming continuously throughout gestation and even after birth, and hence remains susceptible to teratogenic agents throughout infancy.

VALPROIC ACID (VA) AND AUTISM

Some drugs used to control diseases in the pregnant mother may have deleterious effect on a fetus. Valproic acid (VPA), an anticonvulsant used to treat seizures and epilepsy is known to disrupt normal embryo development. Pregnant mothers exposed to VPA bear children with Fetal Valproic Syndrome that include craniofacial abnormalities, such as eepicanthal folds, hypertelorism, flattened nasal bridge, upturned nose, and a long upper lip with a flattened philtrum. Some of the cases were also reported to have strabismus, low-set posteriorly rotated ears, and developmental delays in social and language skills (autism).The increased risk of autism in children born to mothers exposed to VPA has instigated research relating to the causative factors underlying this pervasive developmental disorder. The somatic malformations of those with both thalidomide embryopathy and autism indicate a critical period between days 20 and 24 of gestation, just as the neural tube is closing. VPA exposure at this time has been shown to reproduce some of the anomalies of neuroanatomy reported in histology of human cases of autism, such as reduction of neuron numbers in the cranial nerve motor nuclei. However, the most consistent findings in histology of human autism cases are a reduction in Purkinje cell numbers in the cerebellum. A single exposure to VA *in utero* caused changes in the rat neocortex, including a weaker excitatory synapse response. Further, the cerebella of rats prenatally exposed to VA were significantly smaller in volume than the control cerebella; the treated cerebella also contained fewer Purkinje cells in the vermis. Purkinje cells initiate establishment of the cortical layers during normal brain development.

RETINOIC ACID

Sometimes molecules that are involved in normal development could disrupt development if present in wrong amounts or produced in the wrong time. High intake of vitamin A as well as its chemically similar synthetic retinoids cause birth defects in humans and the defect appears to affect tissues derived from the cranial neural crest. Vitamin A is essential for embryogenesis, growth and epithelial differentiation. Vitamin A refers to retinoid compounds that have the biological activity of retinol. Experiments in animals indicate that retinoids can be teratogenic. In humans, isotretinoin (a 13-*cis*-retinoic acid), a synthetic retinoid used in the treatment of severe acne, causes congenital fetal anomalies. The syndrome, called 'retinoic acid embryopathy'includes a specific group of malformations, including those of craniofacial, cardiac, thymic, and central nervous system structures. The teratogenic mechanism by which retinoids affect the development of cephalic neural crest cells and their derivatives is through their interference with the closure of neural tube. Further, the teratogenic effects of retinoids may come from an effect on the expression of the homeotic gene Hoxb-1 that regulates axial patterning in the embryo. Retinoic acid alters the expression of *Hox* genes to disrupt normal development in the fetus. *Hox* genes are the main determinants of anterior-posterior axis during embryonic development. By altering the expression of *Hox* genes, RA can re-specify portions of the anterior-posterior axis in a more posterior direction and can inhibit neural crest cells from migrating to form the facial cartilage. Furthermore, radioactively labeled RA is seen to bind to the cranial neural crest cells and arrests both their proliferation and their migration.

MECHANISM OF DRUG ACTION

Mechanism of action of both thalidomide and VPA in bringing about teratogenesis is not fully understood. But they along with another teratogen retinoic acid (RA) are known to interfere with pattern formation. RA in particular is a pivotal factor in the anterior/posterior patterning of the mammalian embryo. Artificial altering of RA concentration at critical developmental stages can lead to anteriorization of the developing nervous system if RA concentrations are too low. Increased RA concentration leads to disruption of more posterior parts of the developing hindbrain. RA's action is obviously brought about by their capacity to bind its receptor, retinoic acid receptor (retinoid X receptor) which after activation, binds to its RA responsive element (RAREs) found on the DNA of many developmental genes. One of the genes influenced by RA is the Hox group of homeotic genes. High levels of RA increase the expression of Hoxa1 as well as other Hox genes by binding to RAREs. Hoxa1 is particularly crucial to the development of brainstem: too much or too little Hoxa1 protein or disruption of expression through its RARE is teratogenic. Tight regulation of the genetic expression of this gene is critical to the normal development of the embryo.

While RA exerts its teratogenic action directly on Hoxa1 gene, VPA may have other mechanism of action, although VPA exposure increases the level of retinoic acid in the blood. Clearly, VPA exposed mice showed overexpression of Hoxa1 and altered brainstem development in the embryo. Although VPA is reported to elevate endogenous RA levels, it seems unlikely that they act through RA for Hoxa1 altered expression, because its effect on Hoxa1 is so strong and so rapid. VPA is a potent inhibitor of histone deacetylase (HDAC). Inhibition of HDACs results in increased transcription of many genes by unfolding their chromatin. This is consistent with the fact that VPA exposure alters expression of many genes important for development including other Hox genes and genes related to neural tube formation. Thus, it is possible that inhibition of HDACs is the primary event in VPA's teratogenicity. Another fact that emerges from this VPA teratogenicity on Hox genes is that Hoxa1 is a candidate gene for autism susceptibility. The demonstration that VPA exposure elevates the expression of Hoxa1 during its expression period and can induce expression outside of the normal expression period helps us to understand how VPA exposure disrupts development and also teratogen-induced misregulation of this gene can lead to altered neural patterning and cognitive anomalies seen in autistic children.

INDUSTRIAL MERCURY AND MINAMATA DISEASE

Apart from drugs, industrial chemicals also act as human teratogenic agents. The first report came from Japan, where methylmercury was found to be the cause of neurological deformities in nearly 10% of the children born near the area of Minamata Bay. After world war II, the Mianmata plant of Japan's Chisso Corporation manufactured several organic chemicals such as acetaldehyde, which is an intermediate in the manufacturing of consumer products like plastics, paints and perfumes. Mercuric sulfate is used to oxidize acetylene into acetaldehyde and a by-product in this reaction is mercury, which the company dumped indiscriminately into Minamata Bay. Mercury is converted by the microbes into an organic form, methylemercury, which is consumed and concentrated by shellfish and finfish. This mercury poison finally gets into people who eat these fishes. The villagers who ate the mercury contaminated fish had difficulties in seeing, hearing and swallowing. A startling finding was that pregnant women were inadvertently exposing their fetuses to high degree of this compound, while eating their normal food. Neurological defects in the brain and eye in the developing fetuses could be caused by transmission of mercury across the placenta, which actually concentrates mercury and presents it to the fetus. This disease also affects the axons of neurons that transmit sensory information to the brain. Mercury also could be transmitted through the mother's milk into the infant to cause such defects. Later studies showed that mercury is selectively absorbed by regions of the developing cerebral cortex. Mercury had the same effect on the pregnant mice, fed with mercury in producing small brain and small eyes. Industrial dumping of mercury, cadmium and lead into lakes in USA and Canada have prompted authorities to give warning against eating fish caught from the great lakes.

TERATOGENIC EFFECTS OF ALCOHOL

Recent studies have indicated that the most devastating human teratogen is ethanol. Fetal alcohol syndrome, or FAS, is a syndrome of birth defects in the children of alcoholic mothers. Babies with FAS are characterized by small head size and defective facial features such as narrow upper lip and a low nose bridge. The FAS child has smaller than normal brain, with defects in neuronal and glial migration. There is also abnormal cell death in the frontonasal process and the cranial nerve ganglia. Sadly, FAS is the leading cause of congenital mental retardation, affecting about 1% of the live born infants. Consumption of low quantities of alcohol also leads to 'fetal alcohol effect', a condition that does not cause the distinct facial appearance of FAS, but lowers the functional and intellectual abilities. Children with fetal alcohol syndrome are developmentally and mentally retarded with a mean

IQ of about 68. Experimental studies on the effect of alcohol in the mice also produced same results as in humans. Mouse studies have significantly showed that ethanol induced impairment in the migration of neural crest cells, thus accounting for the defective facial features in the FAS children. In the normally developing fetuses, neural crest cells migrate from the dorsal region of the neural tube to generate the bones of the face (see chapter..). Instead of migrating and dividing, ethanol-treated neural crest cells prematurely initiate their differentiation into facial bones. Ethanol induces apoptosis of cranial neural crest cells by downregulating *sonic hedgehog* expression in the mouse and chick embryos. Ethanol-induced cell death in the nerve cells of forebrain and cranial nerve ganglia is also caused by superoxide radicals, generated by ethanol, oxidizing cell membranes resulting in cytolysis. Since alcohol can pass through the placenta very easily, enormous damage to the developing fetus.

ENDOCRINE DISRUPTORS AND DEVELOPMENTAL ALTERATIONS

Laboratory experiments have demonstrated that exposure of fetuses to endocrine- disrupting chemicals can profoundly disturb organ development by acting as hormone agonists or antagonists. Developmental abnormalities particularly occur in those organs in the fetuses which get naturally exposed to maternal gonadal hormones by their receptors. For example, in the female fetuses, these organs include the mammary glands, fallopian tubes, uterus, cervix, and vagina; in male fetuses, they include prostrate, seminal vesicles, epididymis and testis. In both sexes, the external genitalia, brain, skeleton, thyroid, liver, kidney and immune system are also targets for steroid hormone action and are thus potential targets for endocrine – disrupting chemicals.

The effect of environmental estrogenic compounds on fetus development has been well illustrated by the incidental use of a synthetic estrogen, called diethylstilbestrol (DES), to prevent spontaneous abortions in women in U.S.A. during 1948 – 1971.Thus, DES- exposed humans serve as a model for exposure during early life to any estrogenic chemical, including pollutants in the environment that are estrogen agonists. Interestingly, DES does not bind to estrogen- binding plasma proteins, but enter freely the target cells, thus increasing greatly their biological activity relative to similar concentrations of endogenous estrogen.

Estrogenic activity of any chemical could be determined by their stimulation of mitotic activity in the tissues of the female genital tract in early ontogeny, during puberty and in the adult. Daughters whose mothers took DES suffer reproductive organ dysfunction, abnormal pregnancies, a reduction in fertility, immune system disorders such as high incidence of autoimmune diseases and periods of depression. As young adults, they also suffer increased rates of vaginal clear- cell adenocarcinomas, which is a reproductive tract cancer found only in old women. Hence, the incidence of cancer is much higher in these women than unexposed individuals. DES- treated laboratory mammals adduce further evidence in this regard (see below). Several anthropogenic chemicals not specifically designed to possess hormonal activity, such as some pesticides, herbicides polychlorinated compounds, plasticizers and alkylphenols have been found to mimic sex hormones such as estrogen, affecting human and animal populations including wildlife. In addition to causing teratogenic effect on fetal development and birth defects, they also cause long term effects which in humans may not manifest themselves until adolescence and even much later in life. These effects include, among other things, structural, reproductive, endocrinological, metabolic, immunological, neurological as well as behavioral dysplatic and neoplastic changes.

Environmental estrogenic agents chiefly cause endocrine disruption, leading to developmental as well as functional abnormalities in female reproductive organs. Sex hormones play a crucial role in reproductive neuroendocrine functions. Estrogen or aromatizable androgen plays a significant role in modulating neuronal development and neuronal circuit formation during the perinatal period. Perinatal sex hormone exposure induces permanent sexual dimorphism such as nuclear volume, neuronal number, neuronal membrane organization, synaptic formation and neuronal connectivity in the hypothalamus. Short-term administration of sex hormones, including DES, to rodents during critical period results in persistent changes in the hypothalamo-hypophysio-gonadal system and in reproductive tracts, bringing about infertility .The mechanism of action of xenoestrogen in mammals is by inducing expression of estrogen receptor and then binding to it. In the mouse, the DES induced estrogen receptor (Era) mRNA in uterine and vaginal epithelial cells, 12 hours after injection. Reduced fertility and /or altered sexual behaviour in male rats that received estrogen or DES neonatally. In particular, penile organ contains the estrogen receptor and there is reduction in size and accompanying deviation in sexual behaviour, caused by lowering of testosterone. Abnormally small phalluses in alligators exposed to an excessive spill of estrogenic compounds in Lake Apopka in Florida have also been documented. Neonatal exposure of estrogen to mice causes deregulated expression of oncogenes, c-jun and c-fos mRNAs, resulting in the estrogen- independent persistent proliferation and cornification and hyperplastic downgrowths of the vaginal epithelium. In the mammalian uterus, Abdominal B (*AbdB*) *HOX* gene *HOX*a-10 expression is activated by progesteron and repressed by estrogen. In perinatally DES exposed mice and prenatally DES exposed humans, gene activity of *HOXa-10* is significantly affected resulting in reduced fertility due to defects in implantation. In support, *HOXa-10* knockout mice showed uterine, cervical and oviductal malfunctions resembling those in perinatal DES- exposed mice and prenatally DES- exposed humans. DES exposure also repressed *HOXa-10* gene activity in the Mullerian duct in both *in vitro* and organ culture conditions. DES exposure also affected reproductive tract morphogenesis in the female mice by repressing expression of *Wnt 7a*. Furthermore, fetal exposure to DES results in de-regulation of *Wnt 7a* during uterine morphogenesis.

It is well known that DNA methylation regulates cellular physiology by altering gene expression and is programmed in the growth and differentiation process. In the uterus of the mice, neonatally treated with DES, DNA demethylation occurs in the lactoferrin promoter region upstream of the estrogen response element. Such demethylation requires ovarian hormones to occur in the adults. The demethylation was also maintained in uterine tumours of the neonatally DES- treated mice.

Neonatal exposure to estrogen also induces the permanent and persistent induction of epithelial growth factor (EGF) and lactoferrin that are normally controlled by steroid hormones. Ovary- independent persistent expression of EGF and transforming growth factor-α (TGF-α) mRNAs also occurs in neonatally DES exposed mouse vagina. Neonatal exposure to DES also alters expression of genes such as insulin-like growth factor (IGF)-I, IGF II and keratinocyte growth factor (KGF). Down regulation of Tumour necrosis factor –a (TNF-α) and Fes ligand, associated with apoptotic cell death in mouse reproductive tract also occurs in neonatally DES- exposed uterus and vagina. The up-regulated expression of estrogen- inducible genes, *EGF* and / or *TFG-α*, as well as the down-regulated expression of death factors TNF-α and /or Fas ligand, may play roles in the precocious development of lesions in the female reproductive tracts. Another effect of early exposure to xenoestrogen is the formation of polyovular follicles in the ovary of alligators and mice. These polyovular follicles, after forced ovulation by gonadotropin injection followed by *in vitro* fertilization, showed significantly lower fertilization rate and the arrest of embryonic cell division at 8- cell stage.

Non- genital abnormalities also occurs in immune system, central nervous system, hypothalamo-hypophysial complex and behaviour in mice exposed perinatally to sex hormones and anti-hormones. Estrogen feminizes the bone structure of the pelvis and pubic symphysis in many mammals. Neonatal DES exposure induced persistent reduction of calcium and phosphorus in the pelvis and femur in aged mice, resulting in permanent changes in bone tissues. Another endocrine disruptor that persists in land and water is dioxin which can antagonize the action of estrogen in estrogen target cells. However, this effect does not seem to be due to dioxin binding to estrogen receptors. Dioxin by virtue of its high lipid solubility, bio-accumulates in human breast milk. These xenobiotics are also found in follicular fluid and hence can interfere with reproductive success by disrupting oocyte development.

The ability of 'environmental estrogen' to compete with sex steroid binding proteins in the plasma in human and fish occurs at concentrations about 100 fold greater than estradiol. Hence environmental estrogenic agents are unlikely to produce biological effects by displacing endogenous steroids from plasma steroid binding proteins unless they are present in very high concentrations.

EFFECT OF ENVIRONMETAL CONTAMINANTS ON WILDLIFE

Thousands of synthetic compounds with endocrine disrupting ability are released in the environment allover the globe. Many of the endocrine disruptors are lipophilic and have low vapour pressures, facilitating wide-spread dispersal. They are capable of binding to intracellular estrogen receptors either directly or after in situ conversion to an active metabolite. For example, the pesticide methoxychlor is demethylated in situ to a more estrogenic bisphenol compound. Exposure to endocrine – disrupting chemicals in the environment has been associated with abnormal thyroid function in birds and fish; decreased fertility in birds, fish, shellfish and mammals; decreased hatching success in fish, birds and turtles; feminization of male fish, birds and mammals; defeminization and masculinization of female fish and birds; and alteration of immune function in birds and mammals. In USA, the contamination in the Great Lakes region has lead to several endocrine disrupting instances among aquatic animals living there. According to a report from Gulph University, there is a 100% prevalence of thyroid enlargement in 2-4-year-old salmon in the Great Lakes. In some salmon stocks, there is an extremely high prevalence of precocious sexual maturation in males, poor egg survival and low egg thyroid hormone content. Multiple abnormalities, including behavioral changes, reproductive loss, and early mortality in offspring have been documented in bird species that feed on Great Lakes fish.

The deleterious effect of DDT on embryonic survival in bald eagles due to eggshell thinning and cracking is another example of endocrine disrupting chemicals in these areas. Even after the restrictions in the use of DDT since 1972 in USA, embryonic and chick survival in bald eagles is not adequate to maintain stable populations. In addition to DDT, bald eagles carry elevated levels of other endocrine disruptors such as chlordane, dieldrin, and polychlorinated biphenyls (PCBs). Many PCB residues are either endocrine disruptors or developmental toxicants. In another instance, male rats fed on Lake Ontario fish showed hyper reactivity to stress, and the offspring of females fed with Lake Ontario fish during pregnancy also expressed the same hyper reactive condition, although they were never fed with fish. Many of the effects of endocrine disruptors in wildlife are associated with the presence of a toxic contaminant in the mother due to exposure before egg production in birds and fish or pregnancy and lactation in mammals.

A recent estimate has indicated that approximately 50% decrease in sperm count in man worldwide over the last 50 years. This decrease in sperm count is the result of exposure during the fetal period of testicular differentiation to pollutants that have estrogenic activity. Recent studies have indicated in humans pathological insults, either in the form of environmental biohazards or clinical (anti-cancer) therapies, can lead to premature ovarian failure and infertility by stimulating the intrinsic programme of apoptosis present in ovarian oocytes. For example, polycyclic aromatic hydrocarbons (PAH), released by the incomplete combustion of fossil fuels and through tobacco smoking, have recently been shown to accelerate oocyte loss in mouse and human ovaries by activating apoptosis in oocytes. Treatment of young girls and reproductive stage- women for cancer with either chemotherapy or radiation has also resulted in massive oocyte apoptosis, as a side effect of the treatment. Interestingly, PAH has an intracellular binding protein in the oocytes in the form of aryl hydrocarbon receptor (AHR), a transcriptional factor of the Per-Arnt-Sim gene family. PAH-Ahr binds to the consensus AHR response elements (AHRE) in the promoter region of the pro- apoptotic *Bax* gene. The activation of *Bax* gene promotes apoptosis of the oocyte (for more details see chapter 9 on neural crest cells).

Interestingly, AHR receptor also binds to another xenobiotic compound, dioxin. Dioxins are typical environmental contaminants that exert adverse estrogen- related effects. Dioxin activates the nuclear receptor AhR, which in turn dimerizes with another transcription factor, Arnt. This dioxin activated heterodimer mediates most of the toxic effects of dioxins by binding to xenobiotic response elements (Fig.29.1). This AhR/ Arnt heterodimer also directly associates with estrogen receptors, ER-α and ER-β and then binding to estrogen- responsive gene promoters, leading to activation of transcription in an estrogen- dependent manner (Fig. 29.1 & 29.2).

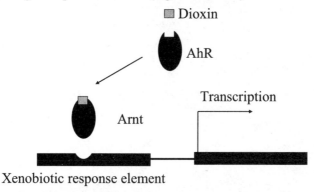

Fig 29.1 The aryl hydrocarbon receptor (AhR), when activated by dioxins, forms a complex with the aryl hydrocarbon nuclear translocator (Arnt) that can also trigger transcription. The complex regulates gene transcription by binding to specific elements in DNA, termed xenobiotic response elements.

Another instance of aquatic pollutants interfering with sexuality of aquatic animals is the herbicide, atrazine. It shows faminization effects on wild leopard frog, *Rana pipiens* with gonadal abnormalities such as retarded development, hermaphroditism and testicular oogenesis. Atrazine affects sex differentiation by inducing aromatase, the enzyme that converts androgens into estrogens, thus causing inappropriate synthesis and secretion of estrogens in males at the expense of androgens. Evidence for this mechanism of toxicity generalizes the possible environmental risk associated with atrazine. In many aquatic forms, the estrogen dependent vitellogenin gene activation has been used as a molecular marker to study the influence of such pollutants with estrogenic effects.

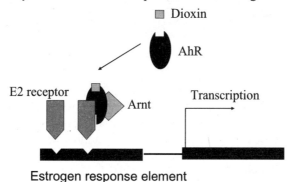

Fig. 29.2 Dioxins mimic the effects of estrogens through a mechanism that involves the activation of estrogen receptors by a transcriptionally active AhR - Arnt complex.

296 Molecular Developmental Biology

Several chemical pollutants, which also occur naturally in the soil, can also cause teratogenic effects in the developing embryos. For example, ingestion of excessive fluoride through food or water causes deleterious effect on embryonic development. Fluoride is known to alter mineralization within bone, although the mechanism for its action is unclear. Experimental studies on the effect of fluoride on chick development have revealed several teratogenic effects such as high bone density and overgrowth of bones in tibiotarsus, tarsometatarsus, and phalanx, affecting bone articulation. Thus, in the fluoride-treated white leghorn chick embryos, the abnormally developed toes showed stiffness and loss of flexibility (Fig. 29.3).

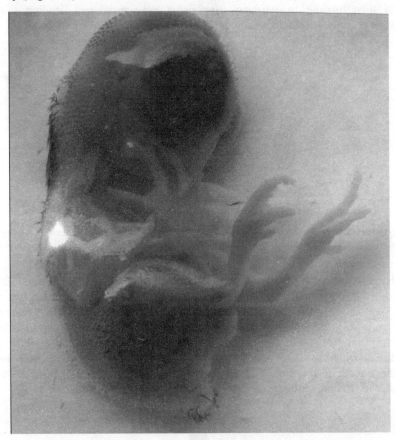

Fig. 29.3 Loss of nestling feathers and abnormal growth of hind limb digits in 12 day incubated chick embryos upon fluoride treatment (Photo courtesy Dr. S. Krupanidhi)

In these chicks, in addition to endoskeletal malformation, the exoskeletal derivatives (feathers) are also severely affected by fluoride toxicity. As seen from the fig.., there is loss of nestling feathers as well as the blackening of adult feathers and their falling due to fluoride toxicity. Similar to the teratogenic agents enumerated above, over-expression of normal developmental genes by way of extraneous growth factors may also cause abnormality in development. For example, chick embryos treated with exogenous fibroblast growth factor (FGF) during gastrulation develop neural abnormalities (Fig.29.4).

Fig. 29.4 Falling and blackening of adult feathers and stiffness of hind limb digits due to fluoride exposure (Photo: courtesy Dr.S. Krupanidhi)

Abnormalities included abnormal neural folds, incomplete closure of neural tube and abnormal somitogenesis. Exogenous FGF at HH stage 4 (gastrulation) of chick embryos caused down-regulation of the *Brachyury* expression as well as enhancement in noggin expression. In frog, noggin codes for a secreted protein which acts as a neural inducer by sequestering BMP4. In the chick, noggin is expressed in the Henson's node and neural plate during axial development and later in the neural folds and somites. Apparently, noggin protein participates in the patterning of the neural tube. Alteration in noggin expression could be one of the causes of the neural abnormalities in the FGF treated embryos (see Fig. 29.5).

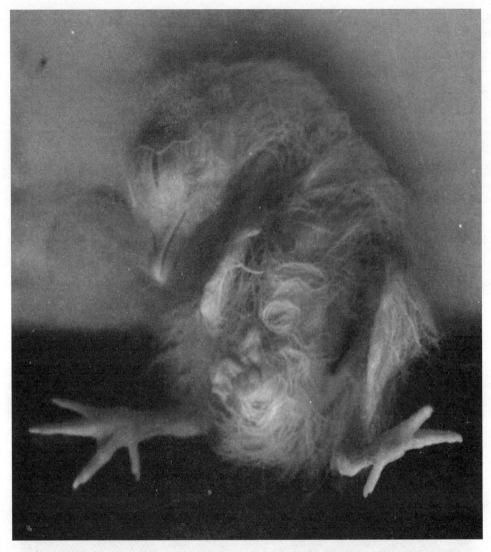

Fig. 29.5 Control chick embryo of 19 days incubation with flexible digits and intact white feathers
(Photo: Courtesy S. Krupanidhi)

TOXICITY STUDIES USING EMBRYONIC STEM CELLS

Mutagenic, embryotoxic or teratogenic substances may exert direct cytotoxic effects and / or induce alterations of embryonic development as a result of mutations at the DNA level. Additionally, developmental defects may be generated by interference of mutagenic or embryotoxic substances with regulatory processes of proliferation and differentiation at the levels of gene and protein expression, respectively. Perturbations of these processes may result in abnormal embryogenesis and malformations. However, reproductive toxicity and teratogenicity studies, under *in vivo* conditions on multigenerational animal models, using pre-conceptional as well as the pre- and postnatal phases, are not only laborious but also time consuming. This difficulty has been now circumvented by the introduction of the embryonic stem cells for such toxicity studies.

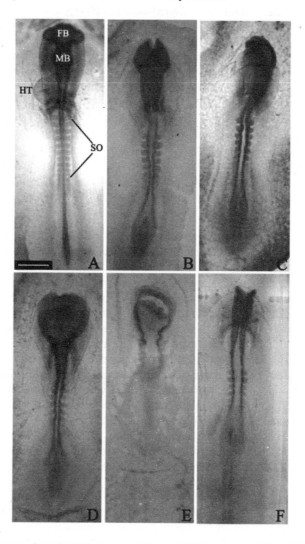

Fig. 29.6 Chick embryos were explanted at HH stage 4 and treated with human recombinant FGF. Excess FGF interferes with endogenous FGF signaling leading to developmental abnormalities after 2 hours. The scale bar is equal to 0.5m.
Courtesy: Drs. Seema Khot and Surendra Ghaskadbi, Agharkar Research Institute, Pune.

(A) Control embryo explanted at HH stage 4 and incubated further for 22 hrs in presence of either saline or bovine serum albumin: Note well developed nervous system with optic vesicles, looped heart and 10 pairs of somites. B-F] Chick embryos cultured *in vitro* at HH stage 4 and treated with exogenous FGF for 22 hrs: Note various abnormalities, such as, open neural tube (B-F), a unusually wide gap between the neural folds in the forebrain region (D), bending of both the neural folds on one side of the body axis (C), incomplete fusion of the neural folds along the body axis (B-D,F), diffused somites (B-D,F), retarded heart development in the form of absence of looping of the heart tube (B-D). A few specimens exhibited poor differentiation of all embryonic structures (E). FB, Forebrain; MB, midbrain; HT, heart; SM, somites.

The undifferentiated pluripotent embryo- derived stem cell lines are capable of developing into differentiated cell types, resembling early embryonic stages. Pluripotent embryonic stem cells with self- renewal capacity and the potential to differentiate into cell types of all three primary germ layers could be derived from embryos and fetal stages of mouse. They are;1) embryonic stem cells (ES), 2) embryonic germ cells (EG) and 3) embryonic carcinoma cells (EC) (for details see chapter on stem cells). *In vitro*, ES and EG cells differentiate spontaneously after cultivation as aggregates, called embryoid bodies (EB). This cultivation systems represent *in vitro* alternatives to animal tests for mutagenicity and embryo toxicity studies by way of investigating the inhibitory or inducing effects on differentiation processes of early embryonic stages, especially on the formation of neuroectodermal and mesodermal cells and tissues. Again, genetically altered ES and EG cell lines are also used to analyze mutagenic effects of chemical substances on embryos. Essentially, cyto-toxicity and embryo toxicity studies are concerned with measuring drug- induced alterations in metabolic pathways or structural integrity of the cells as well as total effect on early embryonic developmental stages. As the stem cell lines recapitulate cellular developmental processes and gene expression patterns of early embryogenesis during *in vitro* differentiation, effects of xenobiotics on specific developmental gene expression could also be perused using the embryonic stem cells.

For example, the influence of retinoic acid, a well known teratogen, displays a gene expression pattern in the EB, very similar to the whole embryos. A strong induction of neuronal differentiation was observed by high retinoic acid (RA) concentrations applied to ES cells during early stages of EB development, at developmental stages characterized by the expression of genes regulating embryonic ectoderm formation. Similarly expression of genes regulating mesoderm differentiation during skeletal muscle cell as well as cardiac and vascular smooth muscle cells differentiation could be achieved by RA treatment during later stages of EB differentiation. The excessive effect of RA on the inducement of differentiation of ectodermal, mesodermal and endodermal cell types is also exemplified by their teratogenic effect on the extranumerary limb formation in anuran amphibian regeneration. Recent studies have indicated that the teratogenicity of retinoic acid lies in their ability to alter the expression of the *HOX* genes and thereby respecifying portions of the anterior-posterior axis and inhibit neural crest cell migration from the cranial region of the neural tube. Interestingly, some of the *HOX* genes have retinoic acid response elements in their promoters. To effect gene regulation, retinoic acid must bind to a group of transcription factors called the retinoic acid receptors. These receptor proteins have the same general structure as the steroid and thyroid hormone receptors. Hence, they can activate these steroid hormones under unnatural conditions.

PROGRAMMED CELL DEATH OR APOPTOSIS.

Apoptosis, a term to describe the suicide of cells, derives its name from the Greek for the fall of autumn leaves. This term is now synonymously used with programmed cell death (PCD). PCD, a usage coined by Lockshin, is a physiological process that leads to the selective elimination of unwanted cells in multicellular animals. PCD thus involves the dying cell's own machinery as an executioner for the cell to commit suicide.

Apoptosis is an evolutionarily conserved form of cell suicide by which an organism eliminates unwanted and damaged cells. It is an essential physiological process that plays a critical role in development and adult tissue homeostasis. This is achieved by controlling the number of cells in development and throughout an organism's life by removal of cells at the appropriate time. Apoptosis plays a pivotal role in controlling cell number and turnover, tissue homeostasis, and specialization, tissue sculpting and pattern formation, oncogenesis and other disease incidents, defence reactions as well as maintaining the overall integrity of the animal.

Dysregulation of apoptosis leads to a variety of human pathologies including cancer, autoimmune diseases (rheumatoid arthritis, type-1 diabetes) and neurodegenerative disorders. Cell death in developing system is not merely a degenerative process, but an active and controlled phenomenon. This involves an intrinsic genetic program that is activated at a controlled time and location to cause the death of the cell. In many instances, this type of cell death is essential for normal morphogenesis.

Typically, cells from multicellular organisms self-destruct when they are no longer needed or have become damaged. To survive, all cells from multicellular animals depend upon the constant repression of this suicide programme by signals from other cells. For example, apoptosis in the free-living nematode, *C. elegans* is initiated by two genes (*ced-3*, *ced-4*). Loss-of-function mutation in either of these two genes allows survival of cells normally programmed to die.

Characteristic features of apoptosis include (1) condensation of the chromatin, (2) cytoplasmic blebbing giving rise to apoptotic bodies and (3) inter-nucleosome cleavage of DNA, as a result of the activation of endonucleases. Another major biochemical feature associated with apoptosis is the proteolytic cleavage of a number of intracellular substances resulting in plasma membrane changes.

Apoptotic cells are rapidly phagocytosed prior to the release of intracellular contents and without induction of an inflammatory response. Hence, apoptotic death does not cause damage to surrounding cells. Apoptosis can be induced in the target cell population, but not in surrounding cells, by a variety of factors such as hormones or growth factors. All in all, the decision to undergo apoptosis in response to a signal is a controlled, differentiative cell process, requiring the activation of a genetic programme.

Apoptosis is different from necrosis, which is characterised by a general cell lysis as a result of membrane damage. In necrosis, there is pycnosis of the nucleus (condensed hyperchromatic nuclei, but no DNA fragmentation), mitochondrial swelling and rupture, and lysosomal rupture and inflammation. Necrosis of the cells is a pathological condition, involving groups of adjoining cells. It is induced by lethal chemicals or physical insult. Necrosis does not require the expression of a specific set of gene products. It is non-physiological and is not encountered in developmental situations.

PROCESSES ASSOCIATED WITH APOPTOSIS

Apoptosis results in the disassembly of cells involving the condensation, shrinkage and fragmentation of the cytoplasm and nuclei into several sealed packets called apoptotic bodies or apoptosomes, which are then phagocytosed by the neighbouring cells or the macrophages. The apoptotic mechanism involves effectors, adaptors, regulators and signals as well as the activity of different endonucleases. The key effector molecules of apoptosis are the cysteine aspartate-specific proteases, called caspases and granzymes. Caspases are the conserved cysteine proteases, while the granzymes are the serine proteases; both specifically cleave after the aspartate residues of many proteins. The activation of caspase precursors is achieved by adaptor proteins that bind to them via shared motifs. TNF receptor superfamily of the apoptogenic cofactors released by the mitochondria is an example of the adaptors. Another adaptor molecule to activate caspase 8 is Fas-associated Death Domain (FADD) (See below). The ability of the adaptors to activate the effector caspases is regulated by other protein molecules that appear to directly interact with the adaptors/ co-factors. In addition, apoptosis involves a number of cell signalling events which may impinge upon the cell death mechanism at any level. The decision of a cell to initiate the act to kill itself is determined by the nature of the signals received from outside the cell, or by the generation of its own signal molecules or a combination of both. Some of the signalling events reported during PCD are (a) an increase in free cytosolic calcium concentration; (b) an oxidative burst manifested as an increase in free radical activity and (c) a collapse in the mitochondrial membrane potential. As mentioned earlier, apoptosis is morphologically expressed through the condensation, shrinkage and fragmentation of the nucleus and the cytoplasm, resulting in the formation of several sealed packets called apoptotic bodies enclosing both cytoplasmic and nuclear materials.

Nuclear fragmentation is preceded by the condensation and marginalization of chromatin in the nucleus. Fragmentation of DNA takes place at the nucleosome-linker sites and effected by endonucleases such as NUC18, DNase I and DNase II, which are present in the nucleus and are activated by Ca^{2+} and Mg^{2+} but inhibited by Zn^{2+}. On electrophoresis, these DNA fragments give a characteristic ladder formation; the rungs of the ladder are multiples of 180 bp. The DNA fragments could be cytochemically determined by the so- called TUNEL staining.

INCIDENCE OF APOPTOSIS

A fertilized egg produces billions of cells to create our body by repeated cell division and differentiation. During this process, many surplus or harmful cells are generated, and they must be removed or killed. Apoptosis is a unique feature of multicellular organisms that enables continuous renewal of tissues by cell division while maintaining the steady-state level of the various histological compartments. Steady-state kinetics is maintained in most healthy adult mammalian tissues by apoptosis occurring at a low rate that complements mitosis. Thus, in cell population control, cell deletion and cell death play an equally important role as that of cell proliferation. Hence, programmed cell death is a constitutive feature during the normal development of mammals.

APOPTOSIS DURING ANIMAL DEVELOPMENT

Apoptosis plays important roles in eliminating unwanted cells during development. They are: (1) sculpting structures; (2) deleting unneeded structures; (3) controlling cell numbers; (4) eliminating abnormal, misplaced, non-functional, or harmful cells; and (5) producing differentiated cells without organelles. The formation of digits in some higher vertebrates is a well-known example, where PCD eliminates the cells between developing digits. If the cell death is inhibited by treatment with a peptide caspase inhibitor, digit formation is blocked. Similarly, PCD is involved in hollowing out solid structures to create lumina, as found during the formation of vertebrate neural tube. During the course of animal development, various structures are formed that are later removed by apoptosis. For example, pronephric tubules form functioning kidneys in fish and amphibian larvae, but they are not used in mammals and are eliminated by PCD. Similarly, Mullerian ducts form the uterus and oviducts in female mammals, but it is not needed in males and is thought to be eliminated by apoptosis. On the other hand, Wolffian duct forms the vas differens, epididymis, and seminal vesicle in males, but it is not needed in females and is removed by PCD.

APOPTOSIS IN METAMORPHOSIS AND MORPHOGENESIS

The best known examples of programmed cell death is involved in metamorphosis of insects and amphibians, where, in the former, the inter-segmental musculature of the larval forms becomes unnecessary in the adult, and in the latter, the tail structures are reabsorbed in the adult. During insect metamorphosis, dramatic changes in body form result in extensive remodelling of the musculature and the nervous system. When the adult emerges from its pupal case, an apoptotic wave removes muscles, motor neurons and inters neurons that are no longer needed by the adult. Like metamorphosis itself, programmed cell death in insects is intimately linked to levels of hormones such as ecdysteroids and juvenile hormone. It is also important in morphogenetic reorganization, such as the elimination of tissue between the digits of the developing amniote limb and in closure of neural tube, heart and kidney morphogenesis. In the limb bud, the dying cells first produce acid phosphatase-rich autophagic vacuoles, then autolyse and fragment, and are then phagocytosed by macrophages or by normal mesenchyme cells in the region which becomes phagocytic.

APOPTOSIS IN *C. ELEGANS*

Another instance of apoptosis occurring during the ontogeny has been found in the free-living nematode, *Caenorhabditis elegans*. This round worm contains a total of 1090 cells in the growing embryo, but precisely 131 of these cells kill themselves during development, leaving 989 cells in an adult worm. Genetic analyses in *C. elegans* led to the identification of four genes namely, *egl-1* (*egl*, for egg laying defective), *ced-9* (ced, for cell-death abnormal), *ced-4* and *ced-3*, that collectively control the death of 131 somatic cells during hermaphrodite development. The protein products of these four genes define a linear pathway. CED-3 is a caspase, which is a cysteine-containing protease that cleaves its substrates after aspartate residues. CED-3 is synthesized as an inactive zymogen. When the cells are programmed to die, the CED-3 zymogen is thought to be activated by the adaptor molecule CED-4. In healthy cells, the pro-apoptotic protein CED-4 is bound to another protein CED-9. The CED-4-CED-9 complex is unable to activate CED-3. However, at the onset of cell death, another pro-apoptotic protein EGL-1 binds to, and triggers a conformational change in CED-9, resulting in the release of CED-4-CED-9 complex. The released CED-4 dimer undergoes further dimerization to form a tetrameric CED-4 complex, which is responsible for the induction of CED-3 autoactivation. The product of *ced-3* acts as a cystein protease during *C. elegans* cell death and is similar to the action of mouse *Nedd-2* gene which induces apoptosis in the brain cells. In addition, genes that are involved in the engulfment of dying and dead cells and the degradation of DNA in these cells have also been identified in *C. elegans*. They are *ced-1* and *ced-2* genes respectively.

APOPTOSIS IN LENS DEVELOPMENT

During lens development, programmed cell death of lens epithelial cells also play a role. In the lens fibre cells, death, by a loss of cellular structure, does not occur; however, lens fibre nuclei degenerate with a strict spatio-temporal pattern. In the chick embryo, loss of DNA begins in the central lens fibres on the 6th day of embryonic development. DNA fragmentation occurring between nucleosomes in the differentiating lens fibres is reminiscent of the classical DNA ladder that accompanies apoptotic cell death. The degeneration of the nuclei including pycnosis is also similar to apoptotic morphology. In the lens cells, apoptotic cell death is mediated by tumour necrosis factor-alpha (TNFα), which induces DNA fragmentation by stimulating endonuclease activity. Nuclear degeneration is accompanied by the breakdown of cell organelles in lens fibres. However, it is not a non-specific cytoplasmic degradation. Whereas organelles like nucleus and mitochondria disappear, the actin cytoskeleton remains intact.

APOPTOSIS IN NEURAL CELLS

Recently, the importance of apoptosis in the development and differentiation of neuronal cells in mammals has been stressed. Unlike other cell types, neuronal cells stop division after they differentiate into adult neurons. Therefore, homeostasis during neuronal development occurs via apoptosis in the neural stem cells (NSC). NSCs have the ability to divide and differentiate into mature neurons and glial cells and are present not only in the developing but also in the adult central nervous system. Neural stem cells undergo apoptotic cell death as an essential component of neural development. In the vertebrate nervous system, both neurons and oligodendrocytes are generated in excess, and up to half or more are eliminated by PCD, to match their numbers to the number of target cells they innervate or the number of axons they myelinate, respectively.

Apoptotic cell death occurs not only during development, but also under pathological conditions as a consequence of cellular damage due to cytotoxic insults, as demonstrated in experimental models using primary culture of NSC obtained from adult rats and a murine- derived multipotent NSC line (C17.2). On exposure to an apoptosis inducer, staurosporine and an oxidative stress inducer, 2,3- dimet*HOX*y-1,4-naphthoquinone (DMNQ), these mammalian neural stem cells are triggered to undergo apoptotic cell death as shown by nuclear shrinkage and chromatin condensation. However, both these cytotoxic stimuli induced only the intrinsic mitochondrial apoptotic pathway (see below), activating the executioner caspase-3. This is further evidenced by the release of cytochrome c from the mitochondrial intermembrane space into the cytosol as well as DNA fragmentation. However, Fas induced extrinsic apoptotic pathway is not induced in the mammalian NSC.

APOPTOSIS DURING VERTEBRATE LIMB DEVELOPMENT

Another example of apoptotic cell death during embryogenesis is found in the development of the vertebrate limb. The embryonic limb buds arise as paddle-shaped outgrowths, which are contoured. The final sculpturing of the limb bud to give rise to the limb with digits etc. is accomplished by the presence of a number of discrete zones of cell death, so that thousands of cells are programmed to die within an 8 to 10 hour period. Among the different zones, the inter-digital necrotic zones, which are responsible for the generation of the individual digits from the solid paddle, are important. Inter-digital zones are found in all amniotes, but are reduced in the inter-digital regions of webbed species. In the amphibians too, they are absent, but here, sculpting of the digits is due to reduced cell proliferation. Several of the areas of cell death in chick limb buds appear to express the homeo-box containing gene *GHOX-7* and *GHOX-8*, suggesting that these genes could be the determinants of cell death. Evidently, programmed cell death is analogous to cell proliferation and differentiation in that it is also involved in morphogenesis and in the regulation of cell population in embryos.

APOPTOSIS IN IMMUNE CELLS

Apoptosis also functions as part of a quality- control process in animal development, eliminating cells that are abnormal, misplaced, non-functional, or potentially dangerous to the organism. Apoptosis of immune cells is a classical example under this category. It has been estimated that more than 95% of thymocytes die in the thymus during maturation. Thymocytes that have failed to rearrange their T cell-receptor gene, or whose T cell receptor may recognise their own tissues, will be eliminated by apoptosis. Certain immune cells release the signalling molecule, TNF-α to initiate apoptosis to kill the infected cells. Steroid hormones of the adrenal cortex such as cortisol and cortisone promote the collective suicide of T lymphocytes in the thymus.

Programmed cell death of lymphocytes plays an important role in the development of immune system to discriminate between self and non-self. Millions of lymphoblasts, the future B cells and T cells of the immune system, rearrange DNA segments at random to produce a large variety of antibodies. This will not only generate lymphocytes whose receptors fit onto foreign antigens, but will also generate self-reactive lymphocytes with receptors that bind self-molecules. This will prove to be fatal if such self- reactive lymphocytes are not eliminated. Apoptosis eliminates in humans all these self-reactive lymphocytes around the time of birth.

APOPTOSIS IN RESPONSE TO DNA DAMAGE AND CELLULAR STRESS

Apoptosis can also be initiated by internal warnings resulting from DNA damage or other such cellular stress. The DNA damage stimulates the ATM pathway by activating ATM or ATR protein kinases (enzymes that phosphorylate certain target genes), and this activation leads to a rise in p53 protein levels. This protein can switch on a number of target genes, including those encoding the BCL-2 family members *BAX*, PUMA and NOXA, each of which contributes to apoptosis in specific cell types.

Cells subjected to hypersensitive response due to pathogenesis and other environmental and abiotic stresses such as osmotic, oxidative, temperature, salt, water, heavy metals, UV, nutrient deprivation, toxins, chemicals etc. often undergo programmed cell death. In this way, cell death is used as a defence against the stress or as a means of producing signals for the other nearby cells to build up defence/immune reactions, thus preventing further infection. In many instances, cells experiencing lethal stress doses often react by activating their PCD mechanisms to commit suicide before they are killed. However, some cells undergo necrosis instead of apoptosis, if the dosage of stress is beyond a particular level.

BIOCHEMICAL AND MOLECULAR MECHANISMS INVOLVED IN APOPTOSIS

The process of programmed cell death involves an epigenetic reprogramming of the cell which results in a cascade of biochemical changes. These changes eventually lead to morphological changes within the cell, resulting in cell death and elimination. The main intracellular effectors of apoptosis are a family of aspartate- specific cysteine proteases, called caspases. These enzymes exist in cells as inactive zymogens and become activated through proteolysis when cells receive apoptotic signals. Caspases are cysteine proteases that cleave their substrates after aspartic acid residue. Because caspases cleave their substrates at Asp residues and are also activated by proteolytic processing at Asp residues, these proteases can collaborate in proteolytic cascades, where caspases cleave themselves and each other.

The apoptotic caspases are divided into two classes: the initiator caspases, which include caspase-2,-8,-9 and –10 in mammals and Dronc and Dredd in fruitflies; and the effector caspases, which include caspases-3,-6 and –7 in mammals and Drice, Decay, Damm, Dcpl and Strica in fruit flies. CED-3 is the only apoptotic caspase in nematodes and functions as both an initiator and an effector caspase (Fig.30.1).

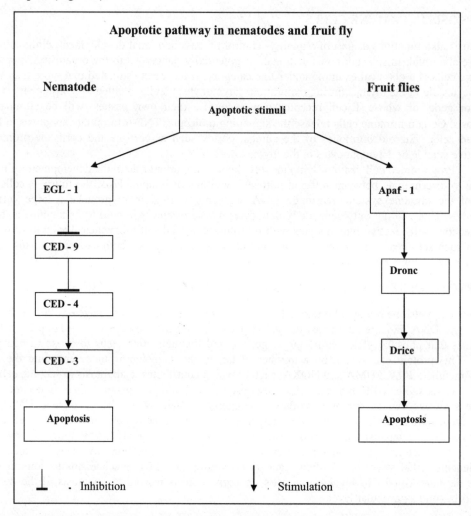

Fig. 30.1 Nematodes: Four genes, *ced-3*, -4 and –9 and egl-1 function sequentially to control the onset of apoptosis. CED-3 is an effector protein and is activated by the adoptor protein CED-4. In the absence of apoptotic signalling, CED-4 is constitutively suppressed by the antiapoptotic protein CED-9 through direct physical interactions. During apoptosis, the negative regulation of CED-4 by CED-9 is removed by EGL-1, which is transcriptionally activated by cell-death stimuli. CED-9 is a functional and structural homologue of the mammalian proteins BCL2 and BCL-XL, whereas EGL-1 is a BH3-only member of the BCL2 family of proteins. Fruit fly: In *D. melanogaster*, Apoptotic-protease-activating factor-1 (Apaf-1) activates the initiator caspase Dronc. Dronc, in turn, cleaves and activates the effector caspase Drice. Dronc and Drice are structural and functional homologues of mammalian caspase-9 and caspase-3 respectively.

BIOCHEMICAL PATHWAYS IN MAMMALIAN CELL APOPTOSIS

Two major biochemical pathways are involved in the process of caspase activation and apoptosis in mammalian cells (Fig.30.2).The first of these depends upon the participation of mitochondria (receptor- independent) and the second involves the interaction of a death receptor with its ligand. As seen from the figure, the first or the intrinsic pathway involves the release of cytochrome c from mitochondria into the cytosol where it binds to apoptotic protease activating factor-1 (Apaf-1), a mammalian homologue of the *C. elegans* cell death protein CED4. The activated Apaf-1 binds to procaspase 9, resulting in the activation of the initiator caspase 9. Caspsae-9, in turn, cleaves and activates procaspase-3, which is an effector protease bringing about cleavage of a set of proteins, including polyribose polymerase, lamin, actin, and gelsolin and causes morphological changes to the cell and nucleus typical of apoptosis.

Recent studies have revealed that cytochrome c is bottled up inside the membrane folds of mitochondria. Sequestering the cytochrome c proteins within the membrane folds protects the organelles' host cells from an untimely end. A protein called OPA1 seals cytochrome c inside the pocket-like folds of the inner mitochondrial membrane. It has also been shown that the actions of cell-death activators might disrupt the OPA1 stopper, releasing the cytochrome c.

ROLE OF BCL-2 FAMILY PROTEINS

Bcl-2 was first identified as a proto - oncogene in follicular β - cell lymphoma, but now recognised as a mammalian homologue to the apoptosis repressor *ced-9* in *C. elegans*. Pro- and anti-apoptotic members of the Bcl-2 family play a major role in governing the mitochondria-dependent apoptotic pathway. Cellular stress induces pro-active Bcl-2 family member like *Bax* to induce the release of cytochrome c from mitochondria into cytosol, while the anti-apoptotic Bcl-2 and Bcl-X_L prevent cytochrome c release from mitochondria, and thereby preserve cell survival. Cytochrome c when inside the mitochondria couples with a heme group to become holocytochrome c, and it is only this form that functions to induce caspase activation.

Evidently, Bcl-2 protects cells from death, whereas other members of Bcl-2 family such as *Bax*, Bad and Bid cause cell to die. Hence for a cell to survive there should be a balance between the activities of antagonistic survival and killer genes. In other words, a misregulation of apoptosis could lead to many human disorders, including lymphomas and other cancers. Another implication of this molecular understanding of the apoptotic pathway is that cancers might be treated by reactivating apoptosis and causing tumour cells to self- destruct. Thus cancer cells can be induced to die either by blocking anti- cell death genes such as Bcl-2, or by activating pro-cell- death genes such as *Bax*.

The second pathway to activate caspase is extrinsic in that the apoptosis involves ligation of the death receptors, like the tumour necrosis factor receptor-1(TNF-R1) and the Fas receptor to their cognate ligands (Fig.30.2).The binding results in an oligomerization of receptors and a subsequent activation of procaspase-8. The active caspase-8 works in two ways. It either cleaves or activates procaspase-3 directly or it cleaves the proapoptotic Bcl-2 protein Bid to tBid, which then activates the mitochondrial apoptotic pathway (not shown in the picture). Although caspase- activated apoptosis is the only known way of cell death in mammalian cells, other pathways could also occur. For example, apoptosis induced by UV light can proceed in animals lacking caspase-9 suggesting the existence of alternative pathways independent of cytochrome c- Apaf-1-procaspase 9 apoptosome.

MACROPHAGE-INDUCED APOPTOSIS

It is well known that macrophages have a critical role in inflammatory and immune responses through their ability to recognize and engulf apoptotic cells. Thus, macrophage involvement in programmed cell death comes only after the apoptotic event, by its specific response to presence of membrane-tethered or soluble 'eat me' signals from the dead or dying cells. However, under certain circumstances, phagocytes (macrophages) can also actively induce programmed cell death. In the mice, macrophages are required for the programmed regression of temporary capillary networks within the developing eyes. Recent studies have indicated that macrophages induce apoptosis in the vascular endothelial cells of the temporary blood vessels in the developing eyes by secreting a short-range paracrine signalling molecule, WNT7b. By cell-cell contact, WNT7b enters into the endothelial cells and activate the canonical *WNT* pathway. The *WNT* signalling pathway has a critical function in developmental cell fate decisions, including cell death. In the vertebrate tissues, *WNT* response requires a receptor complex comprising the co-receptor, low-density lipoprotein receptor-related protein, LRP5 or LRP6 and a multiple-pass transmembrane receptor of the Frizzled (FZD) family. This receptor complex initiates a cascade of events that culminates in the regulation of target genes responsible for the apoptosis of vascular endothelial cells.

Instances of phagocytes promoting programmed cell death are also found in the round worm, *C. elegans*. In *C. elegans*, phagocytes are required for the recognition and engulfment of dead cells. But such phagocyte recognition pathways in disposing dead and superfluous cells can also act as a backup stimulus for the induction of programmed cell death, when the autonomous caspase-driven pathway (such as CED3) is deficient. Taken together, these studies suggest that phagocytes may have a conserved function in signalling cell death in a variety of contexts.

FACTORS CONTROLLING APOPTOSIS

Regulation of apoptosis is mediated by many factors. Viral infection, metabolic derangements such as sudden change in glucose concentrations, heat, irradiation, toxins and drugs are all capable of influencing the transition of a given cell to programmed cell death. On the other hand, growth factors such as epidermal growth factor (EGF) and the insulin- like growth factor (IGF) / insulin family along with their receptors are capable of preventing apoptosis in many cells. In mammalian tissues, many hormones, cytokines and growth factors act as general or tissue- specific survival factors preventing the onset of apoptosis (Tab.30.1). Conversely, several hormonal factors are also known to induce or enhance apoptotic processes. Steroid hormones have a major role in the regulation of growth, development, homeostasis and programmed cell death. Together with other hormones and growth factors, steroids regulate both induction and inhibition of cell death. Glucocorticoid hormones induce apoptotic cell death in immature thymocytes and mature T cells through an active process, characterized by extensive DNA fragmentation. This process is, however, inhibited by interleukins IL-2 and –4.

These adrenal steroids also act on the brain, thus regulating apoptosis in the central nervous system. Particularly, glucocorticoid receptors have a role in containing programmed cell death in the hippocampus and determine the rate of neurogenesis. Several peptide hormones also have a role in programmed cell death. Mullerian inhibiting substance (MIS) is a differentiating and anti-proliferative peptide, playing a crucial role in mammalian sex differentiation. (For details see chapter on sex differentiation). MIS activity is tightly regulated by *SRY* gene products at transcriptional level and by testosterone post- translationally. MIS acts at specific sites through activating a localized programme of cell death that is initiated via receptor-mediated events. Thus, MIS-induced apoptosis constitutes a crucial component of the machinery involved in male gonadal differentiation. The role played by gonadotropins and gonadal sex steroids is well illustrated in germ cell apoptosis in mammalian gonads.

GERM CELL APOPTOSIS DURING SPERMATOGENESIS

Mature sperm cells are products of a precisely regulated developmental sequence in which germ cell proliferation, differentiation, self- renewal and apoptosis are carefully controlled. Germ cell apoptosis during spermatogenesis is mediated by signals derived from the Sertoli cells with which each germ cell is closely associated, as well as by signals originating outside the testis. In particular, the intracellular signalling cascades, which ultimately determine germ cell fate, are best known with reference to Bcl-2 protein family. The complex net work of signals that determine male germ cell survival or death include paracrine signals such as stem cell factor (SCF), leukemia inhibitory factor (LIF) and *Desert hedgehog* (Dhh), as well as endocrine signals such as pituitary gonadotrophins and testosterone. Like other cell types, proteins of the Bcl-2 family provide one signalling pathway, which is essential for male germ cell homeostasis. Some members like Bcl-2, Bcl-xL, Bcl-w, Mcl-1 and A1 promote cell survival; whereas, others antagonise it (eg. *Bax*, Bak and Bim). Because, members of these opposing functions associate and seemingly titrate each other's function, their relative abundance in a particular cell type probably determines the incidence of apoptosis.

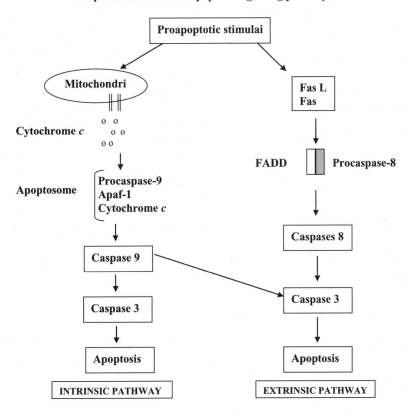

Fig.30.2 Schematic drawing showing the two major apoptotic pathways in mammalian cells. On the left is the intrinsic pathway of apoptosis involving the release of cytochrome c of mitochondria into the cytosol where it binds to Apaf-1 resulting in the activation of caspase-9, which in turn activates the executional caspase-3. On the right side is the extrinsic pathway involving ligation of Fas with FasL resulting in the activation of caspase-8 through interaction involving adoptor molecule such as FADD and the caspases. The initiator caspase 8 activates the executional caspase-3 to bring about apoptosis. Caspase-9 from the intrinsic pathway can also intervene with the extrinsic pathway activating directly caspase-3. (Modified from S.Ceccatelli et al. 2004)

APOPTOSIS IN FEMALE GERM CELLS

It has been estimated that more than 99.9% of the potential germ cell population generated in the human female degenerate rather than survive, in spite of the fact that a woman is capable of ovulating approximately 400 oocytes in her adult lifetime. The first report of apoptosis in any tissue is the description of degenerating granulosa cells in follicles of rabbit ovary in the year 1885. In humans, it has been estimated that about 7 million germ cells in the fetal ovaries at around week 20 of gestation are decimated to 1-2 million viable oocytes in early neonatal life. Those germ cells that survive the pre- and peri- natal waves of apoptosis, now enclosed by somatic cells as follicles, also continue to experience apoptosis, reducing the oocyte pool by a process better known as follicle atresia.

Interestingly, in the primordial, and the early maturational oocytic stages (primary and preantral), apoptosis initiates follicle atresia. In contrast, in the late preantral through antral stage, apoptosis of the granulosa cells initiates follicle degeneration.

MECHANISM OF OOCYTE APOPTOSIS

As for the apoptosis of other cell types, Bcl-2 family members are pivotal regulators of cell death in the ovary. The fate of any follicle, whether to survive or degenerate, depends upon the end result of a complex interaction between multiple Bcl-2 family members. Gene expression studies have implicated *Bax* as a key factor in the initiation of apoptosis in both oocytes and granulosa cells. Recent studies have shown that mice granulosa cells lacking *Bax* are defective in their ability to activate apoptosis. Interestingly, this defect in oocyte death leads to a dramatic prolongation of ovarian lifespan in aged *Bax* mutant females, essentially eliminating the mouse equivalent of menopause in these animals. Similarly, during embryogenesis, *Bax* gene inactivation circumvents excessive fetal oocyte death, caused by Bcl-Lx deficiency in the developing germ line. *Bax*- deficient oocytes are also resistant to the pro- apoptotic effects of the chemotherapeutic drug, doxorubicin. Furthermore, the massive level of oocyte destruction observed in female mice exposed to PAH is completely absent in animals lacking functional *Bax* protein. These results stress the fact that *Bax* gene certainly plays a critical role in most instances of germ cell apoptosis, although *Bax* independent pathway of female germ cell depletion may also exist.

Table 30.1. Involvement of hormones, cytokines and growth factors in apoptosis/ programmed cell death

Known inhibitors of apoptosis
    Testosterone
    Estradiol
    Progesterone
    Growth factors (EGF, IGF-I, NGF, PDGF)
    Interleukins
    Growth hormone
    Prolactin
    Gonadotrophins

Potential inducers of programmed cell death
    Glucocorticoids
    Progesterone
    Thyroid hormones
    Growth factor (IGF-I, EGF, PDGF, NGF) withdrawal
    Transforming growth factor-β
    Tumour necrosis factor
    Fas ligand

In addition to the normal progression of germ cell loss, pathological insults in the form of either environmental biohazards or clinical (anti-cancer) therapies, can lead to premature ovarian failure and infertility by inducing the intrinsic programme of apoptosis present already in the oocytes. For example, polycyclic aromatic hydrocarbons (PAH) released by the incomplete combustion of fossil fuels and through tobacco smoking, have been shown to accelerate oocyte loss in mouse and human ovaries by activating apoptosis in oocytes. PAH acts through aryl hydrocarbon receptor (AHR) to stimulate *Bax* transcriptionally. Premature ovarian failure could also result in young girls and reproductive- age women when they undergo chemotherapy or radiation in the anti-cancer therapy, again by the induction of apoptosis in the oocytes. Thus the competitive interactions of the pro-and anti-survival Bcl 2 family proteins regulate the activation of the proteases (caspases) responsible for initiating and executing apoptosis. Importantly, a proper understanding of the molecular framework responsible for dictating oocyte fate will open new opportunities for the development of apoptosis-based therapeutic strategies to improve the reproductive health and well being of women.

In addition to these intrinsic factors, initiation of granulosa cells apoptosis in response to hormonal deprivation is a well known example. Growth factors and estrogens have been identified as follicle survival factors; however, androgens and gonadotropin releasing hormone potentiate apoptosis of the follicle. Using serum-free cultures of pre-ovulatory follicles, it has been demonstrated that hCG or FSH suppressed follicular apoptosis in a dose dependent manner. Treatment of follicles with IGF-I or insulin also inhibited follicular atresia, whereas IGFBP-3 reversed the inhibitory effect of both hCG and IGF-I on apoptosis. To conclude, sex steroids have an important role in the regulation of apoptotic cell death in the ovary. It appears that estrogens inhibit and androgens enhance endonuclease activity in ovarian granulosa cells.

THEORIES TO EXPLAIN FEMALE GERMLINE CELL DEATH

Three theories are advanced to explain the vast degeneration of female gametes in the ovary described above. The first one deals with the inadequate availability of growth or survival factors, a key component of organogenesis, to the developing oocytes. Thus, in the female mutant mice that lacked germ-cell survival factors such as stem-cell factor (SCF) or interleukin-1α / β, there is gametogenic failure. Mice that are deficient in germ cell survival factors can be rescued from death by the inactivation of the caspase–2 gene. The second theory, also known as 'death by defect' relates to the elimination of oocytes with meiotic pairing or recombination anomalies. A surveillance (or quality control) mechanism detects and removes defective oocytes and retains meiotically competent oocytes for the formation of primordial follicles.

The third theory (death by self sacrifice) implies the coordinated breakdown of germ-line 'cysts'. In *Drosophila*, *Xenopus laevis* and, in mice, female gametogenesis involves the breakdown of small clusters of interconnected germ cells formed by the incomplete cytokinesis of synchronously dividing progenitor cells, which are termed germline cysts. In the typical *Drosophila* ovary, each cyst is composed of 16 cystocytes, surrounded by a single somatic – cell layer of follicular epithelium. The cysts are connected by ring canals. During oogenesis, one germline cell of the cyst develops into an oocyte, whereas the remaining 15 cystocytes serve as nurse cells. Towards the final stage of oogenesis, nurse cells transport most of their cytoplasm into the oocyte and degenerate, and the cyst breaks down to produce a single mature oocyte. The nurse cell degeneration by apoptosis is regulated by *rpr, hid, grim, DIAP1* and *DIAP2* genes. A similar type of altruistic death also occurs in the germ line of *C. elegans* hermaphrodites. Here, large numbers of mitotic and meiotic germ-cell nuclei are partially enclosed by a single plasma membrane to form a syncytium. Over half of these germ – cell nuclei ('oocytes') are eliminated by programmed cell death during meiotic maturation. Cell death genes, CED-4 and CED-3 which are ortholous to Apaf-1 and caspases, respectively in vertebrates are required for oocyte apoptosis in these worms. As in *Drosophila*, cell death during oogenesis in *C. elegans* might have a homeostatic role in eliminating excess germ- cells nuclei, after they have done their job as 'nurse cells' for those germ cells that are selected to become mature oocytes. Such a proposition of death by self sacrifice in the germline cysts in the mammals is attractive; but we have evidence only in mouse ovary.

31

AGEING AND SENESCENCE

Organisms are not immortal; with the passage of time, ageing or senescence occurs resulting in eventual death. Senescence can be defined as the increase of impairment of physiological functions with age, resulting in a decreased ability to deal with a variety of stresses, and an increased susceptibility to disease. In many organisms, ageing and death may be regarded as a programmed event in their ontogeny. The term ontogeny means that new organisms arise from pre-existing organisms. In the annual plants, food is transferred from parent to seeds during the terminal phase of the reproductive cycle, resulting in the rapid death of the parent plant. Reproductive processes culminating in ageing and death also have examples in many animal species. In the salmonoid fish, during the up-river spawning migration, the irreversible utilization of bony tissues such as skeletal muscles takes place to provide amino acids for the formation of eggs and sperm. This kind of depletion of body tissue results in rapid death of the fish after the gametes are shed. In this instance, reproductive process, with its associated endocrine changes is the major cause of death. Relevant to these facts, evolutionists, like J.B.S.Haldane, Peter Medawar and George Williams theorised the ageing process in the following way. Each species comes equipped with a programme of planned obsolescence chosen to suit its expected life span and the age at which it completes breeding.

The genetic control of ageing can be understood in terms of the 'disposable soma theory' which states that natural selection tunes the life history of the organism so that sufficient resources are invested in maintaining the repair mechanisms that prevent ageing at least until the organism has successfully reproduced. Obviously, signs of senescence could appear only after the reproduction is completed. Life seems to cease in an organism, which has successfully performed its duty of procreation, which is vital for speciation and for generations to continue.

For many other organisms, however, ageing does not appear to be directly connected with reproduction in the above way. Unlike the salmon, in which death is linked to a certain stage in the life cycle (spawning), in all other organisms, death comes after a process of gradual ageing. There are two major views concerning senescence: senescence occurring as the organism's post-embryonic developmental programme or the result of wear and tear of body tissues. Organ failure and tissue degeneration are a feature of ageing organisms. As a result of wear and tear, caused to post-mitotic cellular tissues by extended use, exposure to oxidation and the accumulation of toxins, ageing becomes an inevitable event.

Animals have definite life spans characterising the different species. Total life span of each and every organism seems to be genetically determined, as different animal's age at vastly different rates. Table 31.1 gives the life span of some of the mammals. It may be seen that the average age ranges from 3 1/2 years in the house mouse up to 70 years in the Indian elephant. The longest living animal species are however found among the reptiles. For example, there are records of turtles that have lived for longer than 300 years. The phylogenetically important Sphenodon species have lived even longer than the turtles.

Table 31.1 Table showing life span in mammals

Name	Maximum life span (years)
Man	120
Finback whale	80
Indian elephant	70
Horse	62
Chimpanzee	44.5
Indian rhinoceros	40
Gorilla	40
Brown bear	36.8
Dog	34
Cattle	30
Rhesus monkey	29
Cat	28
Pig	27
Squirrel monkey	21
Sheep	20
Gray squirrel	15
European rabbit	13
Guinea-pig	7.5
House rat	4.6
Golden hamster	4
Mouse	3.5

Among the mammals, Homosapiens have the maximum life expectancy and life span. According to records, the actual average age lies between 60 and 70 years in developing countries. In the under developed countries, this could be substantially lower, due to malnutrition, poverty or poor health conditions. In human beings, the average life span has been significantly expanded so that theoretically it is conceivable that a person could live to 140 years (according to a recent report), if we are able to deal with the chronic ailments associated with ageing i.e., heart disease, cancer, Alzheimer's disease, stroke etc. A lot of studies have been made in recent times on the ageing process of human beings. These studies indicate that body functions reach a peak in certain age and then start declining gradually to reach a stage when death is unstoppable. A recent report says that the body reaches peak efficiency at the age 30 and then declines in many ways. Using age 30 as reflective of 100 % performance, the following declines happen: (1) pumping efficiency of the heart is reduced about 20 % when a person reaches 55, (2) kidney function is reduced about 25 % at 55 years of age, (3) maximum breathing capacity declines about 40 % by 55 and 60 % by 75 years, and (4) basal metabolism goes down about 10 %. With normal ageing, there is a decrease in bone mass, muscle strength and lean body mass and an increase in fat body mass. Another important physiological change related to ageing is the decreased immune response to infections. These are the general consequences of ageing in man and can very well apply to other animal species.

CELLULAR AGEING

The process of ageing involves a decline in the efficiency of various cells, tissues and systems. Normal cells can go through only a fixed number of divisions before they die by a process called senescence. During the lifetime, the cells accumulate damage that results in gradual loss of their differentiated function and growth rate.

Various theories have been proposed for ageing at the cellular level. First theory is concerned with the accumulation of inert materials both within the cells and between cells. The best example is the progressive accumulation of lipofuscin granules, which are also called 'age pigments'. The accumulation of lipofuscin in the myocardial cells results in a large displacement of muscle volume, which would result in a loss of efficiency of the contractile elements. The second theory relates to the chemical ageing of macromolecules. The third theory deals with the random metabolic changes that lead to change in cellular phenotype.

REACTIVE OXYGEN AND CELL SENESCENCE

In recent years, much importance is given to accumulation of free radicals within the cells as an index of ageing and senescence. Normally, oxidation involves the transfer of a pair of electrons from one atom to another. When an impaired electron escapes, it can cause damage to the molecules in nearby cell membrane. These single electrons are called the free radicals. They are highly reactive, seeking to capture another electron to complete a pair and in doing so they damage or destroy the function of another molecule. Free radical damage is cumulative, building up with age. In the nerve cells, the cell membrane lipids are oxidised by free radical reaction, thereby impairing cell-to-cell communication and transmission. Dopaminergic neurons are particularly prone to oxidative stress due to the ability of dopamine to be oxidised and produce hydrogen peroxide as a by-product of this reaction. Hydrogen peroxide can react with ferrous (Fe^{2+}) iron to produce highly reactive hydroxyl radicals (OH), which in turn can react with and cause damage to proteins, nucleic acids and membranes eventually leading to cellular degeneration. Parkinson's disease involves such degeneration of dopaminergic neurons of the substantia nigra. Furthermore, free radical damage to immune cell membrane lipids may ultimately impair the ability of immune cells to respond to invasive viruses and bacteria, thereby increasing the vulnerability to infectious diseases in the aged persons.

DIETARY RESTRICTION AND ANTI- AGEING ACTION

McCay in the year 1935 showed for the first time the possibility of increasing the lifespan of rodents by dietary restriction. Since then, several studies have focused attention on this aspect extending lifespan in rodents and other animals. Calorie restriction reduces metabolic rate and lowers the production of harmful molecules called reactive oxygen species, also known as free radicals. In primates, calorie restriction causes a number of physiological changes with positive health benefits. Results of these studies have proposed the following ways by which anti- ageing action of dietary restriction works:

 i. Dietary restriction increases lifespan by retarding ageing.
 ii. Dietary restriction delays or reduces the onset of most age- related diseases and alters most physiological processes that change with age.
 iii. Dietary restriction retards growth and development.
 iv. Reduction in body fat content was the physiological basis for the increased survival of rodents, fed on calorie- restricted diets.
 v. Dietary restriction increased survival by reducing the metabolic rate of the rodents.

However, convincing evidence to support the above concepts could not be obtained in the subsequent studies. Extension of lifespan by dietary restriction has been seen particularly clearly in animals with high reproductive rates (e.g., rodents, worms and flies). In contrast, experiments in rhesus monkeys have yet to show that calorie restriction increases lifespan. The effects of lifelong calorie restriction in humans under controlled conditions will never be studied for reasons of cost and practicality.

Recent studies have now provided evidence that increase in lifespan is associated with enhanced resistance to a variety of stresses. Enhanced resistance to stress thus plays a key biological role in the increased longevity especially in lower organisms, showing the possibility of experimental intervention to retard ageing in higher animals.

RESISTANCE TO OXIDATIVE STRESS

It is well known that free radicals cause deleterious damage that accumulated over time and contributed to the process of ageing and various age- associated diseases. It has also become apparent that many reactive oxygen species (ROS), such as peroxides, that are nor free radicals, also play a role in oxidative damage in the cells. Therefore, the free radical hypothesis of ageing has been modified to the oxidative stress hypothesis of ageing, which hypothesizes that a chronic state of oxidative stress exists in cells of aerobic organisms even under normal physiological conditions because of an imbalance of pro-oxidants and anti-oxidants. Evidence now indicates that the anti- ageing action of dietary restriction might occur through a mechanism that involves increased resistance to oxidative stress. As an example, the activity of one or more of the antioxidant enzymes is increased significantly in the liver of rats fed a calorie- restricted diet compared to rats fed ad libitum. Furthermore, dietary restriction results in a decrease in oxidative damage by decrease in lipofuscin, lipid peroxidation, protein oxidation and DNA oxidation. Resistance to hyperthermia is also an effect of calorie- restricted diet by enhancing the expression of a heat shock protein, hsp70, by protecting the cells against the adverse effects of hyperthermia.

The cellular effects of free radicals represent the most likely contender to explain the ageing process across a wide range of species. When mitochondrial metabolism – fuelled by food and oxygen- increases, more reactive oxygen species are produced as a by-product, with greater adverse effects on cells. Diet restriction seems to retard ageing by its anti-oxidative potentials.

Recent studies have revealed the molecular mechanisms underlying the lifespan extension by calorie restriction in yeast and flies. In the yeast mother cells, calorie restriction extends their replicative lifespan by inducing Sir2 pathway. In *Drosophila*, an increase in dSir2 extends lifespan, whereas a decrease in dSir2 blocks the lifespan extending effect (see below).

TELOMERASE THEORY

Modern techniques in molecular biology and genetic engineering have helped identifying genes that control ageing directly. The genetic basis of cellular senescence lies in a gene called *TEP1* on the human chromosome 14. The product of *TEP1* is a protein forming part of an enzyme called telomerase. At the end of the chromosome there occurs a repeated stretch of nonsense sequence TTAGGG, repeated again and again about 2000 times. This stretch of terminal nucleotide sequence is known as a telomere. Telomeres are DNA-protein complexes that cap chromosomal ends, promoting chromosomal stability. When cells divide the telomere is not fully replicated because of limitations of the DNA polymerases in completing the replication of the ends of the linear molecules leading to telomere shortening with every replication. Thus, every time the chromosome is copied, a little bit of telomere is left off resulting in a gradual reduction in the size of the telomere, as the cells undergo repeated division during growth of the tissue. *In vitro* when telomeres shorten sufficiently, the cell is arrested into senescence. In an 80-year-old person, telomeres are on average about five- eights as long as they were at birth. However, the presence of the telomerase could restore the length of the lost telomere during cell division.

Telomerase, a cellular enzyme, adds the necessary telomeric DNA (T_2AG_3 repeats) onto the 3' ends of the telomere. Its protein component bears a striking resemblance to reverse transcriptase, the enzyme responsible for RNA- DNA transcription. By virtue of the importance attached to the telomerase in the cell division, its gene could be considered close to "genes for youth". Lack of telomerase seems to be the principal reason that cell grow old and die. For example, cells in the walls of human arteries generally have shorter telomeres than cells in the walls of veins. The arteries have to expand and contract with every pulse beat, so they suffer more damage and need more repair. Repair involves cell division that uses up the ends of telomeres. Hence, the arterial cells start to age, which is the reason for death from hardened arteries, and not from hardened veins. People with dyskeratosis congenital, a rare genetic disease that diminishes the ability to synthesize sufficient telomerase, have shortened telomeres and die prematurely from progressive bone marrow failure and vulnerability to infections. Cellular environment also plays an important role in regulating telomere length and telomerase activity. *In vitro* oxidative stress can shorten telomeres and the antioxidants can decelerate chromosomal shortening. Recent experiments have also indicated that chronic psychological stress may lead to telomere shortening and lowered telomerase function in peripheral blood mononuclear cells.

HAYFLICK LIMIT

Non- reproductive mammalian cells (such as human fibroblasts) in culture dishes generally lose their ability to proliferate after a fixed number of doublings- a phenomenon termed the Hayflick limit. With each ensuing division, a greater proportion of the cells enter a state called senescence, characterized by a permanent withdrawal from the cell division cycle. The fixed doubling time of fibroblasts appears to be determined by their species of origin (approximately 20 in mouse and 50 in humans) and has been related to longevity. Thus, animals with longer lives display increased resistance to replicative senescence. An explanation for the senescence of human cells is that every population doubling results in a shortening of telomeres of each chromosome. Reduction of telomere length below a critical threshold can trigger senescence. Another interesting evidence that telomerase could restore loss of telomere during cell division is found in the immortal tumourous HeLa cell line which is used widely in cell culture laboratories. HeLa cells have excellent telomerase. If antisense RNA is added to HeLa cells- that is, RNA containing the exact opposite messenger to the RNA message in telomerase, so that it will stick to the telomerase RNA – then the effect is to block the telomerase and prevent it working. The HeLa cells are then no longer immortal. They senescence and die after about 25 cell divisions. The immortal nature of cancer cells, and of normal stem cells in embryos and adults, might relate to their continuous telomerase activity. It seems then that forced expression of telomerase could potentially reverse or delay the ravages of ageing.

AGEING OF BRAIN CELLS

The telomere theory of ageing cannot, however, be applicable to the ageing of brain, because brain cells do not replace themselves during life. However, brain's supporting cells, called glial cells, do indeed duplicate themselves; with the result that their telomeres do probably shrink. Nevertheless, a distinct decline in the mental capacities could be noticed as one ages. Large populations of networks connecting neuronal groups in many parts of the cerebral cortex and other brain regions do specific tasks of higher mental functions such as abstract thinking, language and memory. As we grow old, we loose large numbers of neurons. Added to that, there is shrinkage of nerve cell bodies; with the result that the dendritic trees become less ramified in the cerebral cortex. Recent studies on rats indicate that their memory impairments are due to both anatomical and physiological changes in the hippocampal region. The hippocampal and surrounding parts of the temporal lobe are necessary for the storing and retrieval of events, facts, names and so forth.

GENETIC CONTROL OF LONGEVITY

C. ELEGANS

A single-gene mutation has recently been identified in the free- living nematode, *C. elegans*, which extends longevity. This mutation however greatly reduces fecundity, suggesting an inverse relationship between fecundity and longevity. In *C. elegans*, *age-1*(Gx546) is a recessive mutant allele that results in an average 40% increase in life expectancy and an average 60% increase in maximum life span at 20°C. *age-1* mutation is associated with a 75% decrease in hermaphrodite self fertility as compared to the *age-1*+ allele at 20°C. This mutation in *age-1* is the only instance of a well-characterized genetic locus in which the mutant form results in lengthened life. In other words, elimination of a normal gene function results in longer life. Thus, life span is shortened by the normal action of a gene whose function may be primarily involved in increasing reproduction. The *age-1* mutation dramatically displays a trade-off wherein reproductive effort is increased by the normal allele at a 'cost' of loss of ½ of post reproductive life as is predicted by the Antagonistic Pleiotropy model for the evolution of senescence. This provides a model in which action of *age-1* in lengthening life results not from the elimination of a programmed ageing function but rather from the lack of the detrimental act of a gene whose primary function is involved in hermaphroditic reproduction. The existence of such genes offers an optimistic view about our ability to intervene in the ageing processes to mimic the action of mutant alleles such as *age-1*.

Environmental conditions may also cause gene mutation, effecting lifespan expansion in *C. elegans*. The first instar larva of *C. elegans* in an uncrowded environment with ample food takes 2 weeks to reach adulthood and then die. In a crowded condition, coupled with shortage of food, the larva enters into a stage called dauer larval state in which it neither eats nor grows until food becomes available again. This condition, however, lengthens the lifespan up to several months. It has been shown that animals with mutation in the gene *daf-2* can also induce the dauer state, thereby lengthening the lifespan. This happens even when the food is plentiful. Possibly, the effect of this mutation is related to the fact that the dauer larvae do not eat. Mutations in another *C. elegans* gene, *clk-1* also slow down the cell cycle, and embryonic and post-embryonic development. These mutant worms have a life span up to 70 % longer than normal. It is not known how the clk genes control senescence, but they may delay it by decreasing metabolic rate and thus reducing exposure to free radicals.

GENETIC CONTROL OF AGEING IN YEAST

A mother yeast cell can readily divide, asymmetrically, to produce a daughter cell early in life. Later on, when the mother cell has divided many times, its capacity to produce progeny diminishes. Numerous environmental and genetic determinants to control the yeast replicative lifespan have been determined. For example, reducing the glucose levels in the growth medium from 2% to 0.5% significantly increased the replicative lifespan of yeast. Such caloric restriction leading to extended lifespan is mediated through the action of a gene termed SIR2. Sir2 belongs to a family of enzymes called sirtuins. The sir2 gene is one of the first longevity genes to be identified. Species from yeast to humans carry variants of it, and extra copies increase lifespan in yeast, worms and flies.

When the yeast is manipulated to produce too much of the protein product of this gene, a similar lengthening of lifespan ensues. Sir2, the protein product of the gene SIR2, extends lifespan in yeast by regulating gene expression or suppressing DNA recombination. The molecular mechanism involves the removal of specific acetyl groups from histones and other proteins that wrap up DNA. This activity of Sir2 in turn depends on the cellular levels of nicotinamide adenine dinucleotide (NAD). Caloric restriction in yeast might increase Sir2 activity by altering either the NAD: NADH ratio or the levels of NAD derivative nicotinamide. Interestingly, caloric restriction in the fruit fly, *Drosophila melanogaster* has also increased the activity of Sir2. Increased Sir2 levels and activity might then dampen gene expression and recombination, leading to an extension of lifespan.

Recent studies have also emphasised the fact that Sir2 protein works with NAD, which is involved in metabolism, so the two together potentially explain the association between calorie restriction and ageing. More recently, a plant-derived molecule, resveratrol, found in grape skins, has also been shown to activate sirtuins. Interestingly, resveratrol is reported to have anti-cancer and neuroprotective effects. Furthermore, a recent study has revealed that this chemical not only extends the maximum lifespan of a short-lived fish, *Nothobranchius furzeri* by 60%, but also seems to protect the fish from neurodegeneration. On the other hand, resveratrol has so far shown no effect on Sir2 function in human cells in culture. Hence, it is quite unlikely that the chemical acts on the Sir2 protein directly, possibly, it might target other biological processes that interact with Sir2.

GENETIC CONTROL OF AGEING IN MAMMALS

Recently, Sir2 proteins have been identified in the mice also. Inactivating a Sir2 family protein in mice causes premature ageing and genome instability. Loss of genome integrity has been implicated in ageing. In the mice, inactivation of SIRT6, a protein related to Sir2 caused not only genomic instability but also premature ageing. Under normal conditions, the genome deterioration is due to the continual production of reactive oxygen species during metabolism. Similarly, SIRT1, a relative molecule of Sir2, regulates the tumour suppressor protein p53 and FOXO3 to suppress apoptosis and promote cell survival. The blood serum of SIRT6 deficient mice contains extremely low levels of IGF-1, compared to normal mice. IGF-1 strongly inhibits apoptosis in lymphocytes, and the age-related reduction in lymphocytes production by the thymus has been ascribed to a decline in IGF-1 levels with age.

AGE- RELATED DISEASES: ALZHEIMER'S DISEASE

In addition to increasing the vulnerability to infectious diseases, ageing also causes certain disease conditions incumbent of advancing ageing and senescence. For example, there are certain chronic degenerative disorders such as Alzheimer's and Parkinson's diseases and the age- related cataracts in the humans. Alzheimer's disease (AD) is a kind of degenerative disease leading to progressive neuronal loss. This disease was first described by Alzheimer as a kind of dementia that characteristically developed before the age of 60 (percentile dementia). He described the 'miliary foci' accumulating extracellularly which are known as senile plaques and the "dense bundles of pathological lesions remain the key diagnostic features of the disease even today. As noted by Alzheimer, this disease causes progressive loss of concentration, memory, learning, abstract thinking, judging and forming of concepts. The affected cells in the brain, which are responsible for producing neurotransmitter substances necessary for signal transmission between nerve cells, fail progressively. Thus communication between nerve cells is impaired. Patients are no longer able to process sensory impressions correctly and to associate these with their existing knowledge. Now it is known that most cases of dementia beginning after the age 60 also are of the AD type. In fact AD is argued that it is an exaggerated form of normal aging, but there is no doubt that it represents a disease entity, both chemically and pathologically.

AD is characterized by several changes in the brain, including the build-up of protein deposits known as amyloid plaques outside nerve cells and the breakdown of neurons. The plaques consists of protein fibres, some 7-10 nanometers thick, that are mixed with small peptides called amyloid-β (Aβ) peptides. Mature plaques also contain degenerating nerve endings, and are surrounded by active astrocytes and microglia- cells that help to clear up debris in the brain. Mutations in genes for β amyloid precursor proteins, presenilin 1 and presenilin 2 have been implicated with the development of AD. Amyloid β-peptide is a fragment of a larger protein, the amyloid precursor protein (APP) that sits across the outer membrane of nerve cells. Fig. 31.1 illustrates the stepwise formation and release of the A β plaque in a AD patient. Essentially, two enzymatic activities are involved in precisely snipping APP to produce Aβ, which is then shed into the brain. In the first step of this enzymatic cleavage process, APP is cleaved at a specific point by β-secretase activity. This is followed by γ- secretase activity, which finally clips off the other end of the amyloid protein, leaving a cytoplasmic fragment that moves inside the nucleus performing important gene regulatory functions. The Aβ peptide then moves out of the cell membrane and aggregates as Aβ plaque (see Fig.31.1). These aggregates of Aβ induce membrane- associated oxidative stress, which along with perturbed calcium balance inside the neuron causes neuronal degeneration.

Electron microscopical studies also show that there is a marked loss of synapses in AD. Many synapses in the basal nucleus are cholinergic and send their axons to the cerebral cortex. Loss of neurons in the basal nucleus may therefore explain why AD patients have severely reduced amounts of acetylcholine in the cortex. The memory impairment can be thus explained by lack of cholinergic input to the hippocamus and nearby cortical regions.

PROTEIN AGGREGATION AS A CAUSATIVE AGENT OF AGE-RELATED DISEASES

Recent studies have revealed that abnormalities of protein aggregation and deposition may play an important role in the pathophysiology of age- related disorders such as AD and cataract. In all of these disorders, the abnormal protein deposits are characterized by aggregated fibrils composed of cross- linked protein with prominent - pleated sheet secondary structure. These deposits are tissue- specific and are found in intracellular compartments, extracellular domains and at the cell membrane. These deposits are highly insoluble and prevent endogenous clearance mechanisms, ultimately leading to progressive loss of tissue function. These protein deposits also aggregate rapidly and cause direct cytotoxicity. Two other age-related neurodegenerative diseases in which protein aggregation is the major cause of cell death are Parkinson's and Huntington's diseases. The aggregation of two proteins, α-synuclein and huntingtin on aggregation in the neurons cause their demise in Parkinson's and Huntingtin's diseases respectively.

In the case of Alzheimer's disease, such abnormal deposits involve a class of amyloid β(Aβ) proteins, causing neuronal demise and cerebral oxidative stress. Among different Aβ species, Aβ-40 is the principal species present in human brain. The deposition of Aβ in the neocortex of transgenic mice over-expressing human Aβ peptides is accompanied by many of the neuropathological features of Aβ, including intraneuronal tau abnormalities and neuronal loss, as well as signs of oxidative damage similar to those seen in AD- affected brain. Recent studies have also indicated that abnormal reaction between amyloid β proteins and redox- active metal ions such as copper and iron, promotes the generation of reactive oxygen species, and protein radicalization. These products then lead to chemical modification of the amyloid proteins, alteration in the protein structure and solubility, and oxidative damage to surrounding tissue. Importantly, dityrosine cross-linking is a major causative factor, conferring insolubility to the proteins. Another important observation in this connection is that the rodent Aβ homologue lacks tyrosine and does not form cross-link after incubation with Cu (II), and hence these animals do not form cerebral amyloid deposits.

320 Molecular Developmental Biology

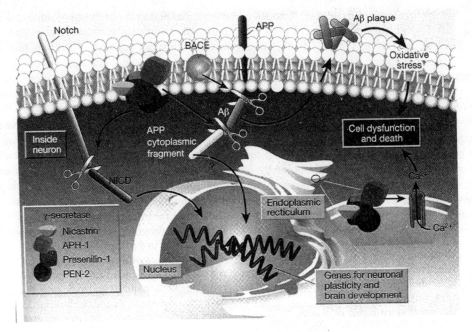

Fig. 31.1 Showing the processing of amyloid β peptide (A β) from amyloid precursor protein (APP). The γ-secretase complex with its protein quartet (shown in insert) is located in the plasma membrane and endoplasmic reticulum of neurons. This complex helps in generation amyloid β peptide. This involves an initial cleavage of APP by an enzyme called BACE (β secretase). The γ secretase librates A β, and as well as an APP cytoplasmic fragments which regulates gene expression. The membrane associated oxidative stress induced by A β aggregation is also shown. (Redrawn from Mattson 2003).

AGGREGATION OF TAU PROTEIN

Alzheimer's disease is characterized by the presence of two aberrant structures, senile plague (as described before) and neurofibrillar tangles (NFT). The main component of the NFT is the paired helical filaments (PHFs), which are mainly comprised of the protein tau in an abnormally phosphorylated status. Tau is a natively unfolded protein that binds to microtubules in normal cells. This microtubules-associated protein gives structural stability to the axons of the neural cells. However, in AD, tau protein loses affinity to microtubules and it self-aggregates into aberrant structures called neurofibrillar tangles. The tau aggregation destabilizes the axonal structure, thereby affecting its functional role in neuronal transmission. Phosphorylation of specific tau sites is accomplished by cAMP-dependent protein kinase activity. The pathological protein aggregation in AD patients involves the formation of β-sheet structure giving rise to the typical amyloid fibrils and neurofibrillar tangles, both inhibiting the normal brain function.

The risk of developing Alzheimer's disease is lower in people who have been intellectually active and the cognitive functions are better preserved in those with higher mental function when in their twenties. Treatment of this disease include agents to modulate the cholinergic system, modulation of other neurotransmitter systems, drugs to reduce oxidative stress that destroys nerve cells and synapses, hormones and drugs to suppress inflammation and ion chelators.

CATARACT

Similar to Alzeimer's disease, age- related cataract, which causes blindness word-wide, alteration of protein structure followed by aggregation, is the main cause of the disorder. In this case, lens protein aggregation leads to the opacification, causing cataract. During catarogenesis, the lens' structural proteins (β crystallin) undergo a conformational transition from long-lived soluble proteins found in the transparent lens to a coloured, insoluble, highly cross- linked, high molecular weight aggregate. This aggregated protein has also undergone protein oxidation to bring about other cytotoxic effects. In addition to protein aggregation, cataract formation is also characterized by oxidative damage. Decreases in the level of antioxidant defence enzymes such as glutathione reductase, glutathione peroxidase and superoxide dismutase, as well as decreases in total glutathione and corresponding increases in oxidized glutathione, have been observed. Cu (II), a cofactor in generating potentially damaging ROS such as H_2O_2 and superoxide may also promote chystalline aggregation in cataract.

WERNER'S SYNDROME

Humans that is homozygous for the recessive gene defect known as Werner's syndrome show effects of premature ageing such as growth retardation at puberty. By their early twenties, those affected by this syndrome have grey hair, and suffer from a variety of illness, such as heart disease, that are typical of old age. Fibroblasts taken from Werner's syndrome patients undergo fewer cell divisions in culture before becoming senescent and dying, compared to fibroblasts taken from normal persons of the same age. The gene affected in Werner's syndrome is thought to encode a protein involved in unwinding DNA, which is required for DNA replication, DNA repair and gene expression. Thus, the inability to carry out DNA repair in Werner's syndrome patients could cause higher level of DNA damage, resulting in premature ageing and death. In particular, a gene WRN is crucial for DNA repair and maintenance of structural integrity of chromosome ends. In humans and mice, WRN mutations and other defects in pathways that protect genome integrity cause premature ageing.

32

PLANT DEVELOPMENT

All animal forms take origin from a single cell, zygote, the fusion product of male and female gametes. The development then starts by repeated division of the zygote, termed as cleavage. After its completion, the resulting mass of cells, the blastula, undergoes cellular movements and displacement resulting in the formation of three concentrically arranged germ layers. This embryonic stage is termed as gastrula. The three germ layers, viz., ectoderm, mesoderm and endoderm, then give rise to all different organ systems found in the adult body by a process of morphogenesis and pattern formation. Both gastrulation and organogenesis involve principally cell migration and interaction. Further growth occurs through a developmental increase in total mass that results in permanent enlargement of the animal.

Although plants and animals are believed to have originated from a common unicellular eukaryote ancestor, they seem to have evolved multi-cellular development separately. There are crucial differences between animal and plant development. One important difference is that plant cells are surrounded by cell walls. Therefore, major changes in shape cannot be achieved by movement of sheets of cells, a common mechanism in animal development. In addition, apoptosis, a major mechanism in maintenance of homeostasis during organogenesis in animal development does not significantly contribute in plant development. Thus, plant development occurs mainly by altered rates and planes of cell division and by cell enlargement. Another significant difference is that post-embryonic development is a continuous process in higher plants and gives rise to all of the adult structures. However, with respect to pattern formation, similar principles seem to be involved in animals and plants. They include axis specification and patterning (both radial and longitudinal), the establishment of developmental compartments and the use of homeotic selector genes to confer positional identities.

In animals, all post-embryonic growth occurs involving all the already formed organs. But in higher plants, post- embryonic growth occurs from localized regions of cell division called meristems, which give rise to all the adult structures of the plant, such as shoots, roots, stems, leaves and flowers. In a way, cells of meristems are like animal stem cells inasmuch as the progeny of these cells can give rise to a variety of tissues, in addition to unlimited cell division potentials. In plants, growth resulting from meristematic activity is possible throughout the life of the organism.

Two meristems are established in the plant embryos; one located at the tip of the radicle (the future root) and the other at the tip of the plumule, the future shoot. Almost all the other meristems of an adult plant are derived from these two meristems. Cell–fate determination in plant development is yet an unanswered question. However, a cell's fate can be altered by a change in its position within the meristem or its place in the differentiating cells derived therefrom, suggesting that signals emanating from other cells must have some influence. Although the thick cell wall is a barrier for the passage of larger molecules like proteins, smaller molecules like plant growth hormones such as auxins, gibberellins, cytokinins, and ethylene could easily penetrate the cell walls. Furthermore, plant cells also communicate through fine cytoplasmic channels namely, plasmodesmata, which link neighbouring cells through cell walls.

Yet another important difference between plants and animals is that a complete and fertile plant can develop from any living somatic cell and not necessarily from a fertilized egg. It is suggestive that the adult differentiated cells of plants retain the capacity for dedifferentiation and totipotency, unlike the animal cells, which have undergone 'terminal differentiation'.

Recent molecular studies have uncovered the mechanisms involved in the transcriptional control of genetic programmes exerted by transcriptional factors during plant development. Additional control mechanisms such as microRNAs have also been discovered in plants. They are complementary to transcription factors with a function in developmental processes. For example, in leaf development, several transcription factors, such as PHAB, PHAV, and CIN like genes are targeted by microRNAs. Temporal and spacial regulation of expression of microRNAs is of utmost importance for the proper destruction of transcription factors during developmental processes and adds an extra level of regulation. In addition, epigenetic control mechanisms are involved in developmental transition (eg. from vegetative to reproductive) as well as in several other morphogenetic processes. Histone modifications, such as acetylation/ deacetylation, are of crucial importance to make DNA available for transcription or to repress transcription. Plant genome actually contains several genes for histone acetylases and deacetylases.

EMBRYONIC DEVELOPMENT IN ANGIOSPERMS

In *Arabidopsis thaliana*, the equivalent of *Drosophila* in the study of plant development, the egg is fertilized by a gamete contained in the pollen produced in turn from the male sex organs, stamens. A pollen grain deposited on the stigma grows into a tube (pollen tube) that penetrates the style and delivers two haploid male gametes into an ovule. One of the sperm fuses with the egg cell, while the other fuses with the polar nuclei or their fusion product, the secondary nucleus. This kind of phenomenon is termed as double fertilization. The latter product goes on to develop into the triploid endosperm, while the former results in the zygote. The endosperm surrounds the developing embryo and provides nutrition as well as regulatory chemicals for its development. The endosperm may either be consumed by the developing embryo, so that the seeds are non-endospermous (pea and beans), or it may persist in mature seeds and continue to support the growth of embryo during germination (cereals, coconut).

Just like animal zygote, the plant zygote undergoes repeated cell divisions, cell growth and differentiation to form a multi-cellular embryo. However, a characteristic early cell division pattern in plants is that the first division of the zygote occurs at right angles to the long axis of the future embryo dividing it into an apical and basal cell. In many species, the first division is unequal and asymmetrical, producing two daughter cells with different identities. Thus, the basal cell divides several times to form a single row of cells constituting the suspensor, which takes no further part in embryo development except having an absorptive function. The suspensor is an ephemeral structure, found at the radicular end of the proembryo. It usually attains its maximum development by the time the embryo reaches the globular stage (see Fig. 32.1). Furthermore, recent studies indicate that the suspensor is a highly specialized, terminally differentiated embryonic structure. The following are some specific functions of suspensor in zygotic embryogenesis: i) orienting the embryo in close proximity to the source of nutrients, ii) involving itself in short-distance transport of metabolites and iii) accumulating hormones necessary for self-regulation as well as for the control of embryo growth.

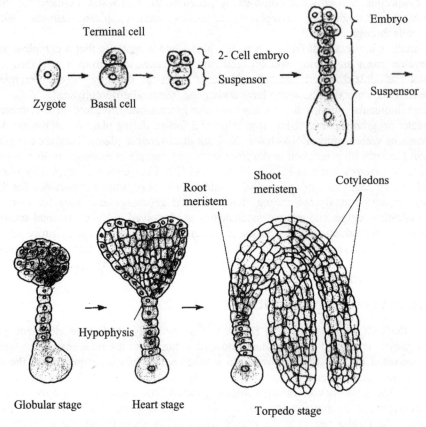

Fig. 32.1 Angiosperm embryogenesis. A representative dicot is shown; a monocot would develop only a single cotyledon. While there are basic patterns of embryogenesis in angiosperms, there is tremendous morphological variation among species. (Redrawn from Gilbert 2000)

Most portions of the future embryo develop from the apical cell which undergoes a series of stereotyped but precise pattern of cleavages in different planes giving rise to the heart shaped embryonic stage, typical of dicotyledons. This develops later on into a mature embryo, which consists of a main axis, with a meristem at either end, and two cotyledons, which are storage organs.

The ovule containing the embryo matures into a seed that remains dormant until suitable conditions trigger germination and growth of the seedling. The hormone abscisic acid is important in maintaining dormancy in many species, whereas another hormone, gibberellin, is responsible for breaking the dormancy. Upon germination, the green shoot and the primary root are formed from the shoot apical meristem and the root apical meristem, respectively.

POST- EMBRYONIC GROWTH

The post embryonic phase of plant development begins with germination. In angiosperms, the embryo has established in it a rudimentary body plan i.e., the apical/basal and radial pattern axis. The apical/basal axis results in the specification of the shoot apical meristem and the root apical meristem at opposite ends. The radial pattern establishes the outer epidermis and the presumptive cortical cells in the center, with presumptive vascular tissue in between.

MERISTEMS

A fundamental difference between animal and plant development is that in animals almost all organs in the adult originate during embryogenesis and they only acquire growth and sometimes maturity during adulthood. On the contrary, the organ systems of the plant embryo (axis and cotyledons) are concerned mainly with the completion of germination. However, the adult organs of the flowering plant are formed post-embryonically and are derived entirely from two small populations of rapidly proliferating cells laid down in the embryo, called shoot and root meristems. They are the growing points of a plant. The shoot apical meristem (SAM) gives rise to all the aerial parts of the plant, whereas the root apical meristem (RAM) gives rise to the root network.

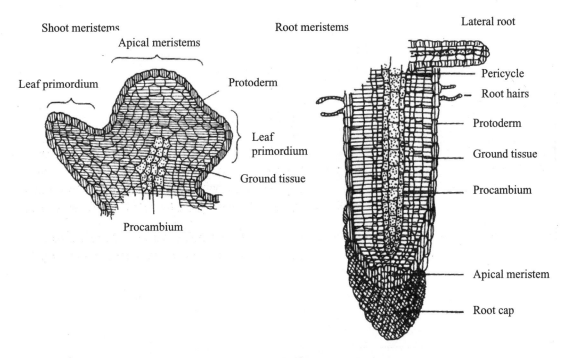

Fig. 32.2 Shoot and root meristems. Both shoots and roots develop from apical meristems, undifferentiated cells clustered at their tips. In roots, a root cap is also produced, which protects the meristem as it grows through the soil. The lateral organs of the shoot (leaves and axillary branches) have a superficial origin in shoot apical meristems. Lateral roots are derived from pericycle cells deep within the root (Modified from Wolpert 1998).

A special meristem, the cambium, is concerned with the secondary thickening. The cambium arises between the primary xylem and phloem and produces secondary tissues. In addition, a cork cambium or phellogen commonly develops the peripheral region of the axis and produces a periderm. The primary growth of an axis is completed in a relatively short period, whereas, the secondary growth persists for a longer period and, in a perennial axis, the secondary growth continues indefinitely and also serve in protection and repair of wounded regions.

The shoot apical meristem produces a repeated basic unit of structure called a module. The vegetative shoot module typically consists of internode, node, leaf and axillary bud (Fig. 32.2). The axillary bud itself contains a meristem, and can form a shoot when the inhibitory influence of the shoot tip is removed. Root growth is not so modular, but new lateral meristems initiated behind the root apical meristem give rise to lateral roots.

Meristems are clusters of cells that allow the basic body pattern established during embryogenesis to be repeated and extended after germination. Just like the animal stem cells, they divide to give rise to one daughter cell that continues to be meristematic and another that differentiates. Thus, most cell divisions in normal plant development occur within the meristems. After the cell leaves the meristem, much of the subsequent growth is due to cell enlargement. Since a meristem remains constant in size during growth, cells are continually leaving it, and as they do so they begin to differentiate. The fate of a cell in the shoot meristem clearly depends upon its position rather than its lineage, because when a cell is displaced from one layer and becomes part of another it adopts the fate of its new layer.

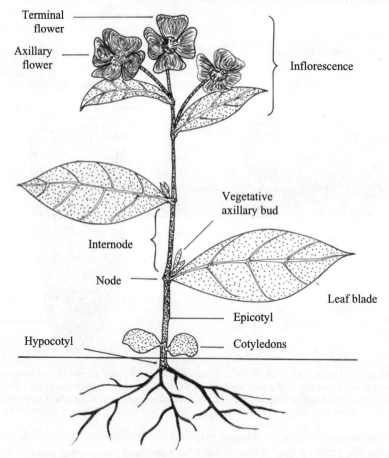

Fig. 32.3 Morphology of a generalized angiosperm sporophyte (Redrawn from Gilbert 2000).

STRUCTURE OF SHOOT MERISTEM

The central region of the shoot or root meristem is called promeristem. The promeristem is the region of new growth in a plant body where the foundation of new organs or parts of organs is initiated.

Sometimes, it is also called primordial meristem and embryonic meristem. The central region of the promeristem contains cells known as initials or organizer that determines the fate of its own derivatives and, in the process, continues to produce leaf or floral meristems repeatedly throughout the life of the plant. The promeristem behaviour is similar to that of animal stem cells. They are self-renewing and give rise to the cells of the meristem. These initials generally divide rather slowly; but their progeny divide more rapidly as they move toward the peripheral parts of the meristem. The shoot meristem of dicotyledons is made up of three layers, namely L1 (dermatogen), L2 (periblem) and L3 (plerome). L1 is the outermost layer and is just one cell thick. L2 is also one cell thick and lies beneath L1 (Fig. 32.4). In both L1 and L2, cell divisions are anticlinal (in a plane perpendicular to the layer), thus maintaining the organization of these layers. The innermost layer is L3, in which the cells can divide in any plane. L1 and L2 are known as the tunica, and L3 as the corpus. In the flowering plants, layer 1 gives rise to the epidermis of all the plant structures, whereas, L2 and L3 contribute to cortex and vascular structures. The L2 layer gets disrupted by periclinal divisions as soon as leaves begin to form.

The stem is terminated by a self-perpetuating population of small, isodiametric, rapidly dividing cells known as the shoot apical meristem. It is this meristem that gives rise to the leaf primordia and produces the tissues that contribute to the increase in length of the stem. The shoot apical meristem is often protected by leaf primordia, and in this configuration, it is known as the bud.

LEAF DEVELOPMENT

Leaves are lateral organs with determinate growth that are initiated at the shoot apical meristem (SAM) at regular time intervals, and at specific positions called phyllotaxis. The first indication of leaf initiation in the meristem is usually a prominence of a region to the side of the shoot apex to form a leaf primordium. It is also known as leaf buttress and is the result of increased localized cell multiplication and altered patterns of cell division. Polarized cell expansion within the meristem also contributes to the formation of leaf primordium. Once specified, leaves differentiate an adaxial (upper) side specialized for light capture, and an abaxial (lower) side specialized for gas exchange.

The arrangement of leaves on a stem is termed as phyllotaxy, which is reflected in the arrangement of leaf primordia in the meristem. Leaves can be arranged singly at each node, in pairs, or in whorls of three or more. The fact that new leaf primordium forms always at the center of the first available space outside the promeristem and above the previous primordium suggests that there occurs a lateral inhibition, in which each primordium inhibits the formation of a new leaf within a given distance from it. Precisely, inhibitory signals originating from recently initiated primordia prevent leaves from forming close to each other.

Leaves have a determinate growth; therefore, leaf initiation involves a switch from indeterminate to determinate growth. The switch from indeterminate growth of the SAM to the determinate growth of the leaf and pattern formation is transcriptionally controlled, involving several genetic interactions. Essentially, *KNOX* genes prevent SAM cells from differentiating; and repression of *KNOX* genes by transcription factors such as YAB3, AS1 and AS2 leads to leaf initiation. The gene ARGONAUTE (AGO), which controls the lateral expansion of leaves, codes for an unidentified soluble protein, which together with another gene product (ROTUNDIFOLIA), is involved in the longitudinal expansion of leaves.

As shown in (Fig. 32.4), three zones could be distinguished within SAM. They are: the central zone that contains stem cells, the peripheral zone that contains the differentiating daughter cells from the stem cells and produces the leaves, and the rib zone that produces the stem tissue of the internodes. The SAM is a self-regulatory unit that keeps the stem cells in an indeterminate mode, maintains equilibrium between the different zones, and initiates leaf formation at specific time intervals and at specific sites at the periphery of the meristem. The equilibrium between the different zones of the

SAM is regulated by the CLAVATA signalling pathway. The *CLV1* and *CLV2* genes encode a receptor that is located at the basal part of the central zone (Fig. 32.4). This receptor recognizes the *CLV3* encoded ligand, produced in the apical part of the central zone. This interaction triggers the signalling cascade to repress the *WUSCHEL* (*WUS*) gene activity. *WUS*, located at the basal part of the central zone, is a homeodomain transcription factor, which prevents the stem cells from differentiation.

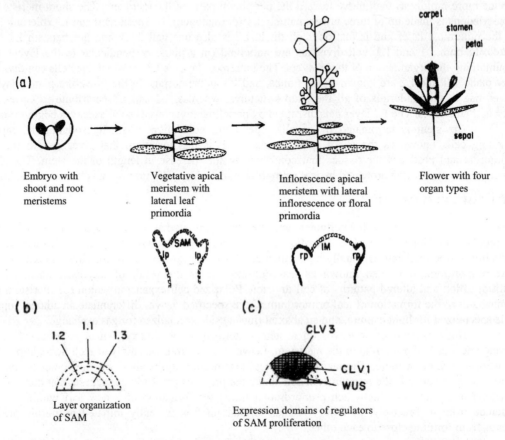

Fig. 32.4 (a) Life cycle of plant showing transition from vegetative to reproductive phase. Vegetative structures an depicted as SAM; shoot apical meristem and lp; leaf primordial. Reproductive structures are depicted as IM; inflorescence meristem and fp; flower primordial. (b) Layer organization of the shoot apical meristem. The SAM is composed of three distinct layers of cells; the epidermal layer (L1), the subepidermal layer (L2) and corpus (L3). (c) The approximate mRNA expression domains of genes (*CLV1*, *CLV3* and *WUS*) regulating the establishment and maintenance of SAM have been shown. *CLV3* is expressed in L1 and L2 cells. *CLV* is expressed in L3 cells, which also include the expression domain of *WUS*, *CLV3* functions as the intercellular signal to regulate *WUS* transcription (Redrawn from Clark, 2001 & Prasad and Vijayaraghavan, 2004).

Another homeodomain transcription factor, SHOOT MERISTEMLESS (*STM*), which is also member of *KNOX* gene family, is also important to keep the stem cells indeterminate. This protein acts independently of *WUS* in preventing premature differentiation of the stem cells and its daughter cells. *STM* represses the activity of two important genes for leaf initiation, *ASYMMETRIC LEAVESI* (*AS1* and *AS*). Hence, leaf initiation requires the downregulation of *STM* at the leaf primordium initiation site that allows *AS1* and *AS2* expression at these sites. In turn, expression of AS genes inhibits expression of the *KNAT* genes which functions like *STM* in preventing cell differentiation in Arabidopsis. The plant hormone auxin, which is produced from neighbouring mature tissues, influences the downregulation of *STM* at the leaf initiation site to facilitate leaf initiation. Auxin is transported through the plant tissues by specific cellular influx and efflux carrier proteins. The polar auxin transport has been shown to be essential for leaf formation, as no lateral organs are formed upon blocking polar auxin transport. Auxin is transported upward to the SAM through the epidermis and that the existing leaf primordia are sinks for auxin, draining the auxin away from the surrounding area, thereby preventing the initiation of new primordia. Thus, the plant hormone auxin plays pivotal role in phyllotactic patterning and also causes lateral inhibition by its withdrawal at a minimal distance away from existing leaf primordia.

PATTERN FORMATION IN LEAF DEVELOPMENT

Upon leaf initiation, the primordium is subject to different patterning processes in order to establish new axis of growth and differentiation. Patterning also results in the formation of a species- specific vascular network and in the regular spacing of specialized cells within a tissue, such as trichomes and the stomata in the leaf epidermis. The sheet-like structure of the foliage leaves is called leaf lamina, which capture light and convert light energy into carbohydrates and oxygen in the process known as photosynthesis.

During initiation, leaves grow along three newly defined axes: length (proximo- distal), width (centro-lateral), and thickness (dorso-ventral) direction. Leaves grow predominantly along the length and width direction, but their growth is restricted along the thickness direction because of early differentiation in a limited number of cell layers. Growth in the leaves is determined and results in a species- specific size and shape that is used as a taxonomic trait.

Axis formation upon leaf initiation leads to polarized (asymmetrical) growth of the leaf along the proximo-distal and the dorso-ventral axes. Several studies based on mutational analysis, genetic interactions, and gene expression analysis revealed the role of transcription factors on axis formation in leaves. In Arabidopsis, a transcription factor of the AP2/EREBP class, LEAFY PETIOLE (LEP), controls the polarity along the proximo-distal axis. The activation-tagged LEP gene converted the petiole into lamina. Another gene, *BLADE ON PETIOLE1 (BOP1)*, modulates cell division activity along the proximo- distal axis. Although restricted cell growth occurs along the dorso-ventral axis, cell differentiation occurs at first at the ventral (lower, adaxial) side of the primordium, in the form of vacuolization of the cells. In contrast, the dorsal (upper, adaxial) side of the primordium remains meristematically visible by the cytoplasm-rich cells. As a result, a fully expanded leaf consists of an upper and lower epidermis that is different, dorsal palisade and ventral spongy parenchyma, and the vascular bundles that consist of dorsal xylem tissue and ventral phloem tissue. The upper and lower surfaces of the leaf are adapted for light capture and photosynthesis and gas exchange, respectively. Several transcription factors thus involved in the determination of adaxial and abaxial identity have provided molecular genetic evidence to this SAM-based signalling model.

PATTERNING IN LEAF VENATION

The shape of the leaf lamina is determined early in the leaf primordium with high mitotic activity which is intimately correlated with the formation of the procambial strands of the major veins. The vascular system is composed of xylem and phloem tissues that transport water and food materials, respectively. The plant hormone cytokinin is important for vascular cell type morphogenesis. The leaf venation in Arabidopsis has a reticulate pattern, the complexity of which increases with developmental age. Procambium is the precursor of vascular tissue and is formed early in the leaf primordium. Xylem and phloem cell types differentiate from the procambium discontinuously. During leaf development, the primary or midvein is formed first, followed by the veinlets or secondary veins that develop as loops that connect to the midvein, resulting in a continuous network. Polar auxin transport determines the venation patterning in the leaves. The leaf primordia are the primary sites of auxin production, which coincides with a high cell division activity. It has been shown that polar auxin flows from leaf margin to the stem vascular tissue, which would create pathways for polar auxin movement that results in the formation of procambial strands.

PATTERNING IN THE EPIDERMIS

Patterning in a plant tissue layer results in specialized cell types with specific functions and with a set position in the tissue layer. As the outer restraining layer of the leaf, the epidermis begins to take shape even before the leaf primordium becomes recognizable as a bulge on the external surface of the shoot apical meristem (L1). The epidermis is composed of a single layer of cells. The leaf epidermis in Arabdopsis consists of a limited number of cell types, such as pavement cells, trichomes and stomata. Trichomes and stomata are spaced at regular positions within the surface according to a pattern. The patterning of trichomes occurs through lateral inhibition with feedback, a mechanism that is commonly used for cell type patterning. In this, neighbouring cells compete to express regulators that specify trichome cell fate as well as repressors that prevent their neighbours from adopting this fate.

Trichomes develop very early at the surface of the leaf primordia and differentiate from the tip to the base, in accordance to the basipetal gradient of differentiation in other tissue layers, such as the mesophyll. In Arabidopsis, the trichome is unicellular and develops into the highly specialized branched structure. The trichome initiation or trichome cell fate acquisition involves a limited number of gene loci. They are: *GLABRA1 (GL1), GLABRA3 (GL3),* and *TRANSPARENT TESTA GLABRA (TTG)*. Although the trichome patterning is genetically controlled, they are also influenced by environmental stimuli. Environmental and hormonal stimuli affect the competence of leaf tissues to form trichomes. Long day photoperiod and addition of gibberellic acids promote the formation of abaxal trichomes on juvenile leaves in which they are usually absent. The stomata are apertures in the epidermis, each bounded by two bean-shaped guard cells. The epidermal cells bordering the guard cells are called accessory cells or subsidiary cells. Stomata occur on both lower and upper leaf epidermis. They are used for the exchange of gases in between the plant and atmosphere.

THE VEGETATIVE TO REPRODUCTIVE TRANSITION

In the majority of the animals, the germ line is set aside during early embryogenesis, i.e., before the germ layer formation. However, in plants, the germ line is established only after the transition from vegetative to reproductive development, namely flowering. The vegetative and reproductive structures of the shoot are all derived from the shoot meristem formed during embryogenesis. In other words, no cells are set aside in the shoot meristem of the embryo to be used solely in the creation of reproductive structures. The flowering process is initiated by a signal that moves from the leaves to the shoot apex, inducing flowering (see below). The change from the vegetative to the flowering state not only affects the apical meristems concerned with flower production but alters physiologically and morphologically other parts of the plant as well.

FLOWER DEVELOPMENT

Flowers contain the reproductive cells in the higher plants. Flower structure ranges from the small simple flowers of self-pollinating species to the large, showy flowers of plants that rely on insects and birds for pollination. However, all flowers share a similar body plan. Typically, flowers contain four types of organs, which are arranged in circles or whorls. The outer whorl contains calyx or sepals, which are the floral organs that closely resemble leaves in many plants. Sepals often contain chloroplasts, trichomes and other cell types found in leaves. Sepals are generally smaller and cup-shaped and serve a protective function by enclosing the interior flower organs as they develop. The whorl immediately interior to the sepals contains the corolla or petals, which are the largest and most colourful part of the flower. Interior to the petals are the stamens, the male reproductive structures that produce pollen. The innermost organs are carpels, the female reproductive organs, which contain the ovules and mature into fruits.

Most flowering plants have a vegetative phase during which the shoot apical meristem generates leaves. The environmental signals such as day length, together with genetic regulators, reprogram the shoot apical meristem to switch over to a reproductive phase and give rise to flowers. Vegetative to reproductive development involves the specification of a determinate floral meristem that gives rise to a series of specialized organs found only in the flower. Flowers develop on inflorescences, which are morphologically distinct from the vegetative shoot that gives rise to leaves. The growth of the inflorescence requires the shoot apical meristem to be respecified as an inflorescence meristem. In species with a single flower, the transition to flowering marks the end of vegetative growth. The inflorescence meristem becomes the floral meristem, generating a stalk with a terminal flower. In other species, the inflorescence meristem gives rise to a succession of indeterminate branches or paraclades which are converted into a floral meristem to generate a flower.

GENETIC PATHWAYS REGULATING FLORAL TRANSITION

The SAM during vegetative phase generates leaves that form a basal rosette in the dicot plants like Arabidopsis. At the onset of the reproductive phase, newly generated primordia at the SAM acquire floral identity; last few leaves (cauline) produced are morphologically different. Secondary shoots (axillary inflorescences) emerge from axils of cauline and rosette leaves. Thus during floral transition, the SAM switches from the production of leaves with axillary inflorescence to the formation of flowers. The genetic pathways regulating the floral transition are well understood in Arabidopsis thaliana. The photoperiod pathway mediates stimulatory signals from long-day (LD) photoperiods, which in combination with the circadian clock, acts via a transcription factor, CONSTANS (CO). While a nuclear protein, GIGANTEA (GI) activates CO transcriptionally, the light sensors phytochrome (phy A) and cryptochrome (Cry2) activate CO post-transcriptionally. The gibberellin pathway has an essential role for activation in promoting floral transition by reducing the levels of floral repressor, FLOWERING LOCUS C (FLC).

The transcription factor, CO promotes flowering by controlling transcriptional upregulation of genes that confer a floral fate to the newly arising lateral meristems. Some genes that mediate the transmission of signal from CO in the leaf to shoot apical meristem, where *LFY* and *AP1* are activated have been identified recently. In fact, the mRNA encoded by one of these, *FT* represents the mobile signal, the florigen, which remained elusive for nearly 70 years.

The length of the day is a major environmental determinant controlling the time of flowering in many plant species. For example, Arabidopsis thaliana is a facultative long-day plant. In this plant, the day-length signal is perceived in the leaf and transduced to the shoot apex to initiate floral development. This long-distance signal is termed as floral stimulus or florigen. A key transcriptional regulator for sensing the day-length signal is known to be CO, which is exclusively expressed in the leaf. CO induces transcription of another gene, *FLOWERING LOCUS T (FT)* in the leaf phloem. Furthermore, recent studies have indicated that the FT mRNA is transported to the shoot apex, thereby forming the major component of the so-called florigen signal that moves from leaf to shoot apex. In the shoot apex, the causation of floral transition and floral morphogenesis by the FT signal involves co-activation of another transcriptional factor, FD. Both FD and FT, through their interaction promotes floral transition and initiate floral development through transcriptional activation of the floral meristem identity genes such as *APETALA1 (AP1)* and possibly *LEAFY (LFY)*.

Mutations in these genes partly transform flowers into shoots. In a LEAFY mutant, the flowers are transformed into spirally arranged sepal-like organs along the stem, whereas expression of LEAFY throughout the plant is sufficient to determine floral fate (floral meristem) in lateral shoot meristems. The transcriptional activation of *LFY* occurs earlier than that of *AP1* and the activation of both genes is early in long day (LD) and delayed in short-day (SD). The third gene activated by CO is *TERMINAL FLOWER1 (TFL1)* that is known to regulate the spatial pattern of *LFY* and *AP1* expression.

HOMEOTIC GENES AND ORGAN IDENTITY IN THE FLOWERS

The floral organ primordia, from which the individual parts of the flower develop, arise in the floral meristem by patterned cell division followed by cell differentiation and enlargement. The flower organ primordia arise at specific positions within the meristem, where they develop into their characteristic structures. In Arabidopsis, the central inflorescence meristem (shoot apical meristem) is surrounded by a series of floral meristems (FM) of varying developmental ages. The inflorescence meristem grows indeterminately, with cell divisions providing new cells for the stem below, and new floral meristems on its flanks. The floral meristem (or primordia) arises one at a time in a spiral pattern. Eventually, such a floral meristem would form sepals, petals, stamens, and carpel primordia.

In the flowering plants such as Arabidopsis, the identity of floral organs is determined by the combination of a series of transcription factors commonly known as the floral homeotic genes. The following genes act in organ identity specification in the developing flower. The gene *APETALA1 (AP1)* is expressed in whorls 1 and 2; *APETALA3 (AP3)* and *PISTILLATA (PI)* in whorls 2 and 3; and *AGAMOUS (AG)* in whorls 3 and 4. Sepals require expression of *APETALA1* and APETALA2 for their differentiation. The homeotic proteins encoded by floral identity genes such as homeotic genes are defined as the ones that specify the identity and developmental pathway of a group of cells. The homeotic proteins encoded by floral identity genes such as *AP1* and AG contain a conserved sequence of 58 amino acids, known as the MADS box, which bind to DNA. It has been shown that different combinations of transcription factors, primarily of the MADS domain class, act to confer specific organ identity to each whorl of the flower. Remarkably, the specific MADS box genes involved in floral identity are conserved in sequence and function in *Arabidopsis, Antirrhinum, Zea mays, Oryza sativa* (rice) and several other species, despite the extreme differences in the appearances of flowers among these species. This suggests that the activities of downstream genes have driven evolutionary change in organ number, position, and morphology.

Furthermore, just like the homeotic genes of *Drosophila* (e.g., engrailed), the MADS box transcription factors of plants are expressed in regions of the flower that exhibit homeotic transformations when one gene is absent. Mutations that alter the pattern of expression of one or more of these homeotic genes result in differing patterns of expression of the others and homeotic transformation of the floral parts. In Arabidopsis, at least five homeotic selector genes (floral organ identity genes) have been identified through the isolation of different types of homeotic mutants (Table.32.1). In the mutant *APETALA2*, the identities of whorls 1 and 2 are changed, so that carpels develop instead of sepals in whorl 1 and stamens develop in place of petals in whorl 2.

The Arabidopsis homeotic mutants fall into three classes according to the whorls that are misidentified, and similar mutants have been identified in *Antirrhinum* (Table. 32.1). In each class, two adjacent whorls are affected. In class A mutants, the identities of whorls 1 and 2 are changed. In class B mutants, the identities of whorls 2 and 3 are changed. In class C mutants, the identities of whorls 3 and4 are changed. Furthermore, the class A genes are expressed in whorls 1 and 2 while the class C genes are expressed in whorls 3 and 4. The domains of the class A and class C genes do not overlap because these genes inhibit each other. The class B genes are expressed in whorls 2 and 3, in a domain that overlaps both the class A and C domains. This divides the flower into four organ-forming zones defined by different combinations of homeotic genes: A = sepals; A+B = petals; B+C = stamens, and C = carpels.

Table 32.1: Floral organ idendity genes (homeotic selector genes) grouped by class in *Arabidopsis thaliana* and *Antirrhinum majus*.

Class (and expression)	*Arabidopsis* genes	*Antirrhinum* genes
Class A (whorls 1 and 2)	*APETALA1 (AP1)* *APETALA2 (AP2)*	*SQUAMOSA (SQA)*
Class B (whorls 2 and 3)	*APETALA3 (AP3)* *PISTILLATA (PI)*	*DEFICIENS (DEF)* *GLOBOSA (GLO)*
Class C (whorls 3 and 4)	*AGAMOUS (AG)*	*PLENA (PLE)*

ROOT DEVELOPMENT

The first visible developmental activity of the germinating seeds is the growth of the embryonic root, or the radicle, which splits the seed coats and emerges outside. The continued growth of the root depends on the formation of cells at its apical meristem and a wave of expansion, elongation, and differentiation of the newly formed cells toward the base. The organization of tissues in the root assumes a radial pattern, comprising single layers of epidermal, cortical, endodermal, and pericycle cells (Fig. 32.5).

The root is set up early and its origin can be identified in the late heart-stage embryo. Radial and axial patterning in roots begins during embryogenesis (Figs. 32.3, 32.5), and continues throughout development as the primary root grows and lateral roots emerge from the pericycle cells, deep within the root. The root hairs are projections of the epidermal cells of the root in the zone of differentiation and extend out tortuously into the soil, greatly enhancing water absorption by the root.

The root meristem is traced back to a set of initial cells that come from a single tier of cells in the heart- stage embryos. Each column of cells in the root has its origin in a specific initial cell in the meristem, and each initial cell has a stereotyped pattern of cell divisions that leads to each column. In a normally growing root, the region of low mitotic activity now known as quiescent centre is lined proximally by the peripheral apical initials, which generate the bulk of the root structure. Cell-cell interactions seem to play a role in determining cell fate even in root development, as removal of some root meristem cells does not disrupt normal tissue development, although altering the pattern of cell divisions. The root apical meristem thus could give rise to daughter cells that produce the three tissue systems of the root. The young root apical meristems are initiated from tissue within the core of the root and emerge through the ground tissue and dermal tissue. The rib-meristem type of growth is characteristic of the elongating root cortex. Root meristems can also be derived secondarily from the stem of the plant.

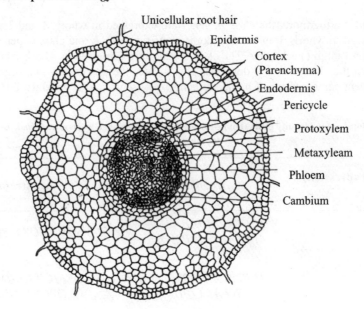

Fig. 32.5 Anatomy of dicot root. T.S. of root of Phaseolus radiatus showing beginning of cambium formation (Modified from Gilbert 2000).

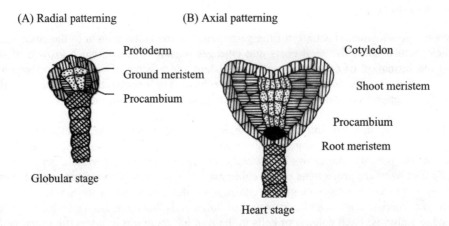

Fig. 32.6 Radial and axial patterning. (A) Radial patterning in angiosperms begins in the globular stage and results in the establishment of three tissue systems. (B) The axial pattern (shoot-root axis) is established by the heart stage. (Modified from Gilbert 2000)

Essentially, root apical meristem resembles shoot apical meristem and gives rise to the root in a similar manner to that in which the shoot is generated. However, unlike the shoot apical meristem, the roor apical meristem is bipolar in nature. As the meristem cuts off cells both distally and proximally, the distal derivatives form a root cap that protects the meristem and aids the root in its drive into the soil (gravitropism). The proximal cells elongate and differentiate into mature tissues of the root. The root branches are usually initiated beyond the region of most active growth and arise endogenously. The root also does not produce nodes and internodes, and therefore grows more uniformly in length than the shoot, in which the internodes elongate much more than the nodes.

REGENERATION AND SOMATIC EMBRYOGENESIS

Unlike animal cells, plant cells show a remarkable capacity for regeneration. Many plants can regenerate new individuals from sections of vegetative tissue containing only differentiated cells. Under appropriate culture conditions, most plant tissues can be coaxed to dedifferentiate and produce a mass of undifferentiated cells called callus. Callus can regenerate entire plants by either organogenesis or somatic embryogenesis. Active cell divisions in the cultured explant results in a callus, in which shoot buds differentiate with the concurrent production of root meristems; these calli are called 'organogenic'. The shoot bud and root meristem are not in direct contact with one another, but are always separated by an intermediate region of callus cells. The two together subsequently give rise to a new plant after establishing vascular connections between them. On the other hand, in somatic embryogenesis, the entire developmental pathway as followed in embryogenesis via sexual reproduction is recapitulated. Some species can even regenerate an entire plant from a single differentiated cell, indicating that differentiated cells in plants are totipotent, and hence can form generative tissue resembling either the early zygote or post- embryonic meristems. In this pathway of whole plant regeneration namely, the somatic embryogenesis, either single cells of the explant give rise to an embryo-like structure (embryoids) directly without an intervening callus stage (direct embryogenesis) or such embryo-like structures are produced from single cells of the callus derived from the explant (indirect somatic embryogenesis); the latter are called 'embryogenic calli'.

Instead of using the embryonic callus derived from differentiated plant tissues/cells, an alternative method employs inflorescence explants to produce somatic embryos, which would give rise to normal plantlets. The immature inflorescence is an excellent source of young meristematic tissue for the induction of direct or indirect somatic embryogenesis. Embryos formed in cultures have been variously designated as accessory embryos, adventive embryos, embryoids, somatic embryos and supernumerary embryos. Of these the term embryoids has been used most widely.

The somatic embryos originating from somatic cells are very similar to zygotic embryos. They pass through the same characteristic developmental stages, and give rise to normal seedlings. However, the pattern of cell divisions in the somatic embryos is distinct from that in the zygotic embryo. Furthermore, the somatic embryo lacks a suspensor. It seems likely that there is a maternal influence on plant embryo development, but this can be mimicked by permissive environmental conditions. Interestingly, a small number of maternal effect embryo patterning mutants have been reported in Arabidopsis. For example, plants homozygous for the short integument mutation produce embryos with axial patterning defects even if they are artificially crossed with plants producing wild-type pollen.

Understandably, every plant cell is primed with the entire developmental program, especially in parenchyma cells. An important component of the molecular pathway is the receptor kinase SERK, which is not expressed generally in somatic cells but is induced under appropriate culture conditions and marks the onset of competence to undergo somatic embryogenesis. SERK continues to be expressed in somatic embryos until the globular stage.

TRANSGENIC PLANTS

The malleable nature of the adult plant cells for regenerating a new plant has facilitated protocol development for plant transgenesis. Transgenic plants have become so popular that many of them find the production of transgenic plants involves transforming individual cells or tissues with foreign DNA significant use in commercial production. As a matter of fact, several transgenic crops whose genes have been modified to be resistant to pest and disease are a permanent part of worldwide agriculture. For example, there are at least four commercial transgenic cotton lines; one was the so-called "Bt cotton" that imparted insecticidal properties to the plant, while the other three were herbicide-tolerant lines.and using these cells or tissues for regeneration. The most common way of producing transgenic plants containing new and modified genes is through infection of plant tissue in culture with the bacterium, Agrobacterium tumefaciens, the causal agent of crown gall tumour (Fig. 32.7).

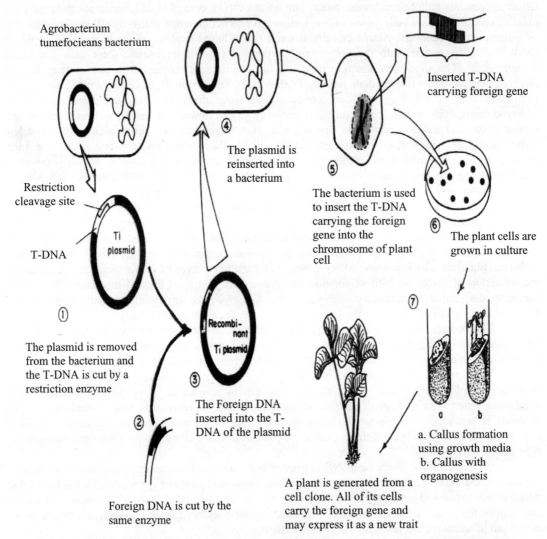

Fig. 32.7 Transgenic Bt cotton production through infection of plant tissue in culture with Agrobacterium (Redrawn from Jamie Pighin Webisite: biotech.ube.ca Biodiversity Transgenic Crops HTML)

This Agrobacterium contains a tumour inducing plasmid (Ti plasmid) that possesses the genes required for the transformation and proliferation of infected cells to form a callus. During infection, a portion of this plasmid, the T-DNA, is transferred into the genome of the plant cell, where it becomes stably integrated. Genes experimentally inserted into the T-DNA will therefore be transferred into the plant cell genome. Ti plasmid, modified so that they do not cause tumours but still retain the ability to transfer T-DNA, are widely used as vectors for gene transfer in dicotyledonous plants. The common method of transfer to plant cell is through a process known as a floral dip. A floral dip involves dipping flowering plants into a solution of Agrobacterium carrying the gene of interest, followed by transgenic seeds being collected directly from the plant. This process is useful in that it is a natural method of transfer and therefore thought of as a more acceptable technique. In addition, Agrobacterium is capable of transferring large fragments of DNA very efficiently without substantial rearrangements, followed by maintaining high stability of the gene that was transferred. The genetically modified plant cells of the callus can then be grown into a complete new transgenic plant that carries the newly introduced gene in all its cells and can transmit it to the next generation (Fig. 32.7).

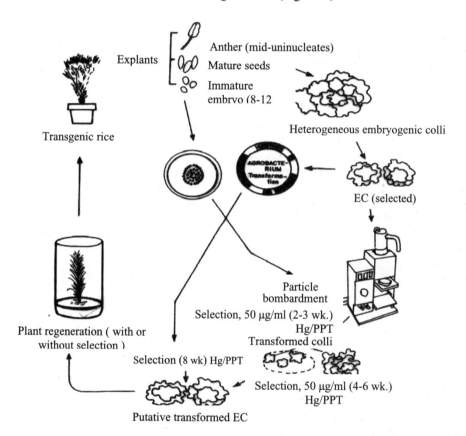

Fig. 32.8 A schematic protocol for production of transgenic rice using biolistic and Agrobacterium methods (Redrawn from Jamie Pighin Webisite: biotech.ube.ca Biodiversity Transgenic Crops HTML)

Other transformation strategies involve direct DNA transfer. In certain species, whole plants can be regenerated from transformed protoplasts. The most common technique of direct DNA transfer is particle bombardment, in which DNA-coated metal particles are fired into the nucleus using gunpowder, a blast of high-pressure gas or an electric discharge. By this technique, multiple genes can be introduced at the same time (Fig. 32.8). The regenerated plants usually contain the transgene in every cell.

REFERENCES

Abzhanov A., Kuo W.P., Hartmann C., Grant B.R., Grant P.R., Tabin C., (2006), The calmodulin pathway and evolution of elongated beak morphology in Darwin's finches. Nature Publishing Group., Nature Vol. 442/3: 563.

Afzelius (1961). Centrioles, cilia and flagella in Biochem. Pharmacol. Physiol. Press Ltd. Great Briton pp. 13-16.

Ahu, J. Miura, K. Li chen and A.S. Raikhal (2003) Cytotoxicity of mosquito vitellogenic ecdysteroid mediated signaling is modulated by alternative dimerization of the RXR homologues ultraspiracle. Proc. Nat. Acad. Sci. USA. 100. 544-549.

Ailles L.E., Weissman I.L., (2007), Cancer stem cells in solid tumours, Current opinion in Biotechnology, 18:460-466.

Bode H.R. (1996). The interstitial cell lineage of hydra: a stem cell system that arose early in invertebrate. J.Cell Sci. 109. 1155-1164.

Browder, L.W. (1980). Developmental Biology. Saunders College, Philadelphia. pp.602.

Chen, T.T. and Hillen, L.J. (1983). Expression of the vitellogenin genes in insects. Gamete Research 7: 179-196.

Carroll S.B. (2005). Evolution at two levels: on genes and form. PLoS Biol. 3, e245. 10.1371/journal.pbio.0030245.

Carroll S.B. (2008). Evo – Devo and an Expanding Evolutionary Synthesis: A Genetic Theory of Morphological Evolution. Cell 134 : 25-36.

Clark. S. E. (2001). Meristems: Start your signaling. Curr. Opin. Plant Biol. 4. 28-32.

Davenport, R. (1979). An outline of animal development. Addison-Wesley Publishing Company Inc, Phillippines. pp.412.

David G. Atwood (1974). Fine structure of spermatozoa of the sea cucumber, *Leptosynapta clarki* (Echinodermata: Holothuriodea) Cell. Tiss. Res. 149, 223-233.

Foltz, K.R. (1995). Gamete recognisation and egg activation in sea urchin. Amer. Zool. 35. 381-390.0

Gerald Karp (1999). Cell and Molecular Biology. 2nd Edition. John Wiley and Sons, New York. pp. 543.

Giese, A.C. and Pearse, J.S. (1974). Reproduction of marine invertebrates. Vol.1. Academic press, New York. pp. 546.

Gilbert, S.F. (1997). Developmental Biology. 5th Edition. Sinauer Associates, Inc. Publishers, Sunderland, Massachusetts, USA. pp.918.

Gilbert, S.F. and Raunio, A.M. (1997). Embryology. Sinauer Associates, Inc. Publishers, Sunderland, Massachusetts, USA. pp.537.

Hand book of stem cells. Vol. I. (2002). Embryonic stem cells Eds. R.Lanza, J. Gearhert, B. Hogam, D.Melton, R. PedeThomson, M. West. Elsevier Academic Press. London.

Hartmann, J.F. (1983). Mechanism and control of animal fertilization. Academic Press Inc, New York, USA. pp.561.

References from which the diagrams were redrawn/adapted/modified

Hopkins, P.M. (2001). Limb regeneration in the fiddler crab, *Uca pugilator*: hormonal and growth factor control. Amer. Zool. 41. 389-390.

Jeffery, W.R. and Martasian, D.P. (1998). Evolution of eye regression in the cavefish *Astyanax* : Apoptosis and the *Pax*-6 gene. Amer. Zool., Vol.No.38, pp.685-696.

Li, A. Murali, S. and Ding J.L. (2003). Receptor-ligand interaction between vitellogenin receptor and vitellogenin : implications on low density lipoprotein receptor and apolipoprotein B/E. The first three ligand binding repeats of VtgR interact with the amino-terminal region of Vtg. J. Biol. Chem. 278.2799-2806.

Li, F., Tiede B., Massague J., Kang Y., (2007), Beyond tumorigenesis: cancer stem cells in met Astasis. Cell Research, 17:3-14.

Mani SA, Guo W, Liao MJ, Eaton EN, Ayyanan A, Zhou AY, Brooks M, Reinhard F, Zhang CC, Shipitsin M, Campbell LL, Polyak K, Brisken C, Yang J, Weinberg RA. The epithelial-mesenchymal transition generates cells with properties of stem cells. Cell 133(4):704-15, 5/2008.

Mark P. Mattson. (2003). The fate of neurons in the developing brain and in Alzheimer's disease may lie with a four-protein complex that regulates the cleavage of two molecules spanning the cell membrane. The role of each protein is now being unveiled. Nature 422, 385–387

McLaren A. (2003). Primordial germ cells in the mouse. Dev. Biol, 262. 1-5

Morrison S. J. and N.M. Shah (1997). Regulatoy mechanisums in stem cell biology. Cell. 88. 287-298.

Morrison, N. Uchida and I.L. Weissman (1995). The biology hematopoietic stem cells. Ann. Rev. Cell Dev. Biol. 11, 35-71.

Prasad T.V Vijayaraghavan (2004). Genetic regulation of flowering: Specification of the floaral meristem and fattening of floral organs. Proc. Indian Nat. Sci. Acad. B70. 413-435

Print C.G. and K.L. Loveland (2000). Germ cell suicide: new insights in apopotosis during spermatogenesis. Bioassays. 22. 423-430.

Quill, T.A. and D.L. Garbres (2002). In fertilization (Ed) D.M. Hardy, Academic Press, New York.

Ramon Pinor Jr. (2002). Biology of Humar Reproduction. University Science Books, Sausalito, California.

Raymond W. Ruddon (2007), Cancer Biology 4th Edition, Oxford University Press Inc., Newyork.

Rohwedel, J. Guan, K. Hegert, C. and A.M. Wobes (2001). Embryonic stem cells as an *in vitro* model for mutagenicitiy cytotoxicity and embroyotoxicity studies. Present status and future prospects Toxicoloyg *in vitro*. 15. 741 –753.

Scott F. Gilbert, David Epel (2009), Ecological Developmental Biology: Integrating Epigenetics, Medicine and Evolution. Sinauer Associates Inc., 23 Plumtree Road, Sunderland, MA 01375, USA.

Sean B. Carroll (2005). The New Science of Evo Devo. Endless forms most beautiful.W.W. Norton & Company Inc. New York.

Shapiro M.D., Marks M.E., Peichel C.L., Blackman B.K., Nereng K.S., Jonsson B., Schliter D., Kingsley D.M., (2004), Geneticand developmental basis of evolutionary pelvic reduction in threespine sticklebacks. Nature Publishing Group., Page 717., Vol 428|15

Sinha Hikim, A. P. Lne, Y, Diuz – Romero, M, Yer, P.H. Wang, C and R.S. Swerdloff (2003). Deciphering the pathways of germ cell apoptosis in the testis. J. Steroid. Mol. Biol. 85. 175 – 182.

Subramoniam, T. (1995). Light and electron microscopic studies on the seminal secretions and the vas

---

References from which the diagrams were redrawn/adapted/modified

References

deferens of the penaeiodean shrimp, *Sicyonia ingentis*. Journal of Biosciences, Vol.No.20(5). pp.691-706.

Subramonium, T. (2002). Crustacean ecdysteroids in reproduction and embryogenesis. Comp. Biochem. Physiol. Part C. 125. 135-156.

Surami, M.A. (2004) How to make eggs and sperm. Nature, 427; 106-107

Surani, M.A (2004) Nature Ceccatelli, S, C. Tamm, E. Sleeper, S. Orrenius 2004. Neural stem cells and cell death. Toxicology letters. 149. 59-66.

Swann K and K.I. Jones (2002). Membrane events of egg activation In fertilization (Ed D.M. Harz) Academic Press New York. Pp 319-341.

Taikhel, A.S. Kokoza, Shu V.A.J. Martin, D. Wang S.F. Li C. Sun, G. Ahmed A, Dittmer, N and G. Attardo (2002). Molecular biology of mosquito vitellogenisis: from basis studies to genetic engineering of anti pathogenic immunity. Insect Biochem. Mol. Biol. 32. 1275 –1286.

Twymab R.N. (2001) Instant notes: Developmental Biology. Bios Scientific Publishers Ltd. U.K.

Wallace, R.A. and Dumont, J.N. (1968). The induced synthesis and transport of yolk proteins and their accumulation by the oocyte in *Xenopus laevis*. Journal of Cellular Physiology, Vol.No.72(2), pp.73-89.

Werver A. Muller (2005). Developmental Biology. Springer Verlag. New York Inc.

Wolpert, L. (1978). Pattern formation in biological development. Scientific American, Vol.No.239, pp.155-164.

Wolpert. L. (1998). Principles of Development. Oxford University Press Inc. New York, USA. pp.484.

Yasno Maeds (2005). Regulation of growth and differentiation in *Dictoyostelium*, Ann. Rev. Cytol. 244. 287 – 332.

References from which the diagrams were redrawn/adapted/modified

INDEX

A

20-hydroxy ecdysone 28
ABC transporters 270
Abscisic acid 256
acquired immunity 130, 131
Acquired Immuno Deficiency Syndrome (AIDS) 138
acrosin 12
acrosomal cap 16
acrosome 12, 13, 14, 15, 18, 32, 35, 37, 38, 40-45
actins 52, 73, 76
active immunity 149
adaptive immunity 149
adenocarcinomas 261
adipocytes 217
AGAMOUS (AG) 344, 345
age-1 mutation 231
ageing 209, 215, 226, 227, 228
age-related diseases 331
agrobacterium 299-301
AhR 32, 245, 263, 264
Alkylphenols 305
allantoic placenta 174
allantois 7, 163-168, 173, 174
allergy 157
allometric growth 178
alzheimer's disease 198, 209, 224, 229-231
ameloblast 99
amnion 147
amniotic cavity 147
amniotic folds 146
amyloid β-peptide 331
amyloid precursor protein- app 234
amplification 225, 228
amyotrophic lateral sclerosis (ALS) 214
anaplastic 225
androgens 299
androgen-binding protein 18
aneuploidy 224, 225, 227, 228
angiosperms 260, 261, 271
angiogenesis 233, 257
animal pole 32-34, 41, 54, 56, 57, 65, 67, 69, 73, 74, 79, 80, 94, 143
Antennapedia 277
antibody dependent cellular cytotoxicity (adcc) 134
antigen 8, 21, 226
antioxidant 141, 229, 230, 333
AP2/EREBP 341
APC 225, 227, 235, 241, 242, 252, 254
APETALA 1 344, 345
APETALA2 344, 345
APETALA3 344, 345
apolipoprotein 24
apoptosis 15, 16, 127, 132, 161, 162, 214 – 225
Araschnia levana 299
archencephalic canal 78

archenteron 65-74
argonaute 258, 264
aromatase 299
ARNT 240, 241
astaxanthin 143
asters 51
astrocytes 191, 197, 210, 331
asymmetric leaves 265
Ataxia telangiectasia 237
Autism 302
autoimmunity 138
autopod 280
auxin 275
axolotl 56, 175, 179

B

B Cells 131-134
B Lymphocytes 137
bafilomycin 141
basophils 135
Beckwith-Wiedemann syndrome 245
beta-carotene 143
bicoid protein 93, 97
Bicyclus anynana 299
bilateral cleavage 55
bindin 12, 39, 41
Bithorax 277
blastocoel 1, 54, 55, 57, 67
blastocyst 77, 148
blastoderm 78, 79, 81, 96-99
blastodisc 57, 58, 75
blastomere 92
blastopore 1, 69-75
blastula 1, 21, 54, 55, 57, 58, 78-81
BODY AXIS FORMATION 277
bone marrow 131-134
boveri 222
Breast cancer 235, 237, 238, 239, 240, 245, 249, 250, 251, 252, 260, 261, 263, 264, 267, 268
Birt-Hogg-Dubé syndrome 244
BRCA1 227, 233, 235, 237, 239, 249, 250, 251
BRCA2 227, 233, 235, 237, 239, 249, 250, 251
broadcast fertilization 36
Bursa of Fabricius 131, 132, 137

C

c-MYC 210
caesin kinase ii 159
calcitonin 86
callus 347
calyx 343

cambium 262, 271
camp 36, 37, 47, 105
cancer 222
Cancer stem cells 261, 263, 265, 266, 267, 268
capacitation 37, 38
capsularis 151, 152
carcinomas 224, 228, 232, 258, 266
carcinogens 223, 229, 231, 235, 259
carotenoid pigments 141
cataract 333
cavitation 78
cell adhesion 223, 268, 269
cell-mediated immunity 131
cells lineage 59
cell-signalling 274
cellular ageing 227
centrolecithal 32
CFU-C (Colony forming unit culture assay) 131
chemotaxis 36
chloroplasts 343
chondrocytes 86, 217
chordamesoderm 159
chordin 87, 101, 102
chorioallantoic membrane 148
chorioallantoic placenta 149, 154, 155
chorion 13, 32
chorion frondosum 152
chorionic villi 152, 155
choriovitelline placenta 154
chronic obstructive pulmonary disease (COPD) 218
c-kit 225, 263
cis-regulatory sequences 274
13-Cis-Retinoic Acid 303
cladogenesis 272
CLAVATA 340
cleavage 3, 5, 25, 51
cleavage nuclei 49, 141
cledoic egg 141
clk-1 231
Cloning 25, 62
CLV1 340
CLV2 340
CLV3 340
CO2 148, 152
colchicin 84
COLORECTAL CANCER 238, 240, 241, 242, 250, 251, 252, 254
compacting 56
Congenital Anomalies 301
corpora allata 166
cortex 21, 45, 132
Corticotrophin-releasing hormone (crh) 152
Cotyledons 336
Cowden syndrome 238, 246, 250, 254
Cranial Neural Crest Cells 303
CREs 274
Cuticle 140-142, 165-167
Cyclin 160
cyclin B2 229
Cysteine 141, 315, 316, 318
Cytochalasin b 84
Cytokines 130, 131, 217

Cytokinesis 28, 52, 57
Cytokinins 259
Cytolysin 134
Cytotrophoblast cells 150

D

2-deoxyglucose 270
daf-2 231
Darwin 272
DARWIN'S FINCHES 289
DDT 239
Decidual basalis 152
deformed 279
Deletions 225, 226
Dendritic Cells (DCs) 134
Desert Hedgehog 222
dermal epithelial cells 217
developmental genes 274
Developmental Synthesis 273
Dextral cleavage 54
Diapedesis 6
Dicotyledons 336, 339
Dietary restriction 228, 229
Diethylstilbestrol-des 305, 335
Differentiative phase 19
Diplotene 7, 20, 23
Distal-less 299
Dityrosine 233
DMEM medium 219
DNA methylation 91
DNA replication 160, 161
DNA repair gene 230, 235
Dobzhansky 273
Dopaminergic neurons 213
Double fertilization 335
Down's syndrome 214
Drosophila melanogaster 274
drug transporters 265
duplications 225
Dynein 13, 14, 36

E

Embryoid bodies 205, 208, 209
E-cadherin 56, 85, 119
Ecdysone 31, 141, 299
Ecdysone Receptor (EcR) 29
Ecdysteroids Response Elements (EcREs) 29
Ecological developmental biology 297
Embryonic carcinoma cells 245
Ectoderm 280, 287, 259
Embryonic germ cells 245
Emys 299
Egg development neurosecretory hormone-ednh 29
Eggshell 141, 331
embryo 288
embryonic stem–cell banks 216
Embryogenesis 6, 10, 15, 19, 21, 29
Embryogenic calli 347
Embryonic meristem 339, 347
Emx (1&2) 248

Endoderm 3, 5, 6, 57
Endometrium 152-155
Endosomes 23, 29
Endosperm 260
Endothelium 115, 151, 155
endothelial progenitor cells (EPCs) 217
Eosinophils 135
Ephrin 90
Epiblast 5, 6, 58, 62, 75-78
Epiboly 66, 71, 78
Epigenesis 2, 3, 4
Epithelial–mesenchymal transition 266
Epithelial growth factor -egf 197, 221, 224, 336, 337
Equatorial plane 33
Eric Wieschans 274
ER ά 240
ER β 240
Ernst Haeckel Theory 4, 248
ES 63, 189-191, 202-214
Estrogens 224, 299, 338, 339
Eukaryotic mitotic cell cycle 160
Eutherian mammals 32, 154, 155
Evo-devo 246, 253, 272
Exocoel 147
Exogastrulation 73
Extra embryonic coelom 147, 149
Extra ovarian organ 139
Extraembryonic cavity 78
Extraembryonic membrane 145-147
Extraembryonic structures 56
eyeless (ey) 289

F

Fab fragments 135
Familial adenomatous polyposis 227, 235, 240, 252, 254
Familial atypical multiple mole melanoma and pancreatic cancer 251
Familial medullary thyroid cancer 253
Familial papillary thyroid cancer 254
fate map 60, 78-81
Fc fragments 135
fertilin 42
fertilization 3, 11, 12, 14, 18, 35-49
fertilization cone 41, 42
fertilizin 35, 37, 41
Fetal Alcohol Syndrome(Fas) 304
FGF fibroblast growth factor-15, 87, 88, 175
Fgf8 291
fibre cells 288
fibroblasts 86, 92, 175, 217
fibronectin 66, 67
filopodia 6, 66, 67
florigen 343
FLOWERING LOCUS C (FLC) 343
FLOWERING LOCUS T (FT) 343
Flt-3 225
follicle stimulating hormone 26, 181, 184
follistatin 87
ford gene 87
FOXO3 232
FOXP2 256, 257

fragilis 6

G

G2 phase 51, 160
gametes 3, 7, 9, 11, 19, 32, 36, 117, 157, 171, 187, 199, 209, 226, 259, 260
gametogenesis 7, 9, 20, 23, 224
ganglion cells 288
Gardner syndrome 241
gastrula 65, 66, 67, 68, 71, 73, 74, 78, 83
gastrulation 1, 6, 32, 55, 57, 65-81
gene conversion 137, 257
gene silencing 90
Geospiza 290
genomic instability 225
genotype 3, 4, 61, 119
germ cells 5-7, 15, 61, 65, 120, 121, 174, 190, 191, 199, 204, 208, 209, 222, 224, 225, 245
germ line 3, 6, 91, 203, 322, 325, 367
germ plasm 5, 6
germarium 26, 28
germination 335-337
gibberellin 335, 336, 343
G Protein 231
GIGANTEA (G1) 343
GLABRA1 342
GLABRA3 342
glycogen 21, 24, 28, 105, 140, 142
glycolipoproteins 139, 140
glycolysis 270
glycoproteins 136, 190, 257
gonadotrophins 322, 324
gonial cells 7, 15, 19, 26
gonocytes 7
gooscoid 91
growth factor 227, 228, 257, 258

H

haplotype 136
Hayflick limit 230
HBOC 237, 239, 249, 251
heavy (H) chains 135
hedgehog 100, 128, 164, 199, 202, 213, 214, 322, 283, 284
HeLa cell line 230
Helper T Cells 134
hematophagus arthropods 143
hematopoiesis 131, 132, 137
hematopoietic stem cells 217
hematopoitic tissues 158
heme-binding protein 155
hemipteran 140
Henson's node 74-76, 102, 256
Hereditary melanoma 246-248
Hereditary pancreatitis 250
Hereditary breast and ovarian cancer 249, 251
Hereditary non-VHL clear cell renal cell carcinoma 243
Hereditary non-polyposis colorectal cancer 240, 251
Hereditary papillary renal cell carcinoma 244
Hereditary leiomyomatosis and renal cell carcinoma 245
Hermann Muller 224

Hereditary non-polyposis colorectal cancer 240, 251
heterodimers 136
heterosynthetic process 25
hippocampal region 231
histone acetylases 260
Histone Deacetylase (Hdac) 304
histone modification 260
histones 17, 47, 49, 21, 232
heterochronic shift 290
H.M.S Beagle 272
holoblastic 54-58, 68, 75, 78, 79
homeobox 6, 75, 99, 100, 108, 174
homeodomain transcription factor 265
homeostatic mechanisms 224
homing factor CXCR4 265
Homo sapiens 257
homology 12, 80, 103, 136, 276, 277, 280
Hox gene 75, 89, 100, 102, 103, 126, 173, 176, 218, 277-287
Hoxa1 Protein 303
Hoxa-10 126, 336
HOXB4 248, 279
Hoxc6 255
Hoxd-10 126, 249, 280
HPC1 239
HPC2 239
HPCX 239
Human Chorionic Gonadotrophin (HCG) 152
Human Placental Growth Hormone (HPGH) 152
Human Placental Lactogen (HPL) 152
Humoral immunity 131
Huntington's disease 233
Hydra Head Organizer 282
Hyla chrysoscelis 298
hypersensitivity 138
hyperplasia 225
hypoblast 5, 58, 75-78, 81

I

IgA 150, 153
IgD 150, 133
IgE 150
IGF-1 181, 256, 262
IGF-2 181, 256, 262
IgG 150-153
IgM 150-153
Immune system 130, 131, 136-138, 152, 191, 195, 218, 305, 339
Immunoglobulins (IG) 135, 257
Immunological rejection 152
Indian Hedgehog 252
INDUCED PLURIPOTENT STEM CELLS (iPS) 210
induction 3, 26, 65, 72-74, 80, 82, 91, 92, 107, 109, 112, 113, 117-119, 128, 132, 162, 164, 166, 173, 180, 189, 199, 211, 214, 216, 321, 323, 338, 345, 372
Industrial Mercury 304
inflorescence 340, 343, 344, 347
ingression 66, 75, 78
innate immunity 130, 134, 137
Insulin-like growth factors 1 and 2 161
interdigitating cells 135
interleukin (IL) 137

invagination 67, 69, 73, 74, 76, 78, 194, 112, 170
involution 66, 71, 73, 74, 78
isoagglutination 35
isolecithal/homolecithal 30
isozeaxanthin 143

J

janus green 60, 95
Julian Huxley 274
Juvenile Hormone (JH) 29, 30, 166, 167
Juvenile polyposis syndrome 242

K

Karagener's triads 14
karyokinesis 52
keratinocytes 158
KIDNEY CANCER 242-246
KLF4 210
KNAT Genes 341
KNOX Genes 340

L

L-dopa 298
L1-Dermatogen 339
L2-Periblem 339
L3-Plerome 339
laminin 7, 66, 88, 119, 198
latency period 223
leaf lamina 266
leaf primordium 339-342
LEAFY (LFY) 344
LEP 341
leptotene 7
Lethal null mutation 292
Lethal regulatory mutation 292
Leukemia 245, 246, 248, 252, 258, 262, 263
leydig cells 18, 119, 122
Li-Fraumeni syndrome 227, 236, 238, 245, 248, 249, 252, 256
LIN28 210
lipophorin 22, 28, 140
lipovitellin 20, 24, 139-144
lutinizing hormone 26
lymph nodes 132, 137
Lymphoma 227, 229, 237, 248, 258
lymphocytes 130, 132-138, 318, 330
lymphocytic leukemia 226
lymphokines 137
lysosomal compartment 141

M

macroevolution 246, 273
macrolecithal 32
macromere 54, 57, 79, 80, 94
macrophages 119, 132, 134-136, 194, 315, 316, 320, 321
Major Histocompatibility Complex (MHC) 135
Mario capecchi 63
marsupials 154
mast cells 135

maternal immunity 137
Medical Embryology 301
Medullary thyroid cancer 253
megalecithal 32
meiotic maturation 15, 19, 26, 225
MELANOMAS 246, 247
Melanoma-astrocytoma syndrome 248
Membrane-Attack Complex (MAC) 137
MEN 2A 253
MEN 2B 253
meridian planes 33
meristem 334-347
meroblastic clevage 57
meroistic 27
Mesenchymal stem cells 196, 212, 215, 217, 219
Mesoangioblast 212
mesenchyme 67, 68, 72, 80, 82, 86, 88, 94, 109, 115, 118, 119, 124, 126, 127, 174-176, 216, 280, 290, 291
mesoderm 1, 6, 57, 65, 68, 70-73
mesolecithal 32
mesomere 54, 79, 80, 117
metastasis 223, 259, 264-269
Methylemercury 304
MHC Molecules 132, 133, 135, 136
Microenvironment 264, 268, 269
microevolution 273
microfilament 67
micromere 54
Microphthalmia 302
micropinocytotic activity 21
micropyles 35
microtubule 14, 51, 69, 84
mid piece 11
midvein 267
Minamata Disease 304
Mitosis Promoting Factor (MTF) 51
mitotic division 1, 7, 23, 51, 52, 157
mitotic spindle 52, 53
MLH1 240, 251
MLH2 240, 251
MLH6 240, 251
Modern synthesis 273, 274
molting hormone 29, 140, 141, 167, 179
monokines 137
morphs 298
morphogenesis 1, 32, 65, 67, 78, 82, 88, 94, 103-110, 124, 157, 162, 173, 175, 312, 316, 217, 318, 334, 341, 343
morula 56, 57, 204
mosaic or determinate development 3
mosaic theory 3
M-Phase 160, 161
mRNA 26, 31, 57, 90, 92, 96, 97, 121, 164, 179, 238, 280, 282, 295
Muir-Torre syndrome 241
multi drug resistance 270
Muller cells 288
Multiple Sclerosis (MS) 138, 208, 209, 213
Multiple endocrine neoplasia 236, 253
MYC 227, 232
MYH 16 256
MYH-associated polyposis 242
myriapods 287

Myelomas 229
myofibril 93, 196
myeloid leukemia 226, 263
myoblasts 217
myosin 51, 53, 66, 84, 90, 92, 176, 256
myotubule 92, 179

N

NANOG 210
natural immunity 130
N-cadherin 85, 88, 124
N-CAM 73, 85, 119, 124, 175
NEK2 229
Neo-Darwinian Synthesis 273
neocortex 233
neonatal immunity 137
neuegulin 88
Neocortex 303
neoplastic cells 225
neural crest cells 84-90
neurectoderm 73, 103
Neurofibromatosis 236
Neurofibrillar Tangles (NFT) 234
neurulation 83, 85, 109, 110
neuwkoop centre 69
NK Cells 134
noggin 87, 102, 242, 243
notch ligand 80
Nusslein-Volhard 274

O

OCT4 210
odontoblast 86, 88
olfactory pits 291
omphalopleure 154
oncogenes 224, 228-234, 253, 255, 258, 262
onychophorans 286
Ontogeny 1, 3, 4, 132, 191, 192, 316, 326, 337, 279, 295, 358
Oogenesis 6-10, 18-20, 92, 98, 139, 225, 240
Organogenesis 65, 82, 109, 112, 116, 119, 142, 149, 168, 224, 237, 274, 287
oscillin 44
osteoblast 86, 191, 194
osteocytes 86, 158, 217
Otx (1 and 2) 280
oviparity 139
ovo-hemerythrin 140
ovule 261, 268
oxidative stress 217, 228-230, 232, 234

P

P53 218, 232
Pachytene 7
PAH 222, 223, 240
Paired Helical Filaments (PHF) 234
PANCREATIC CANCER 242, 247, 251, 252, 268
panoistic 27
Papillary and follicular thyroid cancer 254
paraclades 343

parenchyma 347
Parkinson's disease 198, 209, 214, 217, 228, 232
parthenogenesis 48-50, 216
passive immunity 131
patency 29
pathogens 130, 134, 138, 152
Pax-6 112, 205, 279, 288
PCB 239
PELVIC REDUCTION 291
percutor organ 42
perforin 134
periderm 262
periosteum 158
periplasm 58
personalized medicine 271
Pervasive Developmental Disorder 303
Peutz-Jeghers syndrome 238, 242, 250
PHAB 260
phalanx 241
pharyngeal arch 88, 89
PHAV 335
phellogen 337
phenotype 4, 91, 94, 112, 191, 213, 210, 228, 273, 274, 276, 278, 288, 291, 294
Phocomelia 302
photoreceptors 288
phloem 337, 341, 343
phosphoglycolipoproteins 139
phosphoprotein 21, 23, 25, 26, 28, 140
phosphorylation 21, 37, 44, 47, 140, 234
phosvitin 21, 23-26, 140
photosynthesis 341
phyllotaxis 339
physiological polyspermy 46, 47
phytochrome 343
phylogenetic tree 272
PISTILLATA 334, 345
pituitary gland 18, 26, 161, 202
Pitx1 291
placenta 117, 131, 139, 145, 149, 150-156, 183
Plasticizers 305
plasmodesmata 334
PLENA (PLE) 345
pleiotropic toolkit loci 296
PMS2 240, 251
plumule 334
pluripotency 189, 190, 197, 202, 203, 209
point mutations 224, 225, 232
pollen 138, 335, 343, 347
Polychlorinated Compounds 305
polymorphonuclear leukocytes 133
polyserine domine 140
polytene 28
polytrophic 28
polytrophic ovarioles 28
porphyrin ring 144
Polyphenism 297
post embryonic growth 165
post-embryonic development 167
precursor form 141
premature ageing 232, 235
premeiotic phase 20

Presenilin 1 233
Presenilin 2 233
previtellogenic phase 20, 23
primary lymphoid organs 132
primitive streak regression 157
primordial germ cells 5, 6, 191, 199, 204, 209
primordial meristem 339
primordium 112, 339, 340-342
procambium 341
proembryo 335
progesterone 26, 37, 152, 181, 182, 224
programmed cell death 119, 126, 132, 161, 162, 314-318, 321, 324, 325
PROSTATE CANCER 237, 242, 251
proto-oncogenes 253, 255, 258
proliferative phase 19, 181
promeristem 338, 339
prosthetic groups 140
proteases 28, 41, 137, 141, 142, 315, 318, 323
protein aggregation 233-235
proteolytic processing 30, 141, 318
pseudo placental viviparity 139
pseudopodia 67
PTEN 238, 246, 250, 254, 262
Purkinje Cells 303

R

radial cleavage 53, 54, 79
radicle 338, 339
RAS 227, 229, 232
RA-Retinoic acid 89, 103, 128, 167, 176, 100
Rb 225
Reactive Oxygen Species (ROS) 229, 302
regenerative medicine 261
regulative development 3, 59, 60
regulatory circuits 293
resact 35, 36
retinal pigment epithelial cells 288
Retinoblastoma 227
Retinoic Acid 303
Retinoid X Receptor (RXR) 29, 303
Retroviruses 258
Rhodnius prolixus 140
rhombomere 88, 89, 110
RNA polymerase 90, 91
Rohon Beard Cells 86
root apical meristem 336-338, 345
rotational cleavage 55, 77
ROTUNDIFOLIA 339
rudiments 82, 109, 114, 116, 146, 165

S

Sarcomas 228, 248

SAM 339, 340, 341, 343
Schwann Cells 86, 89
Schistocerca gregaria 298
secondary lymphoid organs 132
secondary nucleus 335

Sean S. Carroll 295
senescence 1, 208, 226, 232
sensory neuromasts 291
sepals 342-345
serine proteases 141, 142, 315
serine residues 140
serosa 147
sertoli cells 7, 15, 18, 119-122, 322
Shh 291
Shh gene 284
shoot apical meristem 336, 337, 339, 343-345
Shoot Meristemless 340
Signal Transduction 227, 229, 231, 232, 261
sinistral cleavage 159
Sir2 229, 232
SIRT1 232
SIRT6 232
slug gene 84, 88
Smad 1 6
Small eye 288
somatic 3, 5, 6, 7, 20, 32, 44, 49, 51, 62, 104, 114, 121, 124, 145, 147, 161, 191, 199, 203, 207, 212, 216, 222, 224, 259, 272
somatic cell nuclear transfer (SCNT) 210
somatic mesoderm 114, 145, 147
somatic mosaicism 226
somatopleure 145
Sonic Hedgehog 128, 198, 201, 210, 211, 283, 284, 305
SOX2 210
specific immunity 134
spectrin 84
spemann's organizer 72, 74
speract 35
spermatids 7, 9, 15, 16
spermatogenesis 7, 8, 15, 16, 18, 222
spermatogonial cells 15
spermatozoa 7, 11, 14, 15, 17, 18, 32, 35, 37, 40, 46, 47, 49, 187
spermiogenesis/spermateleosis 7, 15, 16, 18
S-Phase 179
spindle 51, 55, 57-59
spiral cleavage 59, 60, 104
splanchnic mesoderm 131, 133
splanchnopleure 165
spleen 149, 156, 190, 194
spot CRE 276
stamens 285, 294-296
stem cells (HSCs) 139, 149, 332
stereoblastic 88
stereoblastula 59
stereotyped 336, 346
stigma 334
stimulus 135, 244, 344
stomata 341, 342
subfunctionalism 331
submammalian Vertebrates 139
substantia nigra 231, 251
superficial cleavage 58
suppressor T Cells 151
surface membrane Ig 151, 153
suspensor 336, 339
synapsis 10

Synapta digita 58
syncytioblast 169-171
syncytiotrophoblast 169, 171, 172
Systemic Lupus Erythematosis (SLE) 157

T

T Cells 151-155
T lymphocytes 338
tarsometatarsus 264
T-Cell receptor (TCR) 151, 155
T-cell leukemia virus 258
T-DNA 299
telolecithal 35, 62
telomerase 226, 229, 253
telomerase theory 253
telomere 229, 253
telotrophic 29
tenascin 73, 74, 98
TEP1 253
Teratogenesis 301
Teratogenic Agents 301
terminal differentiation 285
tetrad 10, 11
Thalidomide 302
THREE-SPINE STICKLEBACKS 291
thymidine 91
THYROID CANCER 252
thymus 99, 100, 141, 291
Ti plasmid 348
tibiotarsus 264
ticks 143, 144
Tissue tropisms 264
TNF 306, 315
toolkit 287
Tool kit genes 4
topobiology 3
toolkit genes 274
totipotency 62, 260
totipotent 172, 174, 176, 183, 202, 209
TP53
transgenesis 62, 272
transgenic 62, 185, 331, 349
transgenic crops 349, 350
translocations 225
transparent testa glabra 267
transposable elements 275
transplantation 4, 7, 81, 83, 91, 73, 74, 81, 120, 121, 215, 216, 330-335
trichomes 341-342
trisomy 226, 227
trochophore 104
trophectoderm 61, 68, 205, 226
trophoblast 61, 86, 189, 202, 207
trophoblastic lacunae 169, 170
Trypan blue exclusion 217
Tuberous sclerosis complex 246
tubulins 52, 73
tumor initiation 259, 260, 271
tumor migration 264
Tumor Necrosis Factor (TNF) 153, 172
Tumor promotion 229

tumor suppressor genes 225, 228-232, 241
tunica 289
Turcot syndrome 241
twin of eyeless (toy) 289

U

umbilical cord 166
uric acid 172
uterine endometrium 60

V

Valproic Acid (Va) 302
Vasa gene 6
vectors 351
vegetal pole 36
veratrum californicum 354
vitellarium 29, 30
vitelline artery 167
vitelline vein 132, 167
vitellins 28, 139, 140, 144
vitellogenesis 9, 21, 22, 29, 30, 139+B572
vitellogenic oocytes 28, 139
vitellogenic phase 22, 23
Vitellogenin 264
vitellophages 160
viviparous 34
Von Hippel-Lindau syndrome 243
von Hippel-Lindau disease 236

W

Warburg's hypothesis 223
Werner's Syndrome 259
whorl 290, 294, 295
WING PIGMENTATION 276
Wnt 7a 262
Wnt/ ß-Catenin Signaling 302
Wnt expression 276
Wnt gene 276
Wnt-1 276
Wnt-10 276
Wnt-3 276

Wnt-6 276
Wnt-9 276
WRN 259
WUSCHEL (WUS) 339

X

xbra 81, 116
Xeroderma pigmentosum 228, 235, 248
xylem 341, 342

Y

Yamanaka factors 212
YAB3 339
yolk 165, 169, 172
yolk duct 165
yolk Granules 160, 161, 312
yolk platelets 26, 27, 29, 158
yolk Proteins 23, 149, 159, 163
yolk sac 165
yolk sac placenta 173

Z

zeugopod 280
Zona pellucida 36, 41, 42, 45, 46
Zona radiata 36
ZPA 143, 146, 147, 277
zygote 11, 37, 49-53
zygotene 8, 9
zygotic 9, 51, 55, 110-113
yolk Proteins 23, 149, 159, 163
yolk sac 165
yolk sac placenta 173

Z

Zona pellucida 36, 41, 42, 45, 46
Zona radiata 36
ZPA 143, 146, 147, 277
zygote 11, 37, 49-53
zygotene 8, 9
zygotic 9, 51, 55, 110-113